THE TENANT LEAGUE OF PRINCE
LEASEHOLD TENURE IN THE NEW WORLD

Historical writing about the middle years of the 1860s in British North America has focused almost exclusive'ᵛ ⌐n the Confederation movement and the theme of nation-buildir  ve largely overlooked one of the most sᵗ movements of common people in thᵉ h flourished in Prince Edward Island ᶜ ᵢnt League produced a highly compellir ᵢ- sive role in undermining the leasehola syɔ̄ceⁱⁱ oɪ ɪⁱ.... .ain had imposed a century earlier. Through an exhaustive study of period documents, Ian Ross Robertson examines the origins, the *modus operandi*, and the impact of this organization. In doing so, he has illuminated a rich part of Canadian history.

The Tenant League was a militant grass-roots movement, which, despite its brief life, won the support of thousands of Island farmers. Following league policy, tenants refused to pay further rent until the landlords sold them the land they occupied on terms that they (the tenants) found acceptable. This led inevitably to legal action against league members; in response, the organization practised systematic civil disobedience. In 1865, when British troops were dispatched to assist local authorities, the league, in one of its most daring and unconventional moves, adopted a strategy of inducing soldiers to desert. In doing so, the leaguers intentionally exploited a chronic military problem, and their remarkable success provoked unprecedented counter-measures.

While the league can be compared most obviously with movements in the British Isles, Robertson concludes that circumstances on Prince Edward Island, which combined Old World issues and the New World environment, made it unique. These circumstances included the ethnic mix on the Island, the recent history of pioneering, the institution of responsible government, and the fact that freehold tenure was the norm in North America. Significant chapters in the social, political, legal, and military history of British North America are contained in the dramatic and surprising story of the Tenant League.

IAN ROSS ROBERTSON is an associate professor in the Department of History at Scarborough College, University of Toronto.

IAN ROSS ROBERTSON

The Tenant League of Prince Edward Island, 1864–1867: Leasehold Tenure in the New World

UNIVERSITY OF TORONTO PRESS
Toronto Buffalo London

© University of Toronto Press Incorporated 1996
Toronto Buffalo London
Printed in Canada

ISBN 0-8020-0769-4 (cloth)
ISBN 0-8020-7138-4 (paper)

Printed on acid-free paper

Canadian Cataloguing in Publication Data

Robertson, Ian Ross, 1944–
    The Tenant League of Prince Edward Island,
    1864–1867

    Includes bibliographical references and index.
    ISBN 0-8020-0769-4 (bound)
    ISBN 0-8020-7138-4 (pbk.)

    1. Tenant League of Prince Edward Island.
    2. Land reform – Prince Edward Island – History –
    19th century. 3. Farm tenancy – Prince Edward
    Island – History – 19th century. I. Title.

    HD1333.P75R63 1996   333.33'5   C95-933092-5

The maps reproduced in this book were created by The Cartography Office, Department of Geography, University of Toronto.

This book has been published with the help of a grant from the Humanities and Social Sciences Federation of Canada, using funds provided by the Social Sciences and Humanities Research Council of Canada.

University of Toronto Press acknowledges the financial assistance to its publishing program of the Canada Council and the Ontario Arts Council.

FOR MY MOTHER

Contents

List of Tables

List of Maps

Preface

The purpose of this study is to present, explain, and place in context the history of the Tenant League of Prince Edward Island, an agrarian movement committed to use of direct action to end the system of leasehold land tenure which Great Britain had imposed upon the colony in 1767. The organization flourished in the 1860s, by which time most Islanders believed leasehold to be thoroughly anomalous and anachronistic in New World conditions. The life of the league coincided precisely with the crucial years of the movement for union of the colonies within British North America, an élite-sponsored operation which succeeded in founding a new, although fragile, nation-state. The fact of temporal concurrence with Confederation has tended to divert attention away from the league, an extraordinary body of common people striving to challenge both the imperial power of London and the property rights that London was committed to protecting. Such an undertaking required strategic boldness, imagination, and discipline.

The willingness to defy governments and conventional values, although unusual, was neither ill considered nor fleeting. When soldiers were sent to Prince Edward Island in 1865 as support for the civil authorities, the leaguers attempted to induce desertions. These efforts were rational, for it was a matter of public record that over the previous generation, when there had been a garrison on the Island, the prevalence of desertion had been a chronic problem, sufficiently vexing to the imperial authorities to be a factor in the removal of troops in 1854. The surviving evidence demonstrates that a remarkable number of soldiers forsook their posts within weeks of arrival in 1865. More to the point, it also reveals that Tenant Leaguers played an aggressive part in endeavouring both to encourage the act of desertion and to provide deserters with safe passage off the Island. Sworn evidence at the trials of Islanders convicted of inducing desertion proved the proactive role of the leaguers or their sympathizers. Few deserters were apprehended, and the desertions caused much friction between the civil authorities in Charlottetown and the military authorities in Halifax, leading the latter to press the former to consent to withdrawal of the men. In fact, the leaguers' promotion of desertion had frightened the authorities: the local government resorted to unprecedented measures – such as purchase of civilian clothes for military search parties – to stem the problem; and the military leaders were sufficiently unsettled by the prospect of what might happen if the Island was left without troops during the long winter period when there was no easy means of transportation between it and the mainland that they did not act upon their threats to withdraw. The soldiers remained until 1867, and the alarm surrounding the leaguers' success in encouraging desertion underscored both the strength of the league and how seriously the authorities took its activities.

The innovative tactics of the league are only part of the story. The organization established a phenomenally broad base of support in a brief period. Part of the explanation lies in the aggressive militancy of its tactics and in popular frustration with more moderate approaches. The league was also able to demonstrate to potential supporters concrete achievements in convincing landlords to sell property to tenants. Beyond immediate successes, it is clear that the league played a decisive role in demoralizing the leasehold régime, and hastening its end. For all these reasons – its unique nature, its support among the mass of Island farmers, and its triumphs, for both the short term and the long term – the league deserves scholarly attention.

Nonetheless, scholarly researchers have taken little notice of the Tenant League, and, for the most part, the judgments of those who have commented have been unreflecting and even internally inconsistent. Part of the explanation for the inadequate and distorted perception of its significance is that no writer has made a serious attempt to understand the point of view of the members. Yet that task of taking their perspective into account is essential as a means of interpreting them and measuring them, their organization, and their accomplishments in a reasonable way. Instead, the league has been pictured as, for example, a quaint Prince Edward Island distraction from the real story of the middle 1860s, the Confederation movement. The reasons for the lack of attention to the league can only be a matter of speculation, but the denials, implicit or explicit, of its importance and success are worth consideration.

It is clear from surviving sources that the league received scant sympathy from established contemporary commentators, both Tory and Reform, because of its unconventionality and especially its calculated defiance of the law. Those who were publicly identified as leaders – George F. Adams and Alexander McNeill, for example – were the victims of exceptionally hostile commentary, amounting to vilification. The portraits make the researcher wonder how such people could win a large following, short of mass dementia among their supporters. The leaders had their failings, and may deserve criticism for quailing under pressure. But to flinch in the face of superior armed might is human, and it may be plausible to interpret the tactics of passive resistance into which the league eventually retreated as part of an overall strategy which deserves credit for boldness and success in effecting major social change apparently without a gunshot being fired with the intent to strike another human being. The Tenant League cannot be classified as non-violent, for sticks, stones, and fists were used, bruises were inflicted, blood was shed, and bones were broken. Nonetheless, despite all the fury and tension, both sides – the authorities and the leaguers – seem to have had a sense of proper limits to their strategy and tactics.

Yet the importance of the Tenant League is more than process or method. It is also rooted in substance, for the organization was part of the broad and passionate struggle for human liberty and individual – or perhaps more accurately, household – 'independence' which marked so much of the nineteenth century. The ideal was that of the freehold farmer, the small property-owner, who had no master or lord. In locating the leaguers ideologically, we must note that they did not reject property, and that they were willing to pay their landlords what they considered a fair price for freehold title. It was as an agrarian movement committed to the freeholder ideal that the league won mass support for a program which contemplated, as a method, resort to illegal actions. The social and political conditions – the predominance of leasehold tenure that spawned the league and the governmental system of ministerial responsibility based on a broad male franchise that gave its supporters exceptional leverage – formed a combination unique to the Island.

Beyond its intrinsic qualities, the history of the league is important because of its success, a success which will be explained in the course of this study. In the view of Island observers of all political stripes, including rivals and bitter enemies, the league rapidly gained the backing of most producers within the colony. Its level of support on the Island constituted a remarkable achievement. But beyond that, I will argue that the league played a vital role in hastening the end of the system of leasehold tenure. It was not simply an interesting organization with broad public sympathy; it was effective, and its flourishing marks a turning-point in the entire history of what was known locally as the land question. The reverberations of league activities appear to have caused leading land-lords to reconsider the future of leasehold tenure in the colony. Once the will to continue expansion of the system had been shaken, abolition was only a matter of time.

The significance of the Tenant League of Prince Edward Island is multifold. This book will argue that its importance extends into Canadian history as a whole, and into the general history of agrarian and popular movements.

Acknowledgments

This work was not conceived and executed in isolation. Harry Baglole, Director of the Institute of Island Studies, University of Prince Edward Island (UPEI), did pioneering work in rethinking the Prince Edward Island land question in the 1970s. From the beginning of the present project he has been generous with his insights, encouragement, and assistance. J.M. Bumsted of St John's College, University of Manitoba, has made major contributions in changing and deepening our understanding of Prince Edward Island history in the first several decades after the establishment of leasehold tenure in 1767. For many years he was an exciting and inspiring fellow-researcher with a stream of fresh perspectives, and I have benefited from his energetic and energizing comradeship.

In the course of doing the research, I spent much time in the search rooms and basements of the Public Archives and Records Office of Prince Edward Island (PAPEI) at their present and previous locations. I owe much to the people who have worked there. Nicolas J. de Jong, the first full-time provincial archivist in Prince Edward Island, created a congenial and professional working environment. Harry T. Holman, his successor, has followed through with an outstanding sense of commitment to Prince Edward Island history. Kevin MacDonald, who has worked there throughout the period when the research has been done, has been a never-failing source of good cheer, good sense, and efficient service. F.L. Pigot, creator of the Prince Edward Island Collection at the Robertson Library, UPEI, has provided much assistance over many years. Orlo Jones, formerly at the Prince Edward Island Museum and Heritage Foundation (PEIM), has given guidance in areas where genealogical research or information has been relevant.

Two former associates at Scarborough College, University of Toronto, deserve special mention. Once a student of mine at the undergraduate level, M. Brook Taylor of Mount Saint Vincent University has taken an active interest in this project, prodding me to see it in as broad a perspective as possible. He read critically an earlier version of the manuscript, and I hope his advice has had the desired effect. John S. Moir, as my senior colleague in Canadian history, encouraged scholarship in various ways over many years. He also read every chapter of a previous draft, and made constructive suggestions; he is responsible for the idea behind map 9. In addition, I must acknowledge the influence of students in my Atlantic Canada course at Scarborough College, whose probing questions and astute observations have provided stimulation and food for thought.

At crucial times I have had the benefit of advice and encouragement from two outstanding scholars in Irish and Scottish history, where significant parallels with aspects of the Prince Edward Island land question can be found, and where the historical literature is much more developed. Samuel Clark of the Department of Sociology, University of Western Ontario, has alerted me to works concerning the Irish land question in the nineteenth century, on which he is an authority. An exemplar of scholarly generosity, he read several chapters with a critical eye and made invaluable suggestions for sharpening the analysis. T.C. Smout of the University of St Andrews, Historiographer Royal in Scotland, is someone whose encouragement of this study extends back more than two decades. He has also assisted by keeping me abreast of pertinent developments in Scottish history.

Byron Moldofsky of The Cartography Office, Department of Geography, University of Toronto created the maps, and I have profited from his expertise and commitment to precision. Thanks are due to the University of Toronto Department of History and the Principal's Office at Scarborough College for financially supporting the map work, which is essential for a study like this, whose focus is land. Harry Holman and Marilyn Bell at PAPEI and G. Edward MacDonald, curator of history at PEIM, provided expert advice in selecting the illustrations.

Robert A. Ferguson of the University of Toronto Press has efficiently guided the manuscript through its many stages, and both he and Gerry Hallowell have given me editorial advice on improving its structure and style. The anonymous readers who assessed the manuscript for the University of Toronto Press and for the Aid to Scholarly Publications Programme in Ottawa contributed helpful criticism and suggestions which have made this a better work. Judy Williams has done an exceptionally fine job as copy-editor.

In addition, there are many other people who have assisted in a variety of ways, by encouraging the project in the earliest stages, by facilitating access to particular materials, by drawing my attention to a source or an angle I might otherwise have overlooked, by advising on a legal or linguistic matter, by checking a reference for me in a distant archive, by sharing information gleaned from painstaking genealogical research, or by some other act of kindness. I have attempted to acknowledge some of these debts in a number of specific places in the endnotes, but among those not mentioned elsewhere are Phillip A. Buckner, of the University of New Brunswick; Allan Greer, of the University of Toronto; and Frank MacKinnon, formerly of the University of Calgary.

Financial support from the Social Sciences and Humanities Research Council of Canada facilitated the research which has made this book possible. Parts have appeared in different form in the following: 'Political Realignment in Pre-Confederation Prince Edward Island, 1863–1870, Acadiensis, 15/1 (Autumn 1985), 35–58; 'The Posse Comitatus Incident of 1865,' Island Magazine, no. 24 (Fall–Winter 1988), 3–10; and 'Introduction' to Ian Ross Robertson, ed., The Prince Edward Island Land Commission of 1860 (Fredericton: Acadiensis Press, 1988), ix–xxx. On 24 April 1980 I presented a paper on the Tenant League at the opening session of the Atlantic Canada Studies Conference in Halifax. Revised versions were delivered to the Canadian History Colloquium, at Massey College, University of Toronto; the Atlantic Canada Institute, a travelling summer school, at UPEI; and the History Department at the University of Manitoba.

Alexander McNeill. Schoolmaster, tenant farmer, auctioneer, and the secretary of the delegation who called the founding convention of the Tenant League on 19 May 1864, McNeill became general secretary of the league's Central Committee.

John Balderston. Balderston was present at the founding convention of the Tenant League in 1864. He was dismissed as commissioner for the recovery of small debts in 1865, and was elected to the Legislative Council in 1866.

Robert Poore Haythorne. A resident proprietor who sold his estate according to the prescription of the Tenant League in 1864, Haythorne was the landlord of Alexander McNeill.

Grafton Street, Charlottetown. This streetscape faced the north side of Queen Square. At the far left is the corner where the deputy sheriff of Queens County attempted to arrest Tenant Leaguer Samuel Fletcher on 17 March 1865.

Benjamin Davies. Davies was a veteran radical in Charlottetown. On 17 March 1865 Tenant League demonstrators reportedly stopped in front of his business establishment and applauded. He was elected to the House of Assembly in 1867.

Thomas W. Dodd. On 19 July 1865, Dodd, the sheriff of Queens County, informed the government that he was 'completely powerless to execute any Writs placed in [his] hands.'

James Colledge Pope. Pope was premier during the Tenant League crisis in 1865. He eventually used soldiers to assist the Queens County sheriff.

William Henry Pope. The older brother of the premier, William Pope was colonial secretary, editor of the *Islander* newspaper, and chief Conservative strategist.

William Campbell. A justice of the peace, Campbell gave a defiant answer when the Pope government questioned him about his connection with the Tenant League in 1865, and he was fired as a result.

George Coles. Coles was Liberal opposition leader in 1865, and premier three times in the 1850s and 1860s. Like other established political leaders, he opposed the league.

Edward Whelan. Liberal assembly-man, editor of the *Examiner* newspaper, and the leading journalist in the colony, Whelan criticized the Tenant League forcefully, antagonizing members.

Robert Hodgson. Chief justice of the Prince Edward Island Supreme Court, Hodgson served as administrator for several months in 1865, when the crisis was at its peak.

Colonel Bentinck Harry Cumberland. Cumberland was an absentee landlord whose writ led to riots on 18 and 19 September 1865. The presence of his land agent on the grand jury in January 1866 led to the quashing of its indictment of the rioters.

A drawing of William Graham's house, in New London, by Henry Jones Cundall, 1856. It is one of the few drawings of a rural settler's home from this period. Cundall, the artist, was a land agent, surveyor, and diarist.

Benjamin Balderston Jr. Brother of John Balderston, district schoolteacher, and Tenant League activist, Balderston was appointed registrar of deeds and keeper of plans in 1867, but the appointment was annulled. He was named reporter to the House of Assembly in 1868.

Bagnall's Inn. The inn was located in Hazel Grove, Lot 22, a focus of Tenant League strength. It became the site of more than one confrontation between Tenant Leaguers and the forces of the Queens County sheriff.

Prince Edward Island legislature and Supreme Court. The Colonial Building, constructed in the 1840s and situated on Queen Square, was the seat of political and judicial power in Charlottetown.

Charles and Eliza Dickieson. Charles Dickieson was the most famous Tenant League prisoner. During his capture in a confrontation at the Curtisdale Hotel in Milton on 18 July 1865, the deputy sheriff struck him on the mouth with a billy, and afterwards he wore a beard to conceal the scar. He is pictured here in later life, with his wife, Eliza.

Charlottetown city hall *cum* police court. This site was the focus of the demonstration and riot of 26 July 1865, when Charles Dickieson was taken from the jail to the court and back to the jail (see map 7).

James Horsfield Peters. Peters was the Supreme Court judge who quashed the indictment of the accused rioters on the Cumberland estate and sentenced Charles Dickieson.

Roy Dickieson, 4 September 1994. Farm activist and grandson of Charles Dickieson, Roy heard Tenant League stories first-hand in the 1920s and 1930s. He is pictured here next to Charles's tombstone, New Glasgow, Lot 23.

Herb Dickieson, 4 September 1994. Political activist and son of Roy Dickieson, Herb was elected leader of the Prince Edward Island New Democratic Party in 1995. He is pictured here near the spot where his great-grandfather was arrested in 1865.

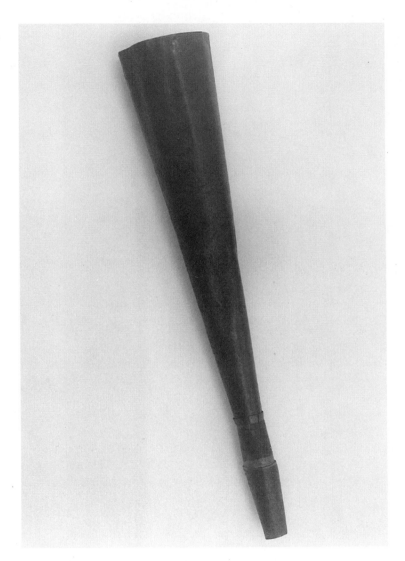

The tin trumpet was a symbol of Tenant League solidarity. Instruments like this, which were ordinarily used to summon workers from the fields at meal-times, became the means by which Tenant Leaguers warned one another of the approach of the sheriff. In some instances at least, leaguers were required to have trumpets, bugles, horns, or conch shells 'to sound the note of alarm on the approach of *rent-leeches.'*

THE TENANT LEAGUE OF
PRINCE EDWARD ISLAND

Introduction

In the spring of 1865 the barn and stable of a landlord at Glenaladale, eastern Queens County, in rural Prince Edward Island burned, likely as the result of arson. Less than two months later a group of men set upon a deputy sheriff and three assistants near the Curtisdale Hotel, in Milton, seven miles north-west of Charlottetown, the capital. The deputy and the bailiffs were returning from a country district with distrained property they had seized on behalf of a landlord for rent; the deputy's arm was broken by a roundhouse blow from a fence post, and most of the property was liberated. The blowing of tin trumpets had summoned the assailants, who were associated with an organization known as the Tenant League. Exactly two months after that violent incident, tin trumpets called together a crowd of angry men on horseback to oppose the serving of writs by the same deputy and one of the same constables. They forced the two to decamp by boat from Rocky Point, two miles south-west of the capital, across the Charlottetown harbour. The deputy and the constable had to leave a horse behind, and when they returned the next day its mane and perhaps its tail had been shaven; although certainty is impossible, some reports indicated that a dog left behind on the wharf had also been mutilated or killed. A few days later, ten miles west of Charlottetown at Strathgartney, the estate house of the largest resident landlord on the Island, what was apparently intended to be a threatening letter was discovered nailed to the front gate, with a heading in English, 'Take Warning,' and an indecipherable text.[1] Soon soldiers were being used to assist the sheriff when serving legal processes for rent.

These incidents seemed greatly out of place in British North America in the 1860s, when the other colonies were fretting over railways and inter-colonial union. Yet not all aspects of the situation were unprecedented. Large tracts had been granted to great landowners in other areas of North America at various times in the past and had been farmed on a leasehold basis. In fact there was an obvious echo from recent American history in one tactic of the rent resistors which attracted much attention: an anti-rent movement between 1839 and 1845 in leasehold areas of eastern New York state had used dinner horns for warning that a sheriff was approaching with writs, and indeed in some instances members had resolved to use the horns for no other purpose. The anti-renters of New York had received widespread publicity, in part because of the flamboyance of their disguises. Men painted their faces like Amerindians to assert priority on the land, which formed part of aristocratic estates of pre-revolutionary and Old World origin, and dressed in women's clothing to accentuate the humiliation of the authorities when preventing them from serving writs.[2]

But even when large estates had been established elsewhere, they did not collectively monopolize the agricultural land of an entire colony or state. What made Prince Edward Island different was the predominance of leasehold tenure, the fact that most farmers had to rent the land they worked from landlords or 'proprietors.' On the Island virtually all the land had been granted in large estates to favourites of the British court in the previous century. There was almost no crown or public land to be granted free or on easy terms to farmers, as there was elsewhere. This predominance of leasehold tenure was absolutely untypical of North America.

The people, of course, were as important as the specific forms of land tenure, and they brought with them cultural characteristics and histories, and formed their own history in the colony. In terms of ethnicity, the largest group within the Prince Edward Island population was Scottish in origin, constituting perhaps 45 per cent of the 80,857 counted at the 1861 census. The next-largest groups, each probably between 20 and 25 per cent of the whole, were English and Irish in origin. Acadians formed approximately 10 per cent.[3] Of the four national groups, the English were in the best social and economic position. According to historical geographer Andrew Hill Clark,

In general those of ultimate English origin (and these included descendants of people from more than half the counties of England, Loyalists, New Englanders, and disbanded soldiers) were situated where agriculture was most intensive and productive or in the best locations for ship-building or fishing. On average they had more capital and more applicable agricultural skills. With the English should be grouped some, at least, of the Lowland Scots. But the great majority of Highland Scots and Southern Irish had come as poverty-stricken immigrants and had advanced their circumstances very slowly over the years; they were yet in the 1850's (and many of them still in the eighties) as close to the level of a European peasant tenantry as one would be likely to find in the New World. Through the middle decades of the nineteenth century, where the land was poorer, rougher, swampier, or less accessible, there Gaelic, the Acadian patois, or a distinctly Caledonian or Hibernian inflection of English was likely to be heard.[4]

During the years immediately preceding the rise of the Tenant League, religious differences had seemed much more important than ethnic origin in public life. The previous three general elections had featured increasing polarization of the voters between Protestants, 55 per cent of the population consisting of Scots and English, and Roman Catholics, 45 per cent, made up from Irish, Scots, and Acadians. By 1863 all Protestant constituencies were voting Conservative, and all Roman Catholic ones were electing Liberals. Denominational divisions seemed, therefore, to dominate the political culture of the colony, and to be hardening.

In the story of the Tenant League there are parallels with Old World experiences, and indeed the parallels with Irish history in the nineteenth century were striking. Arson, animal maiming, and the threatening letter fitted patterns of protest known in Ireland under the rubric 'agrarian outrage.' Even the name 'Tenant League,' applied to the organization behind rent resistance in Prince Edward Island in the 1860s, was rooted in Ireland. But events on the Island were not an exact reflection of events anywhere else, just as the experiences of Ireland and Scotland did not mirror each other. In Scotland the great agrarian struggles and controversies of the nineteenth century revolved around attempts to remove cultivators, full-time or part-time (these pursuing a second occupation out of necessity), from the land, usually in order to make way for large-scale sheep grazing.[5] These were the 'Highland Clearances' of legend, which derive their name from the efforts by landlords, as a matter of policy, to 'clear' crofters from the land as expeditiously as possible.

One of the major questions historians have asked concerns the apparent docility or passivity of Scottish crofters, who did not as a rule offer physical resistance, although they were undoubtedly distressed by the landlord initiatives which meant disruption of their lives and, not infrequently, emigration. Such outbreaks of resistance as occurred tended to be spontaneous, uncoordinated, and bereft of a mobilizing ideology. There was virtually no significant political manifestation of resistance. In fact politics became important in Scottish agrarian conflict only at the time of the 'Highland Land War' late in the ninteenth century, when a combination of direct action, political agitation, sympathetic public opinion in the south, and the authorities' determination that the situation not deteriorate into something similar to that in Ireland led to a royal commission and reform legislation. On a smaller scale and somewhat later than the movement towards grazing, the expansion of deer forests put crofters under further pressure.

In Ireland the issues, the nature of the controversies, and the intensity of resistance were quite different from the Scottish experience.[6] The fear of the working farmer in typical circumstances was not sheep or deer, but the possibility that he might be displaced by someone who would pay a higher rent. For the sake of convenience, the agenda of the tenants can be summarized as 'the three Fs': fixity of tenure, fair rent, and free sale. Since most tenants occupied their land on a yearly basis, 'fixity of tenure,' the legal right to remain on the farm one worked as long as the rent was paid and other reasonable conditions were met, was the central demand of tenant movements for decades. In the province of Ulster, something close to it existed on a common-law or customary basis and was known as 'Ulster custom.' 'Fair rent' was a related demand, reflecting the concern that rents might increase beyond the means to pay, or at least to a level which might despoil the tenant of the benefits of increasing output. The establishment of a fair rent involved taking into account many factors, and there were inevitably elements of subjectivity. 'Free sale' was a third demand, and it referred to the right of a tenant to sell his interest as a tenant upon vacating a property voluntarily or involuntarily. It was what some contemporaries meant when they used the term 'tenant right,' and although tenant right existed in Ulster and was referred to as 'the Ulster custom of tenant right,' it was neither universal in that province nor limited to it.[7] Where there was tenant right, including freedom of sale, it could involve the transfer from the new tenant to the outgoing tenant of a large sum of money clearly worth more than improvements made by the latter. Interpretation of the significance of such payments has a bearing on many other matters. Why, for example, would new tenants spend substantial sums, entirely aside from the rent to be paid to the landlord, for the right to occupy farms held on yearly tenancies, if they faced insecurity and unreasonable rents?

The Irish land question was entangled in politics, and movements of resistance might or might not take a political form. The Irish Tenant League, founded at mid-century, had entered politics at the British general election of 1852, and had faded by the late 1850s, but the long-term trends in the ninteenth century were for the Irish land question to become a factor of growing importance in politics within the United Kingdom, and for political parties and legislators to play an increasing role in the land question. There was also a tradition of violent resistance, which could go as far as assassination, although uncritical use of available data and existing labels could lead to a great exaggeration of the actual extent of landlord-tenant violence. Police in Ireland kept a separate record of agrarian crimes or outrages, but in the words of historian W.E. Vaughan, 'analysis of actual cases ... shows that the police defined all crimes arising out of disputes about land as agrarian, including family rows and disputes between neighbours. ... only a fraction of agrarian crime was caused by landlord-tenant disputes.'[8] Thus most of what the police classified as agrarian crime in Ireland was wreaked not upon landlords, land agents, or bailiffs, but upon neighbours, family members, and other persons similar in social position to the perpetrators.

The Prince Edward Island land question was a unique historical phenomenon, with a mixture of Old World issues and New World environment and expectations.[9] Events surrounding the Tenant League of the Island did not follow either the Scottish or the Irish pattern in any exact way. Indeed they could not, for several reasons. The population was more 'mixed' than that of England, Ireland, Scotland, or Wales, and the traditions that it drew upon could not be those of one nationality. The largest single group of people came out of Scottish traditions; the points of friction resemble most closely those of Ireland. Moreover, by the 1860s the Island had established a history of its own, and even a system of 'responsible government,' or ministerial responsibility, based on virtual manhood suffrage. Beyond their sharing in a liberalized version of British political institutions, Islanders breathed the air of North America, where freehold tenure was the norm. Rent-paying was anathema, identified with the Old World, and believed to be inconsistent with a spirit of independence. But in addition there was the fact that farmers, or their parents, had had, in most instances, to create farms and homes out of forests, a process sometimes summed up with the significant phrase 'making land.'[10] In such circumstances, there was a personal identification with one's farm and dwelling place which transcended what the tenant in the British Isles felt for the property he leased from a landlord, and which, when he entered upon it, was a fully operational farm. Most farms in Prince Edward Island had been created out of thickly wooded wilderness by working farmers who were still living in 1865. Therefore, the sentiments surrounding such issues as rent levels, distraint for non-payment of arrears, eviction, and compensation for improvements had to be profoundly different from what they were in Ireland. They had also to be different from the feelings surrounding the Highland Clearances of the early ninteenth century in Scotland, when landlords sought to remove crofters in order to replace cultivation and cattle raising with sheep grazing, or surrounding the movement later in the century to turn the land into deer forests. There was no such tension between types of farming and types of land use in Prince Edward Island.

Throughout this study, parallels with other agrarian movements or conflicts or modes of protest, especially in England, Ireland, Scotland, and Wales, will be noted particularly. Islanders themselves saw their problems through the prisms of both America and the United Kingdom; and in London, where there was a potential veto on all land reform legislation, and where there was the authority and capacity to send military force to put down resistance to the law, there was vigilance lest precedents be set which could undermine the rights of property in the British Isles. But the reader has to keep in mind that some of the similarities were superficial, since identical forms in widely differing contexts may have quite different meanings for the persons involved. From the point of view of an Irish tenant, most Prince Edward Island tenants, with 999-year or 'perpetual' leases, rents of 1s. per acre, and freedom to sell improvements, could be seen as having 'the three Fs' already, although matters were not that simple from the perspective of the Island tenant.

In summary, there was nothing precisely similar to the Prince Edward Island land question in either the Old World or the New – because of its combination of the predominance of leasehold tenure, representative political institutions based on a broad male franchise, New World expectations of freehold tenure, and recent history of pioneering – and therefore no patterns established elsewhere could provide a exact guide to its development or to the nature of the Tenant League. As well as being distinctive, the league is a clear demonstration that more than Confederation, railway building, and school disputes was going on in British North America, the future Dominion of Canada, in the middle years of the 1860s.[11] One of the earliest organized movements of farmers in Canadian history was taking shape as an extra-parliamentary movement for radical change, willing to resort to civil disobedience and to defy the law in an organized, calculated, and disciplined way in pursuit of its objectives. It would be exceptionally successful.

This book will examine the background, origins, development, and impact of the Tenant League of Prince Edward Island. The league deserves to be better known also as a part of agrarian history combining elements from the Old World of the British Isles and the New World of North America, with their different circumstances and values. It deserves to be better known also as a part of Canadian history, both as a forerunner of later agrarian insurgencies and as a distinctive and robust example of the power of civil disobedience when used in a disciplined, intelligent way by a popular movement with widespread support. Its record of achievement demonstrates forcefully the effectiveness of its approach to solution of the land question, which amounted to advocacy of a rent strike as a means of pressuring landlords to sell their properties. When the local government sent the sheriff to enforce the legal rights of the proprietors, the league organized resistance at the community or district level. Indeed, its defiance of law would be extraordinarily audacious, extending even to attempts to induce desertion among British soldiers sent to support the civil authorities. At the same time, it avoided an escalation in violence which could have resulted in discrediting itself, in ruinous consequences for its supporters, and in catastrophe for its society – unlike, for example, the rebels of 1837 in Upper Canada.[12]

Significant chapters in the social, political, legal, and military history of British North America will be found in the dramatic and surprising story of the Tenant League. In the course of making this history, it achieved its objective: driving the leasehold system on Prince Edward Island much closer to extinction.

1

Social and Political Background

[T]he Owners have as good a right to their Property as you or I ... have to our Estates; and it would be as unjust and of as bad an example to *extinguish* the Rights of these owners, as it would be to *extinguish* our Rights.
Prime Minister Viscount Palmerston to British Colonial Secretary Henry Labouchere, 19 December 1855[1]

It is difficult for an European to understand why almost every man in America considers it a personal degradation to pay rent. ... The prejudice in favor of a freehold tenure, if it is one, is beyond the power of reason.
Royal commissioners John Hamilton Gray, Joseph Howe, and J.W. Ritchie, 18 July 1861[2]

I have a near connection who has lands in the island and is much interested in the matter.
Palmerston, prime minister, to the Duke of Newcastle, colonial secretary in the British government, 30 June 1863[3]

Leasehold Tenure on Prince Edward Island

For more than a century the predominance of leasehold tenure as a form of land holding made Prince Edward Island unique among English-speaking colonies in British North America. The hegemony of leasehold in the overwhelmingly rural and agricultural colony lasted until the 1860s. The subject of this study is the Tenant Union of Prince Edward Island, whose activities undermined the leasehold system decisively. Yet the life of this organization, usually known to contemporaries as the Tenant League, was short; it did not exist at the beginning of the decade, and was only a memory at the end.

The Prince Edward Island census of 1861 revealed that almost 61 per

cent of occupiers of land were either tenants or squatters.[4] Still more significantly, leasehold tenure was continuing to expand: since the previous
census, in 1855, the number of leaseholders (although not their per centage among occupiers of land) had risen. The balance shifted definitively
against leasehold as a form of tenure in the 1860s. Because the census of
1871 did not record the numbers of persons in different types of tenurial
relationships, it is impossible to provide precise figures on the shifting
proportions of tenants, squatters, and freeholders during the decade. But
the best available measure – differences in total acreage held by freeholders and by non-freeholders (that is, tenants and squatters) – indicates
dramatic change, which is apparent in table 1.1. Between 1861 and 1871,
a ten-year period when total occupied land increased by only 6.4 per
cent, the land held by freeholders increased by 53 per cent and that held
by tenants and squatters decreased by 35.2 per cent. These differential
changes meant that the proportion of occupied land held by freeholders
increased from 47.2 to 67.8 per cent. Clearly, large numbers of farmers
who had been tenants or squatters became freeholders.[5]

Thus the 1860s were the epochal decade in the Island's passage from
its distinctive leasehold régime to the North American norm of freehold
tenure. The failure to include in the census report of 1871 some of the
statistics on land tenure – the numbers of freeholders, tenants, and
squatters – which had been collected systematically for a generation is
itself evidence that the Island had passed a critical turning-point. The
superintendent of the census returns was John McNeill, the same person
who had been in charge ten years earlier, but public concern over a system whose end was approaching had diminished. Islanders could get
on with the same affairs as other British North Americans, such as
building railways. This transition represented an enormous change in
sensibility since 1861, when issues surrounding landlord-tenant relations still dominated the landscape.

The Tenant League, founded on 19 May 1864, played the crucial role
in subverting the leasehold system which had been imposed in 1767. Its
objective was elimination of leasehold tenure, and its strategy involved
refusal to pay rent or arrears of rent, and support for those who refused.
Rent was classified as a legally collectible debt, and in the event of non-
payment, eventually the landlord could call upon the assistance of the
law. Supporting those who refused to pay rent could and did lead logically to resistance to sheriffs and constables. Although the Tenant
League had a brief history when measured against the 108 years
between the establishment of leasehold and its effective abolition by law
in 1875, it had an extraordinary impact.

TABLE 1.1
Changes in total acreage and in percentage of total occupied land held by different categories of land occupiers, 1861–71

	1861		1871		Increase/Decrease		Percentage change in acreage
	Acres	% of total	Acres	% of total	Acres	Percentage points	
By freeholders	455,943	47.2	697,598	67.8	+241,655	+20.6	+53.0
By leaseholders or holders of agreements to lease	407,169	42.1	279,601	27.2	–127,568	–14.9	–31.3
By tenants with verbal agreements	38,440	4.0	20,931	2.0	–17,509	–2.0	–45.5
By squatters	64,636	6.7	30,110	2.9	–34,526	–3.8	–53.4
By all non-freeholders	510,245	52.8	330,642	32.2[a]	–179,603	–20.6	–35.2
Totals	966,188	100.0	1,028,240	100.0	+62,052		+6.4

[a]This figure does not match the sum of the percentages of the three constituent subcategories because of rounding

In order to understand the advent, nature, and appeal of the league, and the opposition to it, one must know, in general outline, the history of the Prince Edward Island 'land question,' a term which encompasses a range of controversies surrounding not only the rightness of leasehold tenure, but also intra-élite attempts to alter ownership patterns without changing tenurial relationships. Without this knowledge, it is impossible to comprehend the sense of grievance, of entitlement, and of frustration which both landlords and tenants carried into the middle 1860s and which made the situation explosive. This opening chapter will accordingly offer a capsule account of the land question.[6] It must commence with an explanation of the perspectives of landlords and tenants.

1767 and the Landlords

The Island had become British territory as a result of the provisions of the Treaty of Paris, 1763, which concluded the Seven Years War. It was then known as the Island of St John, and would remain so until 1799, when it was renamed Prince Edward Island, after the fourth son of King George III, then commander of the forces in Nova Scotia and New Brunswick. Following a survey, conducted 1764–6, it was divided into sixty-seven numbered lots or townships of approximately twenty thousand acres. Almost all were distributed by means of a lottery in London on a single day, 23 July 1767, to individuals or groups, mostly resident in Great Britain. Typical grantees were leading political or military figures, or merchants with already established interests in the region. Among the townships, there were three exceptions, all in Kings County: two, Lots 40 and 59, had already been granted to groups of merchants who had actively sought land on the Island since 1763, and the third, Lot 66 – the smallest township, and less than one-third the size of the others – was reserved for the crown. Land was also set aside for three county towns and, in each case, a surrounding 'Royalty,' to be used for pasture lots to sustain the townsfolk. Together, Lot 66 and the 'town and royalty' areas totalled approximately 21,600 acres,[7] representing 1.5 per cent of the land mass; aside from this, all of the Island had been alienated. Effectively, there was no crown land. For common settlers, it meant that freehold tenure was probably out of the question, since the owners of the townships had no intention of selling to small holders. By 1769 a group of proprietors, as the owners came to be known, persuaded the British government to detach the Island from Nova Scotia

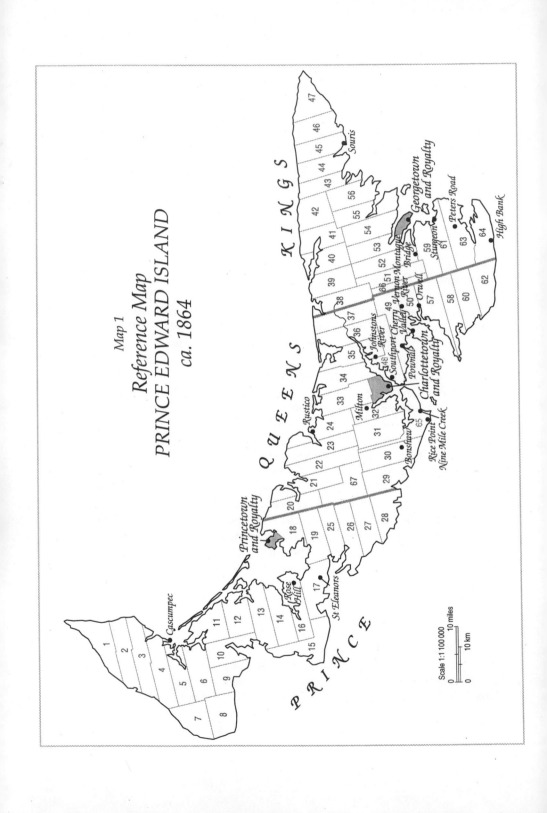

Map 1
Reference Map
PRINCE EDWARD ISLAND
ca. 1864

KINGS

QUEENS

PRINCE

Souris

Georgetown and Royalty

Peters Road

High Bank

Sturgeon

Vernon River
Cherry Valley
Montague Bridge
Orwell

Johnstons River
Southport
Pownal
Charlottetown and Royalty

Rustico

Milton

Bonshaw
Rice Point
Nine Mile Creek

Princetown and Royalty

Cascumpec

Rose Hill

St Eleanors

Scale 1:1 100 000

10 miles

10 km

politically and make it into a separate colony with its own government. The first governor was a proprietor from Ireland.

The new owners received their townships upon certain conditions. They were to settle them with one person per two hundred acres within ten years; the settlers were to be Protestants from places other than Britain; and the proprietors were to pay annual quitrents (fees of feudal origin) to the crown, the amount to be determined by the general quality of their land. The two important conditions for the future of the land question concerned quitrents and the number of settlers. In arguments over many years concerning the performance or non-performance of the proprietors, no one of consequence took seriously the requirement that the settlers be non-British Protestants. Quitrents were, in theory, to be the basis of public finance in the colony in the early years. Once the Island developed sufficiently, revenues could be raised in other ways as well. Settlers, who should have numbered at least 6,600 by 1777, would make it productive and enable the proprietors to pay their quitrents with ease. Governance should, if the plan worked out, cost the British taxpayer nothing.

There are two ways of explaining why the system never worked: either because it was unworkable or because the proprietors failed to honour the bargain they had made with the British government. Historian J.M. Bumsted has examined closely the record of an original grantee, Sir James Montgomery, a capable, energetic, and politically influential man who was genuinely interested in developing his Island property. If there was an early proprietor who fitted the ideal, a developer rather than a speculator, it was Montgomery. Although he never visited the Island, he poured a great deal of money into settlement efforts over many years.[8]

The problems for a 'developing' proprietor like Montgomery were legion. There were, in the first place, the legitimate expenses of attempting to settle a wilderness area – usually thickly wooded – and the difficulties inherent in trying to supervise effectively business operations an ocean away without a modern system of communications. The predicament of the conscientious proprietor was compounded by dog-eat-dog political factionalism on the Island, the biased and unprofessional administration of justice, and the near-impossibility of finding honest, able, and dedicated agents in the colony. In order to steer a path through the political factions, one needed a local helper; and if control over the latter was to be effective, there had to be a system of legal redress available to the proprietor. But for more than half a century the courts simply functioned as tentacles of the local élite.

The morass of pitfalls and dangers surrounding involvement in Prince Edward Island, which cost a proprietor like Montgomery dearly, worsened during the American War of Independence. Communications with the Island broke down almost entirely, cargoes could be lost to the enemy, and, given the climate of utter uncertainty, proprietary activity came to a virtual halt. This meant that settlers ceased arriving in any organized fashion. Many who were already there concluded that the proprietors had abandoned them, and as many as one-half of the approximately 1,500 who had been present in 1775 left for other colonies. The proprietors, for the most part, also ceased paying quitrents; from their point of view, a war whose causes had nothing to do with them was preventing development of their properties. Consequently, in addition to the crisis for individual settlers there was a financial emergency for the colonial government and those who depended on it for salaries. The pecuniary plight of the government officials had been bad enough before the war that the death in 1774 of the first chief justice had been attributed, in part, to starvation.[9] With the war, assistance from the British Treasury became an absolute necessity, and by 1777 the local government was receiving an annual grant.

Hardships tend to create a sense of grievance which can lessen inhibitions. Such an axiom partly explains a clumsy attempt in 1781 by the governor, Walter Patterson, and other officials to appropriate for themselves a number of townships by means of an improper land auction. The rationale was apparently to compensate themselves for salary arrears, suffering, and general inconvenience. This auction set in motion a train of events that cost Patterson his job several years later.

The proprietors got their lands back, but the events surrounding the auction and its undoing resulted in deep and lasting hostility between them and the local officials.[10] The proprietors distrusted the officials, who they believed were bent upon despoiling them of their lands. As for the officials, they tended to lump the proprietors together as a parasitic element whose inactivity held back the development of the colony. While the details are not germane to the main thrust of the present story, which centres on landlord-tenant relations, this abiding mutual antagonism between proprietors and officials helps to explain the intense feeling of grievance which some proprietors brought to the Tenant League conflict in the 1860s. Such men and women believed they could not trust any politicians or officials on the Island, regardless of ostensible political or ideological orientation – that is, in spite of whatever pro-property rhetoric such Islanders might mouth.

Part of the reason for the proprietors' disillusion with local officials and politicians stemmed from their own experiences with some of the same people when they hired them as land agents. These experiences could frequently be summarized in one word: bad. The difficulty in finding honest and competent personnel has already been mentioned. Somehow, even when a proprietor did find someone who seemed suitable, matters just did not work out. Expenses often exceeded the revenues collected, and the probability was great that the agent would be applying to his employer for money, rather than remitting funds. Thus the property became a drain rather than a source of gain. There was no easy answer to the conundrum of how to manage an estate from the other side of the Atlantic Ocean.

Although an unwritten assumption of the land lottery of 1767 appears to have been that proprietors would take up residence on Prince Edward Island, few actually moved there. Nor were they likely to send relatives, who would presumably be the most trustworthy and committed stewards, to tend their properties. The Island was simply too underdeveloped and lacking in the amenities desired by gentlefolk to be an attractive place to spend a career. Proprietors therefore relied on the local élite for agents, and there was no satisfactory formula. That colonial élite consisted, for the most part, of opportunistic individuals who had come to the New World to make their fortunes by hook or by crook. One could choose among professional people, elected politicians, non-elected public officials, resident proprietors looking for opportunities to make extra money, or, occasionally, persons who were agents more or less on a full-time basis.

An example of a possible land agent was the Irish lawyer James Bardin Palmer. He arrived in 1802, played an intensely controversial role in local politics, and made a legion of enemies. But he was a man of some talent and plausibility, and he had sponsors. Through one of these, in 1815 he became agent to George Francis Seymour, an overseas proprietor who had never seen the Island. The relationship lasted for more than eight years. In 1820, at Palmer's request, Seymour gave him a power of attorney granting considerable discretion to sell land from his estate without prior authorization. By the time an alarmed Seymour fired Palmer in 1824, the agent had sold some 1,400 acres for a total price approximating £1,200, without remitting anything. Palmer died in 1833, and it was only in 1840, after gaining a judgment for debt against his estate, that Seymour recorded 'having rec[d] 500 from Messrs. Palmer [the heirs] in compromise for about 1100£.'[11]

By the early nineteenth century there had been so many changes in ownership that only a few of the original grantees, notably the Montgomery, Sulivan, and Townshend families, remained. These transactions opened up opportunities for individuals and groups (usually families) to accumulate more than one township. Several large-scale landholders, such as the Montgomerys, the Sulivans, and the fifth Earl of Selkirk, emerged. The reasons for the turnover included the general lack of development on the Island and the difficulty regarding agents. Some proprietors gave up any pretence of actively managing their estates, and simply hoped for an appreciation in value which would allow a profitable sale sometime in the future. A proportion of the new proprietors seem to have acquired their lands as speculative investments, and to have determined to risk as little as possible beyond the purchase price. This lack of interest in development may explain in part the fact that some large estates consisted of areas geographically isolated from each other. In any event, the theory of the speculative proprietors was that the efforts and risks of others, whether proprietors or settlers, would be their own means of profiting from the Island. From the perspective of economic rationality, it is difficult to fault their judgment. The example of active proprietors like Montgomery was a caution to others. He spent a generation trying to make his Island estate, which included land of high quality, remunerative, and he died in 1803 without success and without even an adequate accounting of where the money had gone.

The experiences of the few proprietors who moved to the Island were not especially heartening for others who might have considered following suit. John MacDonald of Glenaladale, an important colonizer in the 1770s, lost many of his tenants during the years of the American War, when he was absent on active military service. Two of his sons and one of his grandsons became legendary for their poor relations with tenants.[12]

Charles and Edward Worrell, brothers, moved to the Island in the early years of the nineteenth century and invested large sums in their extensive estate, which was geographically cohesive, but they saw few returns.[13] Explanations were various. Their lands were not of high quality. There are references in the sources to the 'eccentricity' of Charles, who stayed on the Island long after his brother had left for London, and also to his kindliness. The eccentricity referred to his rigid policy of charging high rents and giving leases which were, by Island standards, short. Such behaviour was bound to make tenants who had choices think twice before settling on the Worrell estate. But then there was his kindly side; when he was unable to collect rent from his tenants, for whatever reason, he was unwilling to use all the sanctions at his disposal. Doubtless this fact became well known among his tenants, and militated against their paying any rent at all, especially if they had other obligations. Although he was reputed to have a rent roll totalling £2,000 in 1840, he stated that he was not receiving £100 annually 'as Rent in any form.'[14] When he left the Island in 1848 in his late seventies he was financially ruined.

Robert Bruce Stewart was a landlord who arrived in 1846 to manage his father's property and his own, and eventually he became sole owner. He was less inhibited than Worrell in dealing with his tenants, but although hard financial evidence is lacking, it is crystal clear that he found many aspects of his situation unsatisfactory. Regardless of his reputation as a severe and grasping landlord, he was convinced that he had extended much leniency to his tenants and received little in return.[15]

The proprietors who did nothing but sit on their land titles provided the best ammunition for critics of the leasehold system. Sometimes – even with large-scale proprietors like the Sulivan family, whose property consisted of four townships spread widely over three counties – there was no local agent. This meant there was no one to whom the self-starting and well-intentioned prospective tenant could apply for a lease. The result, if he persisted in settling on the estate, could be years of work without a legal entitlement to remain should the proprietor or his agent show up. The uncertainty and insecurity inherent in the situation set many nerves on edge. The sight of a surveyor was most unwelcome, and a conflict between a squatter and a proprietor leading to eviction of the former could be absolutely ruinous, since there was no legal provision for compensation for improvements, including the settler's home. He could lose everything. Squatters could obtain freehold title if they proved twenty years of undisturbed possession; that was a ray of hope.[16]

The Tenants and the Terms

Between the extreme cases of eviction and gaining freehold tenure through squatter's title there were several possibilities for the farmer who had settled on a piece of land without bothering with any formalities. Since many proprietors did not have local agents, the settler might not hear from a landlord or land agent for many years. If eventually one showed up and was willing to let the settler remain as a tenant, then the crucial next step was *attorning*, or accepting a lease or an agreement to lease. By this act, the occupier acknowledged the proprietor's title to the land, and since attorning was irrevocable,[17] the decision to attorn was exceptionally important. Attorning meant signing a document which would govern such matters as a farmer's access to timber growing on his farm, his access to mill sites, and his liability for arrears of rent owing on the land. On the other hand, the implications of having to depart after having established himself, cleared some acres of land, and built a dwelling were daunting. Consequently the pressure to attorn was great.

The usual rent was one shilling per acre per year, and the terms of the lease were almost certain to provide that the tenant was also responsible for paying the equivalent of the quitrents due, plus any land taxes levied. In other words, the terms, which were often non-negotiable, allowed the landlord to pass on to the tenant the annual public charges on the land. These were in addition to such initial fees as 'lease money' or 'plan money,' totalling perhaps two pounds or more, which the new tenant would probably be required to pay. The standard farm consisted of one hundred acres (although there were many farms of other sizes), and thus, by the usual terms, the tenant would owe five pounds annual rent plus several odd shillings and pence for quitrents and taxes.

In most cases the lease specified that the rent was to be one shilling *sterling*, as opposed to *currency*, which meant local currency. Sterling, or British money, was scarce on the Island, and hence it was conventional practice for landlords or their agents to accept payment in currency, with a calculation being made for exchange. Here custom, as it evolved in the nineteenth century, favoured the tenant over the landlord. For ordinary commercial transactions, that is for all dealings other than the payment of rent, it was accepted that there was a 50 per cent rate of exchange; a debt of £5 (British) sterling could be paid with £7 10s. (PEI) currency. The tenant, nonetheless, was customarily allowed to pay his rent at a one-ninth rate of exchange, and the statistical evidence from the 1861 census drives home the point that this was an issue directly involving the material interests of most tenants: 74.2 per cent of reported leases and agreements to lease provided for sterling rents.[18] By mid-century it had become rare for a landlord to insist on the full 50 per cent, although he retained the legal right to do so. The exercise of this prerogative appears to have been capricious, and could most frequently be accounted for by the tyrannical disposition of an unusually harsh landlord or agent or by a specific conflict with a particular tenant. In normal

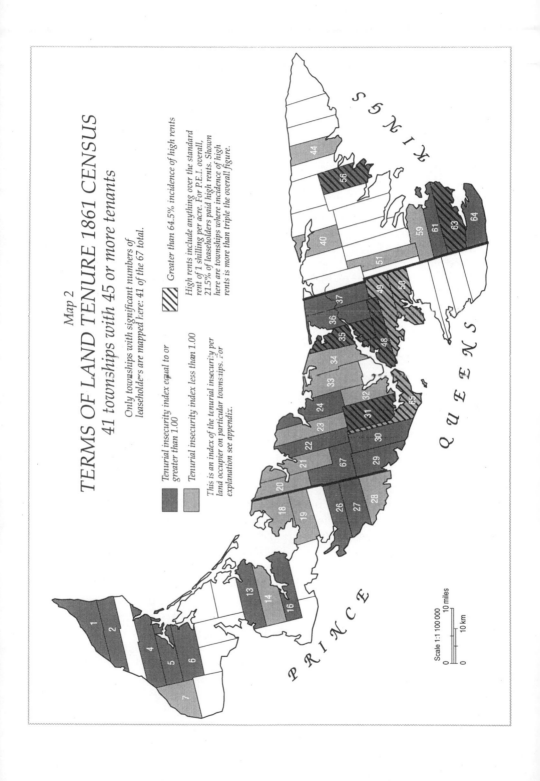

Map 2

TERMS OF LAND TENURE 1861 CENSUS
41 townships with 45 or more tenants

Only townships with significant numbers of
leaseholders are mapped here: 41 of the 67 total.

Tenurial insecurity index equal to or
greater than 1.00

Tenurial insecurity index less than 1.00

This is an index of the tenurial insecurity per
land occupier on particular townships. For
explanation see appendix.

Greater than 64.5% incidence of high rents

High rents include anything over the standard
rent of 1 shilling per acre. For P.E.I. overall,
21.5% of leaseholders paid high rents. Shown
here are townships where incidence of high
rents is more than triple the overall figure.

PRINCE

QUEENS

KINGS

Scale 1:1 100 000

0   10 miles

0   10 km

circumstances, for the tenant who was owing £5 rent, if his lease stipulated 'British sterling,' it meant that he would be paying £5 11s. 1½d. currency.

Common Prince Edward Islanders had their own perspective, whether concerning overseas or resident proprietors. Despite much rhetoric over the years about 'absentee' landlordism, the experiences of the McDonalds, the Worrells, and Stewart suggest that resident proprietorship was no more palatable to Islanders. The reasons for popular opposition to landlordism are not difficult to discern. In the first place, most settlers had paid their own passage to the Island, or at least had received little or no assistance from the landlord. Secondly, the settler had usually been faced with dense woods which he had to clear – again, with no assistance from the landlord.

George Wightman, a researcher working for a royal commission on landlord-tenant relations in 1860–1, estimated that a new settler operating without capital would be fortunate to clear two acres a year. It would be many years before such a settler would be able to clear enough land to support a family and pay rent. Some leases took account of this to a certain extent, allowing minimal or no rent to be paid for several years, with the amount slowly escalating until the full annual rent was due. Even so, it was difficult for a settler to avoid falling into and remaining in arrears. It is reasonable to infer from available evidence generated over many years that virtually all tenants were in arrears at one time or another; that at any specific time a large number were in arrears, some being several years behind; and that some never escaped being in arrears. For the record, it should be noted that a standard written lease provided the tenant who observed its terms with absolute security of tenure for its duration, which was in most cases set at 999 years, often referred to as 'perpetual tenure.' But regardless of the generous length of time specified in the leases, if a settler fell into arrears on his rent and was unable to pay upon demand, a long term could become meaningless if the landlord or agent decided to sue.

The tenant could be sued at any time for back rent. If he could not pay the sum and the legal costs of the suit, which might exceed the debt, he was ultimately liable to eviction. An evicted tenant had no more legal right than an evicted squatter to compensation for improvements, in effect for the results of years of labour. Although there was no systematic collection of statistics on eviction over the years, it must be emphasized that eviction and loss of all improvements were real possibilities. Wightman related that 'several' informants estimated that 'at least one half of the new settlers lose their farms, or hold them only until the land-lord can get a tenant who will pay a part of the back rent.' His report reflected a common belief among many Islanders: 'the tenant in arrears finds it to his interest to improve no more than he can help; improve-ment would only invite purchasers.'[19] In summary, the leasehold sys-tem left the working farmer in a state of insecurity and, if in arrears, with little incentive to improve his farm.

Yet not all emigrants arrived without capital and lived on the brink of disaster for decades. In some instances settlers were able to purchase freehold, either on arrival or later, from a consenting proprietor. But although possible, this transition to freehold tenure was not at all the norm. As a matter of policy, many proprietors refused to sell freehold title to tenants or prospective settlers. Part of the conventional wisdom of estate management on the Island was that one should either sell large blocks of land or entire estates or not sell at all. The reasoning was clear. The cost of administering an estate remained more or less constant as it was reduced in size on a piecemeal basis, farm by farm, since the mile-age involved, and therefore many of the expenses, did not change sig-nificantly; and if the management was done through an agent, his retainer was unlikely to change as the estate shrank. The process of piecemeal sale also reduced the potential revenue from the estate, since those likeliest to purchase were almost certain to be the most productive tenants on the best lands, and therefore probably also the ablest to pay their rent on a regular basis. Hence it was best not to sell freehold title to tenants one by one – and consequently by the middle decades of the nineteenth century there were a significant number of tenants who could afford to pay a reasonable purchase price for their farms, but whose landlords were adamant that they would not sell to individual buyers under any circumstances. Many tenants believed that other pro-prietors set purchase prices unrealistically high so as to discourage sales.

The refusal to sell even to tenants who could guarantee immediate payment in full was especially galling when Island farmers considered the rest of North America, where freehold tenure was the norm. They believed proprietors who had a policy of refusing to sell were guilty of perverse behaviour. For their part, many landlords had parallel thoughts about their tenantry. They believed that some tenants who could pay their rent did not do so, and they were certain that a tenant faced with a choice between paying a debt to a merchant and paying rent would, as a matter of course, choose the former.[20] Indeed, one financially rational reason for a tenant to prefer the option of paying other creditors first was another custom on Prince Edward Island that, like the acceptance of a one-ninth rate of exchange on sterling rents, favoured the tenant, namely, the practice of not charging interest on rent arrears. In such circumstances, arrears functioned as interest-free loans from proprietors to settlers, although to put matters into a realistic perspective, it appears that the waiving of interest was not the result of proprietary indulgence so much as tenant insistence. An absentee proprietor visiting his estate in 1840 recorded that when examining his rent roll books with his land agent he had yielded on the issue because the agent advised that it 'was not the customary usage for tenants in P.E. Island to allow' interest on rent arrears; the wording is significant.[21] Distrust permeated landlord-tenant relations, just as it permeated relations between the local élite and the proprietors in the United Kingdom.

Each party nursed a sense of grievance against the other. The landlord found it difficult to extract rent, and the tenant found conversion to freehold tenure difficult or impossible. Essentially these were private matters between the two parties, with the land agent, if any, probably being the sole significant intermediary. But by the 1830s relations between tenant and landlord were at the centre of Island politics, and by the 1860s the respective parties had experienced a full generation of intense political debate about the fundamentals of tenurial relationships.

Island Politics and the Colonial Office, 1830–1860

It is necessary to inquire into Prince Edward Island politics and particularly the limits of conventional political action in order to account for the emergence of the Tenant League, which would explicitly reject parliamentary endeavour as a means of resolving the land question. All those who spoke for or professed to understand the viewpoint of the tenantry in the period of the league believed that the failures of the official political system lay close to the root of the anger among ordinary farmers. There was by the middle of the 1860s a sense of betrayal which would result in rejection of parliamentarism and distrust of those who had parliamentary links. But despite renunciation by the leaguers of political action, official politics was relevant to what actually happened and why it happened. In fact, without comprehending the frustrations associated with political action, it is impossible to explain the sudden rise, the nature, and the strategy of the league.

Although in a broad sense the land question had always been central to Island politics, before the 1830s the political struggle had really been confined within the numerically tiny élite and its various factions, with overseas proprietors becoming involved when they perceived threats to their interests. Occasionally there would also be a restraining intervention from the Colonial Office in London, in the form of a veto or a dismissal. From 1769 until the early 1830s the aim of almost everyone challenging the *status quo* had been to gain personal possession of particular lands or even entire estates, or to manipulate the system in some way that benefited the challenger while leaving leasehold tenure intact. J.B. Palmer had been portrayed at one stage in his career as a dangerous, even seditious radical; the record suggests that he was simply seeking self-enrichment. Over the next generation several different approaches to the issues at stake emerged.

The first major political break with the tradition of accepting the leasehold system as the norm on Prince Edward Island came with the Escheat movement of the 1830s.[22] Escheat was a term derived from the feudal traditions of the Old World. If a landlord holding his land from the crown on condition of performing certain duties failed to meet his obligations, his land was subject to being *escheated*, or returned to the crown. The applicability to the Island situation was manifest: none of the original proprietors and their successors had fulfilled the granting conditions of 1767 with respect to quitrents and promotion of settlement, and therefore their lands should revert to the crown. QED.

The advocates of escheat, led by William Cooper, a sailor and former land agent, sought to use the logic of feudalism to undo the neo-feudal system on the Island. After the crown resumed title, the next step would be to regrant the land in freehold tenure to the actual occupiers. This movement, with its advocacy of a neat legalistic solution to the land question, won widespread support in the 1830s. One reason for its electoral success was the enfranchisement of the large and impoverished Roman Catholic minority in 1830. Roman Catholic Islanders were overwhelmingly tenants or squatters, and their sudden acquisition of the franchise combined with the low property qualification for voting to create a mass base for Cooper and his colleagues.

In 1838 the Escheat wave crested with a stunning 18 to 6 victory in the general election for the House of Assembly. Cooper went to London in the following year as a one-man delegation to the Colonial Office with three possible options in the hope that Downing Street would choose one. The most radical proposal was establishment of a court of escheat. This was the movement's panacea, for such a court would 'investigate' the titles of proprietors, with the foregone conclusion that the result would be expropriation without compensation – or 'confiscation,' to use the word the process conjured up in the minds of the propertied. As an alternative, the Escheators suggested that the crown could purchase the rights of proprietors at a cost of £200,000, that is, a buy-out. This would be the most costly solution for the British government in a financial sense, and the rationale would be that, since the imperial government had created the problem in the previous century, it had an obligation to finance its resolution. The third and most moderate proposal was a heavy tax on wilderness lands, which Escheators interpreted as being held for speculative purposes.

Lord John Russell refused even to grant Cooper an audience. The fact that the Escheators had a fresh and overwhelming mandate from the electorate of Prince Edward Island did not seem to matter. Why did London not respond to Escheat?

Let us examine their proposals, from the most moderate to the most extreme. There was already a wilderness tax, and thus this measure would not entail a radical change in policy. But Downing Street had only confirmed the tax in December 1838 after a proprietary lobby against it which had lasted more than a year. Consequently, the British government believed a drastic increase so soon afterwards would be premature at best.

The purchase proposal would have entailed a huge expenditure by the British Treasury. That very fact doomed the suggestion. But there was another reason for spurning the notion of a British-financed purchase. Although a less radical solution than a court of escheat, it was rooted in the assumption of 'imperial error': that the imposition of leasehold tenure in 1767 had been a fundamental mistake in policy, which required redress. Such ideas were not to be encouraged – a point related to the rejection of escheat.

The most obvious reason for refusing even to consider establishing a court of escheat was respect for property rights. Every British government of the period included a significant number of members who were, as private individuals, owners of estates with many tenants paying rent to them in a system of feudal origin. Mass dispossession of overseas landlords could raise awkward questions closer to home. This consideration was especially likely to make an impression in London because so many Island proprietors resided in the United Kingdom and were well connected. Laurence Sulivan, for example, was deputy secretary in the War Office and a brother-in-law to and close friend of Viscount Palmerston. Such proprietors would be sure to point out that it was British landowners who faced seizure of their property by a British colonial government if such a proposal were accepted, and that only an ocean separated owners of land in the United Kingdom itself from a like fate. Russell termed the proposal for establishment of a court of escheat 'inadmissible.'[23]

Many years later, Palmerston, as prime minister, would remind his colonial secretary that 'the Owners [of the Island estates] have as good a right to their Property as you or I ... have to our Estates; and it would be as unjust and of as bad an example to *extinguish* the Rights of these owners, as it would be to *extinguish* our Rights.' This inescapable mental linkage between land issues in Prince Edward Island and the United Kingdom was a significant factor in the attitude of the British governing class to the land question. In a political system based on custom, precedent, and convention, only the foolhardy would overlook the danger of creating precedents which could be turned against them. Palmerston emphasized that 'We must not be led ... by the unprincipled Scramblers of the Island ... to establish a principle of authorised spoliation even in a remote corner of the Queen's Dominions.'[24]

Before returning to the political situation in Prince Edward Island, it is worth touching upon another issue concerning the Colonial Office. Over many years on the Island there was much talk of 'proprietary influence,' that is, an assumption that the proprietors could generally get their way at Downing Street. Things were not as simple as that, at least not from a proprietary point of view. Some proprietors found the British government exasperating at times, and some officials at the Colonial Office occasionally indicated that they found some of the proprietors difficult, if not unreasonable. On 24 October 1865 at the height of the Tenant League troubles, Sir Frederic Rogers, permanent undersecretary of state for the colonies, alluded, with some impatience, to 'the no surrender party ... who would like the Imperial Gov^t to carry on war with the Tenants.'[25]

But if one regards the exercise of 'influence' as including behaviour more subtle than crude wire-pulling, the Islanders who complained of proprietary influence at the Colonial Office surely had a point. 'Access' and the right to provide information are important in the exercise of influence, and there is no doubt whatever that proprietors enjoyed an access to the corridors of power in London that tenants could never dream of. For example, Seymour, a proprietor intensely interested in his Island estate, was sergeant at arms to the House of Lords from 1818 to

1841, master of robes to King William IV (1830–7), and an admiralty lord in the 1840s. On the basis of a reading of his diaries and correspondence, it is a fair inference that for more than a generation he knew everyone who was anyone in the British military, political, and court establishments. Samuel Cunard, owner of one-sixth of all the Island by the end of the 1830s, had founded Cunard steamship lines, and the Colonial Office treated him as an authority on affairs in the Maritime colonies – which, with his multiplicity of contacts, he probably was.[26] Men like Sulivan, Seymour, and Cunard were not patronized or dismissed as cranks with dangerous ideas.

In 1866 Thomas William Clinton Murdoch, chairman of the Emigration Board, who had long-time intimate knowledge of the way the British government operated with respect to colonial affairs, and who was, in the words of historian P.A. Buckner, an 'exceptionally able public servant,'[27] wrote to Thomas Frederick Elliot, deputy undersecretary of state for colonies, that

it has been a continual, and not unfounded, complaint on the part of the Legislature and Inhabitants of Prince Edward Island, that the Laws passed in the Colony to rectify the evils caused by the manner in which the Island was originally portioned out by the British government, have been constantly defeated by the representations of absentee proprietors resident in this Country. The irritation which this has created is natural.[28]

In other words, there was significance in the documented fact that the Colonial Office would listen to one side and not the other and that they treated one side as credible and the other as not. Perhaps knowledge of this unbalanced approach was one reason Murdoch had declined the lieutenant governorship of the Island in 1850.[29]

Cooper's failure in London was the beginning of the end of Escheat as the dominant political force among the Island tenantry. The London trip made it clear that the British government would not countenance a policy of confiscation of estates. To put it baldly, the Escheat program was impracticable within the British empire. Such astute Island reformers as George Dalrymple had predicted failure because of the ingrained attitudes of the British government; and some Islanders must have recalled that it had rejected escheat in the first decade of the century by withholding assent from enabling legislation, and that it had also stopped Lieutenant Governor Charles Douglass Smith from taking further action in 1818 after he had, on his own authority, ordered the escheat of two thinly populated townships for whom no owners could be identified.[30]

There were troubling aspects of Cooper's leadership in any event. Cooper was a radical who continued to pay his own rent while counselling other tenants to withhold theirs. His personal probity had been in question since his dismissal as land agent for the Townshend estate shortly before he became an enthusiast for escheat. Historian Harry Baglole has pointed out that 'his political enemies frequently alleged that Cooper had misappropriated funds, a charge ... which Cooper never seemed quite able to refute convincingly.'[31] Doubts about Cooper's integrity may well be in order, for Baglole has also observed that he 'purchased several farms on nearby Lot 55 [one of the two lots escheated by Smith] in the 1830s and '40s, but refrained from having the sales registered until the 1850s, possibly for political reasons.'[32] Cooper conforms in a general way to one type of leader Eric Richards, the scholarly authority on the Highland Clearances, has identified in incidents of disturbance or resistance to landlord power in Scotland, namely, 'a failed land-agent.'[33]

The Escheat movement faded in the early 1840s. Some individual leaders, like Cooper, remained active in politics off and on over many years, and formed a sort of radical rump within the more moderate Reform or Liberal[34] party which replaced Escheat as the political group with the greatest support among the tenantry. Politically impotent, and beset by a leadership with faulty judgment, Escheat moved to the margins of Island politics, providing little more than nuisance value. In the end, the major significance of the movement for the history of the Tenant League was that Cooper's prolonged agitation had placed the land question at the centre of popular politics in the colony, and had done so, in the words of Baglole, 'so forcefully [that] ... it could never again be ignored.'[35] There was also a continuity of certain patterns of reasoning and behaviour linked to Escheat. Many tenants continued to think of the leasehold system as illegitimate, and even in the 1860s Escheat-inspired jargon about 'land claimants' – Escheators' term for the proprietors – would show up in resolutions passed at local meetings, in the wording of petitions, and so on. Finally, there was a tradition of direct action attached to the Escheat movement, and this too would resurface in the 1860s.

It is worth emphasizing lessons emerging from the Escheat experience: that earth-shaking radical change which directly threatened established concepts of property rights would not succeed because the Island did not exist in a political vacuum; that limited reform, such as the wilderness tax of 1837–8, could succeed despite the lobbying of influential proprietors; and that the British governing class would not spend the money of British taxpayers on land reform in Prince Edward Island, especially when such reform was predicated on rejection of fundamental assumptions rooted in the political structure of the United Kingdom. There was real hostility on Downing Street towards radical solutions to the land question. This was a fact any reformers on the Island would have to take into account as long as the Colonial Office had the right to reject local legislation, and as long as governors were specifically enjoined to ensure that all acts affecting property rights include suspending clauses which provided that they not take effect until they had received royal assent.

The Reform party which surfaced after the Escheat débâcle demonstrated that it had learned from the failure of the Escheat movement. The most eloquent advocate of the new orientation was Edward Whelan, a Roman Catholic native of Ireland who arrived in Prince Edward Island in 1843, one year after a decisive Tory victory at a general election. Although still a teenager, Whelan, a brilliant writer, had edited a newspaper in Halifax. He commenced a paper in Charlottetown, and carefully avoided becoming identified with escheat as a solution or condemning it outright. He allowed Escheators access to his columns and admitted the justice of a partial escheat,[36] but claimed that advocacy of a general escheat was impractical until the Island had obtained responsible government. This monumental reform of the political apparatus, he argued, would provide the necessary leverage to force an end to leasehold tenure. From the beginning, Whelan tried to unify Island reformers around the constitutional issue. Although accepting the prevailing assumption that the land system was the root of the Island's problems, he would not commit himself firmly on matters over which reformers might disagree, such as escheat or annexation to Nova Scotia as solutions to the land question.[37]

The struggle over responsible government dominated the later 1840s, and by 1849 George Coles, a former Tory, a native Islander, and an Anglican, had emerged as the forceful leader of the reformers in the assembly. He was a businessman, a brewer and distiller, and, compared with most reformers, wealthy. He represented a different kind of capitalism from that which characterized the local élite, who made their livings as middlemen. Land agents, lawyers, and merchants, they feasted off both landlords and tenants. They were at least as opposed as landlords to abolition, for if abolition came with compensation, the owners could walk away with cash in their hands, but the middlemen would lose their function and hence their livelihood. Coles more than once stated that land agents were more resistant than landlords to change when confronted with reform proposals. In contrast to the middlemen, Coles was a manufacturer of goods for common consumption, and as such his prosperity depended upon the widest possible diffusion of purchasing power throughout the population at large. On the other hand, his class interests would suffer if the leasehold system exported capital *en masse* to the United Kingdom. The best safeguard for his interests was abolition.

Led by Coles and nourished journalistically by Whelan, the reform movement won a decisive victory at the polls in 1850 on the issue of responsible government. By the next year the British government conceded the constitutional change. It also authorized the lieutenant governor to use force if necessary to maintain the property rights of the landlords. The Reformers remained in office until 1859, with the exception of a brief interval in 1853–4. They had the most comprehensive reform agenda of any of the parties which came to power in the Maritime colonies in this era. Their Free Education Act of 1852, through which they made primary education 'free' by paying the entire salaries of district teachers from the colonial treasury, was particularly important. Previously the districts had been required to raise part of teachers' stipends by local assessment and tuition fees. After 1852 local taxes were to be used only for the erection and maintenance of school buildings, and tuition fees were abolished. The Reformers also extended the franchise so as to make it almost universal for adult males.[38]

Given the situation on Prince Edward Island, the land question had to be the major focus of the Reformers. They pursued a dual policy. One thrust emphasized ameliorative measures designed to reduce the insecurity and arbitrariness to which the working of the system might expose the land occupier, and taxation to make leasehold tenure less financially attractive to the owners. The Reformers had limited success. London at first refused assent to a bill to entrench legally the right of certain tenants to pay their sterling rents at an exchange rate of one-ninth, but later assented to an amended version. A narrowly circumscribed tenant compensation bill failed to receive royal assent. A bill imposing a tax of 5 per cent on the nominal rentals of landlords[39] met the same fate.[40]

The other major element in the Reformers' land policy was the Land

Purchase Act of 1853, which authorized the government to make large-scale purchases of land on a voluntary basis for resale to their occupiers. The noncompulsory aspect was necessary for acceptance by London; but it also left the Reformers at the mercy of landlords, who, as a group, were not interested in selling. The only major property purchased in the 1850s was the Worrell estate. By this time Charles Worrell was in his eighties and living in poverty in London. Unfortunately for the Reform government, for the colonial treasury, and for the proprietor, Worrell's local representatives were able to twist the sale in such a way as to make the purchase price more than 70 per cent greater than the amount Worrell received, skimming off some £10,000 in a manner that amounted to breach of trust. The government paid the exorbitant price, probably because they believed they had little choice. To abandon the sale on account of the impending swindle would undermine the credibility of their land reform program, for no other large estate was on the market; and proceeding with the sale would compromise the same program by identifying the Land Purchase Act with a scandal. In the end they were probably swayed by the belief that the piecemeal abolition of leasehold tenure had to be commenced, regardless of the embarrassment of being victimized by the 'Worrell Job.'

The Liberal government left office in 1859 after an election which turned on denominational differences over the role of Bible reading in the school system. Island politics were taking a new direction. Divisions centring on the relationship of church and state in education would emerge in several different forms over the succeeding twenty years.[41] Ironically, the Conservatives, led by Edward Palmer (a son of J.B.), changed almost nothing in terms of practice with respect to Bible reading in the district schools when they had the power to do so. In 1860 they essentially gave a statutory basis to permissive daily reading of the Bible. They had used the 'religion and education' issue and the fear of 'Catholic domination' to split the Liberal voters along denominational lines, and to unite the Protestant majority behind them, leaving Coles's Liberals dependent, in terms of mass electoral support, on the Roman Catholic minority, which constituted approximately 45 per cent of the population. As a political strategy for attaining office, it had been a striking success, and it left the opposition Liberals demoralized. After many years of preaching that persons of different ethnic backgrounds and religions must work together to break the leasehold system, the Liberals saw the elemental force of denominationalism fracture the political party system they had shaped.

During the election campaign, the Tories had treated the land question as a secondary matter. But once in office they had to face the reality that they had an unresolved and potentially explosive issue on their hands. Throughout the 1850s Palmer had argued that the Liberal land reform program was ineffective, without explaining precisely what his party proposed to do. The years in opposition had provided him with opportunities for embarrassing the Liberals, and on one occasion when objecting to the Land Purchase Act he, a land agent and proprietor, had even found himself allied with Cooper and other diehard Escheators, who were upset by the statute's implicit recognition of proprietary titles as legitimate. Once in office, Palmer did not have the luxury of flirting with Escheators to score opportunistic political points; something more concrete was required. In effect, he was now on the spot, and it is clear that he did not have a comprehensive program of land reform. Given his own conservatism and the make-up of his government, which included several landlords and land agents, it would have been unrealistic to expect drastic reforms from him. Thus, with little apparent sense of direction on the issue and perhaps with the idea of buying time, Palmer established the land commission of 1860.

The Land Commission of 1860 and the Fifteen Years Purchase Bill

This royal commission would disagreeably surprise those who were initially its most enthusiastic sponsors. Composed of three members, none of them Islanders, it was created to investigate differences relative to the rights of landowners and tenants, and make recommendations for their resolution. A group of six prominent absentee landlords, including Cunard, who together owned more than one-quarter of the Island, had suggested the formula for choosing the commissioners: one would be chosen by the proprietors, one by the Island government (acting on behalf of the tenantry), and the third by the imperial government.[42] In announcing their appointment, the British colonial secretary, the Duke of Newcastle, stated that despite the method of their selection, they were to act in a quasi-judicial manner, not as advocates for the respective parties who had designated them. Their recommendations were to be directed towards equity and the common good. Both landlords and tenants approached the commission with a sense of the justice of their respective cases. Given the events that followed, this belief in the evident rightness of their cause must be emphasized with respect to the proprietors. Cunard's group gave the appearance of submitting in advance to the recommendations of the commissioners. They insisted that the decision be binding, and from subsequent comments they and their associates made, it would appear that they were confident the commissioners would uphold their fundamental rights and were hoping the process would 'settle' or 'quiet' the land question once and for all time. One has to conclude that they expected confirmation of the *status quo*.

The commissioners – John Hamilton Gray of New Brunswick and Joseph Howe and J.W. Ritchie of Nova Scotia – held public meetings between 5 September and 1 October 1860. They heard from just about every type of Islander affected. A delegation from Lot 30 personalized for their benefit the potential for harshness in the system by having a squatter being sued by Robert Bruce Stewart brought before them. In the custody of the sheriff and, in the words of the official reporters, 'having his head bound round with a handkerchief, and looking very unwell,' he was on his way to jail. According to Stewart, he was a non-paying tenant. It appears that, as a means of effecting removal, Stewart had taken the roof off his house, which the sheriff, Henry Longworth, described as 'the most miserable abode that I ever beheld.' The presentation, near the end of the parade of witnesses, had significant impact, for it provoked Ritchie, the nominee of the proprietors, to remark that 'We have been made to believe all along by the proprietors that there were no cases of hardship under the tenant system, and here is a case that has come up, which distinctly shows that there are cases of hardship caused by the proprietors.'[43] Premier Palmer's name arose more than once as an example of a landlord or land agent, and always in a context certain to be politically embarrassing. His brother Charles, lawyer for the proprietors, was reminded by a witness of *his* role as a proprietor. Charles Palmer immediately claimed to be indulgent towards his tenants.

There were the predictable differences in perspective, but, beyond that, in some instances witnesses denied specific statements made by others. On 26 September, Howe remarked: 'I am afraid that let our decision be ever so wise, it will not calm the elements. The statements are so conflicting, that we scarcely know how to find out the truth.'[44] They received some three hundred memorials and other documents, including protests from several proprietors (not members of Cunard's group) against being affected by the proceedings.[45] A major difficulty was the disorganized and inadequate state of the public records available to them on the Island, and in order to assist them in analysing some of the contradictory evidence they hired George Wightman as a researcher. A native of Nova Scotia and unknown on the Island, he spent several months gathering information as discreetly as possible. He was submitting material as late as 20 June 1861.[46]

The commissioners issued a unanimous report on 18 July 1861, and the thoroughness and length of their labours would eventually prompt the imperial government to double their pecuniary compensation. Newcastle released the report officially on 7 February 1862. A model of lucidity, it included an historical summary of the land question since 1767 and an analysis of several specific sub-questions, as well as recommendations. The 'award,' as it became known on the Island, may not have been entirely satisfactory to the tenantry, but it came as a bitter shock to the proprietors. The commissioners recommended that tenants, with two categories of exceptions, be given the right to purchase freehold from their landlords, with arrears prior to 1 May 1858 being forgiven.

That is where simplicity stopped. In addition to the conflicting claims and counter-claims, and the chaotic state of public records on the Island, the commissioners had to cope with another level of complexity which would have complicated their task even if everyone had agreed on the facts and if the public records had been well organized and readily accessible. They stated in a draft interim report that 'we soon discovered that all parties ... had underestimated the delicacy and difficulty of this enquiry. If one landlord owned the Island, if one family had controlled its territory, one system would be apparent to the eye and one remedy might be applied.' Such, however, was not the case. Each township had begun with a different owner or group of owners and many had been subsequently subdivided among several proprietors. As the commissioners pointed out, 'even where several estates have fallen, at a later period, into the hands of one proprietor, their past treatment affects the length of leases, the value of the property, the amount of arrears and the equitable considerations which must ultimately control our decision.'[47]

Recognizing the tremendous variation in circumstances among individual cases, the commissioners did not presume to set a purchase price or even a scale of possible purchase prices. Instead they recommended that in each case the landlord and tenant attempt to reach agreement, and proposed rules to give both sides an incentive to be reasonable. If they could not agree, then the price should be determined by arbitration. In effect, the commissioners advised a form of compulsory conversion to freehold. They described this as 'a compulsory compromise' because, while making conversion ultimately *required*, it also involved general recognition of existing titles and leases.[48] The Palmer government professed satisfaction with the award, and passed two pieces of legislation to implement it. Since the acts dealt with property rights, they included suspending clauses.

The award of the commissioners was absolutely unacceptable to the proprietors for two reasons, one of principle and the other of practicality. In the first instance the award was giving the tenant's *claim* to freehold priority over the proprietor's legal *right* to his property. This was a serious objection, and few knowledgeable observers could have expected the proprietors, no matter how anxious they were to see the land question quieted, to consent readily to what they regarded as a breach in their fundamental rights. They considered the right to decide on whether to sell freehold title to tenants or prospective settlers to be part of their basic entitlement. The second reason they found the award unacceptable was that, as explained earlier, any prospect of selling on a piecemeal basis, one farm at a time, raised the spectre of estates shrinking steadily in size over several years, with declining revenues, fairly fixed expenses, and rising proportions of uncollectable rents and of acreage relatively poor in quality.

Thus the proprietors had two strong reasons to object to the award of the commissioners, one founded in principle and the other grounded in the practical considerations of rational estate management. But in the end the objection which apparently provided them with the argument the British government found most convenient to use in rejecting the award was procedural. The commissioners had been empowered to recommend a means of settling differences between landlord and tenant, and it was to that empowerment that Cunard's group of proprietors had consented. The commissioners had not been authorized to delegate the power to settle differences between landlord and tenant to an unknown third party, that is, the proposed arbitrators. Hence the proprietors argued that in delegating the right to resolve impasses the commissioners had exceeded their authority.[49] On 9 August 1862 Newcastle informed the Island government that royal assent would be withheld. In the spring of 1863 he obtained the opinion of the crown law officers, which supported the decision of the government; and on 30 June 1863 his prime minister, Palmerston, reminded him that more than principle was at stake, for he had 'a near connection who has lands in the island and is much interested in the matter.'[50]

The award was dead – Newcastle and the crown law officers even objected to the term 'award'[51] – but its rejection set in motion a train of events which would shake the system of leasehold tenure on the Island to its very roots. The Conservatives survived the next general election after the report of the land commission, on 21 January 1863, by contesting it exclusively on the supposed threat to Protestants from the Roman Catholic minority. This was a variation on the theme that had been used successfully in 1859, and it worked: all Protestant constituencies voted Tory, and all Roman Catholic ones voted Liberal. If anything, it was more cynical than the campaign four years earlier.

Yet the land question would not go away. The Conservatives, after March 1863 led by Colonel John Hamilton Gray (no relation of the commissioner bearing the same name), realized that they must take some initiative. In the late summer of 1863 they sent a delegation overseas to negotiate a compromise with a group of proprietors led by Cunard. The delegation consisted of Edward Palmer, now the attorney general, and William Henry Pope, the Island's colonial secretary. Pope, who was also editor of the leading Conservative newspaper, the *Islander*, and the chief beneficiary of the Worrell Job, was the more active and aggressive of the two delegates. He remained in London into the new year following the departure of Palmer in the autumn. But he failed to reach an agreement with Cunard that he believed would satisfy Island tenants.

The only palliative the Conservative government provided at the next session of the legislature was a statute that became known as the Fifteen Years Purchase Bill. It applied to the estates of twelve consenting landlords. Tenants who held leases with unexpired terms longer than forty years would for a period of ten years be allowed to purchase their lands for the equivalent of fifteen years' rent plus arrears owing since 1 May 1858. The bill also adopted the suggestion of the land commission regarding remission of arrears. In fact, as a sort of olive branch to all tenants on the estates of the consenting proprietors, those arrears were to be remitted whether the tenant purchased under the terms of the bill or not.[52]

Attorney General Palmer, one of the consenting proprietors, stated in the Legislative Council that this 'measure ... is the first invasion of the rights of the proprietors on this Island – it is inserting the thin end of the wedge.'[53] Palmer, with his lawyer's sense of the importance of precedent, had a point: any proprietor named in the attached 'Schedule (A)' was required by law to sell to a tenant who offered the terms specified in the bill. This was absolutely new in Prince Edward Island. Many small proprietors and one of the largest, Robert Bruce Stewart, were not on the list, but the remaining five of the original six non-residents who had suggested the composition and mandate for the commission were there.[54] The other seven were resident proprietors who probably realized that the mood of the tenantry had become volatile.[55]

The bill was obnoxious to Island tenants for a number of reasons. In the first place there was its origin: it grew out of the failure of the land commissioners and the Island government delegates, and the ability of the proprietors to revise to their own advantage the outcome of any process. The bill was modelled closely on a written proposal made by Cunard to the Colonial Office on 5 December 1863, and it had been published in the *Islander* on 26 February 1864.[56] So it was Cunard's bill, and that fact was public knowledge. By 1864 Island tenants were inclined to be intolerant of it for that very reason.

Secondly, there was the inflexibility of the formula for purchase. The royal commissioners had recognized the range in the situations of tenants from estate to estate and within a single estate, and had despaired of producing a blanket formula. That, they believed, was a recipe for injustice, and therefore they had recommended that the parties negotiate purchase prices on a case-by-case basis.[57] The flexibility of the Tenant League agenda which would emerge later in 1864, with its emphasis on estate-by-estate negotiations, was in fact closer than the Fifteen Years Purchase Bill to the approach of the land commissioners.

Thirdly, the price set was high: few tenants thought fifteen years' rent plus arrears dating back to 1 May 1858 was a fair price for land they had improved and made into a farm. Moreover, the price had to be paid in full if the landlord was to be compelled to sell – a provision that constituted an impossible barrier for many tenants.[58] In the words of an anonymous poet who contributed *The Tenant's Song* to *Ross's Weekly*, an Island newspaper, in the spring of 1864:

Of all the terrible bills,
With which a poor man is curs't,
For clothing, or food, or parchment or pills,
Sir Samuel's bill is the worst. ...
With thirty odd pounds, from those in arrears,
To be paid in ready cash.
Ye members who take your ease,
In the cut-stone house so grand,
Who oft at the hustings vow'd to release
The 'serfs' who till the land;
Beware of this mons'trous bill,
It rivets the tyrant's chain,[59]

The Fifteen Years Purchase Bill proved to be a vain hope for Cunard, the Island Tories, and anyone else who may have counted upon it to prevent disorder – which was the rationale Cunard cited more than once in writing to the Colonial Office in its support.[60] Over the years 1864–8 a total of forty-five tenants used the act to become freeholders, resulting in only 2,911 acres, or 0.82 per cent of the 353,537 acres to which it applied, being converted to freehold.[61] Even before Pope arrived with news of the failure of the delegation (but after the return of Palmer), indications of tenant restlessness had begun to appear in the press.

The events of 1864 and 1865 would indicate that many, probably most, Prince Edward Island tenants were absolutely frustrated and disillusioned with the conventional processes of politics, law, and government. It seemed as though they or their representatives had tried almost everything. The tenantry had voted for Cooper and his promise of escheat; that would never be acceptable to the British government, which held the key to the proposed process. They had voted for the more moderate solutions of the Reformers led by Coles and Whelan, and in the end Downing Street and the proprietors had stymied those gradualist policies. They had voted for the Tories, who had good connections with the proprietors, so it was said; and the Tories had produced even less than the Reformers, namely, the Fifteen Years Purchase Bill, which became a despised and irritating symbol of futility and impotence in the face of landlord power. Again, the nameless author of *The Tenant's Song* summed it up:

> The great Commission's at rest,
> It seemed a plausable [sic] scheme, ...
> And prov'd a fanciful dream.
> The Land Purchase Bill has freed
> Two or three townships or so,
> Where the settlers hold the government creed[62] ...
> The Delegates have performed
> Their journey to Downing street,
> And Pope with a goose quill has boldly storm'd
> Sir Samuel Cunard's retreat.
> But the lion holds his prey,
> Despite the Delegate's quill,
> Each farthing of rent the tenant must pay,
> Or purchase under his bill.

The Escheat movement, Reform governments, Tory governments, a royal commission, and a delegation had all failed. *The Tenant's Song* conveyed the resultant dismissive feeling towards politicians of all stripes, and also the tendency to lump them together with the proprietors:

> Bad luck to the landlords' crew,
> Bad luck to them great and small,
> Bad luck to the members of Parliament, too,
> Lib'rals and Tories and all.
> Quit-rents, reserves[63] and escheat,
> For thirty long years and more,
> With hopes that were false, the tenant did cheat, ...

It was time for the tenants and their allies to take matters into their own hands.

2

A Tenantry in Ferment

The Tenantry of this Island, heretofore loyal, sympathetic and patriotic, have ultimately been driven by Proprietary oppression into extreme measures, if not desperation.
Preamble to resolution passed at meeting held in Mount Albion, Queens County, 7 March 1864[1]

The Tenantry in 1864

Even before William Henry Pope returned to Prince Edward Island from London in February of 1864, it was apparent that the tenants of the colony were 'disturbed,' to use the contemporary term. Understanding the volatility of the situation requires an attempt to empathize with the tenantry.

The individual circumstances of tenants varied widely, but for the sake of simplicity it is possible to divide them into three groups: those who were relatively secure on their farms, those who were insecure because of tenurial status, and those who were insecure because of rent arrears. Those who were relatively secure had long leases and were up to date with their rent and other payments. But they were not necessarily content, for they were often anxious to obtain their 'independence' by gaining freehold tenure, yet unable to do so because their landlords would not sell under any conditions. There were two types of insecurity, and the first was the vulnerability inherent in being a tenant without a written lease or a tenant with a short lease; at the far end of the spectrum of this sort of insecurity were those the census referred to as 'Occupants being neither freeholders nor tenants,' the squatters. This category of insecurity was independent of the financial status of the individual land occupier.[2] But there was another sort of insecurity for Prince Edward Island tenants. Large numbers, as noted in chapter 1, were in arrears on their rent, and for them there was no security of tenure. They were constantly vulnerable to legal action which could result in eviction. All three groups, the secure, those insecure because of the terms of their tenure, and those insecure because of back-rents, wanted the leasehold system either abolished outright or modified so that tenants had security on their farms and the right to purchase when they had the financial capacity to do so.

In the name of the tenants, the Island government had consented to a royal commission on the land question. Six non-resident proprietors – Sir Samuel and Edward Cunard, Sir Graham and James Montgomery, the Earl of Selkirk, and Laurence Sulivan – had suggested the means of selecting the commissioners, and had requested that the recommendations be binding. The British government had indicated in 1860 that unanimity would give 'double weight' to any decision the commissioners reached.[3] They had proceeded to recommend unanimously in favour of the tenant's right to purchase freehold tenure. Joseph Howe recorded in his diary that he and his fellow commissioners 'never had an unkind word. Discussed every point manfully, and came to conclusions without ever deciding any thing by a vote. We were in fact unanimous throughout.'[4] Yet, despite all the apparent advantages the landlords had in determining the make-up and mandate of the commission, they petitioned London against the 'award' of the commissioners. Downing Street responded to the proprietors' wishes, the unanimity of the land commissioners notwithstanding. Island tenants had received hope from an unexpected source, and then had seen some of the very people who had said that they should put their trust there snatch that hope away.

With their report, the commissioners had responded to the tenants' aspirations; but the British government had negated that response. This was the state of the land question, as understood by the tenantry, after disallowance of the award and the refusal of leading proprietors like Sir Samuel Cunard to make some equivalent concession. Tenants believed that the political system, as well as the landlords, had betrayed them. The land commission had turned out to be, as the opposition Liberals had charged from the beginning, an exercise in futility. Even when chosen as the proprietors wished, given the powers the proprietors wished, and acting unanimously, as Downing Street wished, if the commissioners came to a conclusion inconvenient to the proprietary interest then their deliberations would have no effect. Edward Whelan had written to

Howe, for whom he had worked as an apprentice, in an attempt to dissuade him from serving as a commissioner on the grounds that he would be outvoted by the nominees of the imperial government and the proprietors, which he was not, and that in the end the proprietors would determine what was acceptable, which turned out to be correct.[5]

Rightly or wrongly, tenant farmers on Prince Edward Island were convinced that freehold tenure was necessary for their future prosperity. One of the most striking features of the testimony before the commission was the prevalence of talk about out-migration because of pessimism about the future. The factor witness after witness cited was leasehold tenure, which meant that their sons would never be able to own their farms.[6] No one who reads the proceedings of the commission can fail to be impressed with how frequently the leasehold system was linked with lack of hope for the future. This was, it must be noted, before out-migration had become a popular choice in the Atlantic region. The 1850s had been a good decade economically for the Maritimes, including Prince Edward Island. In fact, by conventional measures the Island had done extraordinarily well: the annual growth rate of visible exports between 1850 and 1859 was 11.5 per cent.[7]

The purpose of this outline is to explain the point of view of Island tenants as a whole on the eve of the formation of the Tenant League. One may argue with the justice or accuracy of their viewpoint. One may say that they did not understand the importance of the technical or procedural point on which the proprietors pegged their rejection of the award. One may object that their talk of out-migration in the presence of the commissioners should be interpreted as a device to dramatize their sense of grievance. One may emphasize the importance of solemn contracts (leases in this case) and property rights. One may believe that they over-valued freehold status, and point to the rapid economic growth in the 1850s, when most farmers were tenants. But, regardless of how one may discount their views, their comprehension, their rhetoric, or their faith in freehold tenure, Island tenants clearly believed they had been deceived once too often, and they were determined that the leasehold system must be abolished. Without understanding this perception, it is impossible to account for the events of 1864 and 1865 in rural Prince Edward Island. Although the tenant who was far in arrears and the leaseholder who was debt-free and prosperous approached the situation from different places on the economic spectrum, there was a remarkable congruence of opinion. Leasehold tenure had to end, and only the tenants and their friends could or would do the job.

Reports of Disturbances on Lots 29 and 63

By December of 1863 the initial signs of a new tenant insurgency were beginning to appear. Since the Tenant League was a movement of grassroots revolt against established political ways of doing things, it is not surprising that its precise origins are obscure and difficult to pin down with certitude. Between 1880 and 1992, at least five authors suggested in published works that the league originated with a meeting held in Sturgeon, Lot 61, in southern Kings County, in December of 1863, although it should be noted that each of the five acknowledged that formal foundation of the league took place subsequently in Charlottetown.[8] Since four gave no sources, and the only source the fifth cited was one of the first four, it is necessary to examine whatever relevant evidence survives in order to evaluate this claim.

According to the official report of debates in the House of Assembly for 1866, when the Tenant League was the topic of the hour, James Duncan, Conservative member for 4th Kings, stated that 'measures were concocted for the organization of the League in the school-house at Sturgeon. A very few of those who attended ... were from Peters's [sic] Road.'[9] Both Sturgeon and Peters Road, Lot 63, were in 4th Kings, and although not adjacent, are only a few miles apart. Exactly a month after the end of the legislative session, a letter writer in a local newspaper, signing as 'A Farmer,' purportedly from Sturgeon, wrote to refute Duncan; he claimed that the first resolution on the subject of the formation of the league was passed at Peters Road, and not the Sturgeon schoolhouse.[10] Nine years later Manoah Rowe, a former Tenant Leaguer who was then a Liberal assemblyman for 4th Kings, stated in the house that he was representing 'the district in which the Tenant Union took its rise.'[11] That could mean either Peters Road or Sturgeon. Neither Duncan nor 'A Farmer' nor Rowe gave a date for the meeting.

In light of the limited and contradictory nature of the available information, the significance of the December 1863 meeting in Sturgeon alluded to by the five authors has not been conclusively established. Indeed, its very occurrence is not certain, since the first evidence which survives is dated 1866. Although it is possible that such a meeting was held, and that it manifested a will to create a tenant organization, very few of the sources which survive from the era can be interpreted as supporting this hypothesis.

For observers in December 1863 who received their information through Island newspapers, the first evidence of a new tenant movement surfaced in western Queens County, approximately fifty-five miles to the east of Sturgeon and Peters Road. Whelan's *Examiner* reported under the heading 'Tenant Outbreak' that 'We are informed that an attempt to enforce old arrears of rent on the Melville Estate, Lot 29 (near Crapaud), has been met with armed resistance, and that something in the form of a Tenant League has been organized there, to keep aloof the Bailiffs of the Agent.' The issue of arrears was particularly explosive after the report of the land commission, dated 18 July 1861, had recommended that all arrears prior to 1 May 1858 be forgiven. The date 1 May 1858 was arbitrary, but this proposed remission, once known, became the minimum demand of tenant spokesmen. Although over the preceding twenty years Whelan had frequently advised tenants to unite against their landlords, and only a few years earlier had come very close to advocating formation of a tenant league, on this occasion he displayed little sympathy with the tenants involved.

It used to be slanderously said that no class of people in the Island resisted the payment of rent but Irish Catholics. In this case, the 'rebels' are not Catholics or Irish, but good Protestants and strong supporters of the Government. We are not at all surprised at their acting in the manner described to us, for *they*, at least – having served the Government at the elections – had reason to believe that they would be relieved from the payment of the back rents, which had been so often promised to them. ... It is not unpleasant to hear of a rebellion arising amongst the slaves of the Tory party – they deserve to have the screws tightened upon them by their ungrateful task-masters. ... if they don't pay, well! – let the screws go on, and let them get out of their troubles as best they can.[12]

The next number of the *Examiner* featured a letter from John Roach Bourke, the land agent of Henry Dundas, the 3rd Viscount Melville. Bourke was also a proprietor, and one of the twelve listed as consenting to sell under the terms of the Fifteen Years Purchase Bill. According to 'Schedule (A),' annexed to the statute, he owned one-half of Lot 37. While not denying that there had been a confrontation of some sort on Melville's half of Lot 29, he stated that the incident had occurred on the Melville Road (now known as South Melville),[13] an area populated mostly by Roman Catholics of Irish origin, and that only one family involved supported the government. They were the Beers, headed by John Beer Sr, who was, according to Bourke, almost £90 in arrears on two hundred acres. Assuming the annual rental to have been close to the usual one shilling plus taxes per acre, this meant that Beer was

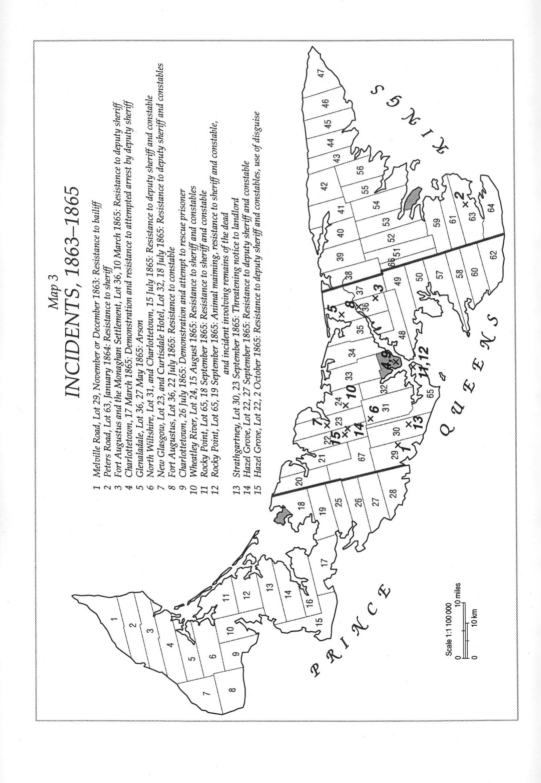

Map 3

INCIDENTS, 1863–1865

1 Melville Road, Lot 29, November or December 1863: Resistance to bailiff
2 Peters Road, Lot 63, January 1864: Resistance to sheriff
3 Fort Augustus and the Monaghan Settlement, Lot 36, 10 March 1865: Resistance to deputy sheriff
4 Charlottetown, 17 March 1865: Demonstration and resistance to attempted arrest by deputy sheriff
5 Glenaladale, Lot 36, 27 May 1865: Arson
6 North Wiltshire, Lot 31, and Charlottetown, 15 July 1865: Resistance to deputy sheriff and constable
7 New Glasgow, Lot 23, and Curtisdale Hotel, Lot 32, 18 July 1865: Resistance to deputy sheriff and constables
8 Fort Augustus, Lot 36, 22 July 1865: Resistance to constable
9 Charlottetown, 26 July 1865: Demonstration and attempt to rescue prisoner
10 Wheatley River, Lot 24, 15 August 1865: Resistance to sheriff and constables
11 Rocky Point, Lot 65, 18 September 1865: Resistance to sheriff and constable
12 Rocky Point, Lot 65, 19 September 1865: Animal maiming, resistance to sheriff and constable,
     and incident involving remains of the dead
13 Strathgartney, Lot 30, 23 September 1865: Threatening notice to landlord
14 Hazel Grove, Lot 22, 27 September 1865: Resistance to deputy sheriff and constable
15 Hazel Grove, Lot 22, 2 October 1865: Resistance to deputy sheriff and constables, use of disguise

Scale 1:1 100 000

0   10 miles
0   10 km

several years behind in his payments. Distraint, also known as distress, is the legal act of taking movable property from a person, for example, in satisfaction of a debt such as rent; it was in the course of an attempt to distrain upon Beer that the bailiffs encountered resistance.

Bourke gave no information as to whether firearms were actually present or to what degree the resistance was successful. Yet he went on to declare that 'the ringleaders shall be dealt with as the law directs; and ... no Tenant League, country or politics will prevent me from doing my duty as an agent; but if they persist, and to use your own words, let the screws go on, and let them get out of their troubles as best they can.'[14] He closed with a reference to William Hughes, a tenant from southern Kings County. Hughes was still in jail after being sentenced in 1860 to four years' imprisonment for his part in the shooting of a horse from under a sheriff who was acting in a case to recover arrears of rent.[15] In reply, Whelan pointed out that, despite the corrections in matters of detail, 'the Agent's letter confirms the most material part ... namely, that distraints WERE made for arrears of rent – that the distraints were resisted by an organized party of the tenantry.'[16]

The incident on Lot 29 went virtually unreported in the rest of the Island press. The Tory *Islander*, normally edited by Pope, who was still overseas on the delegation to the Colonial Office, professed ignorance, but warned tenants to heed the history of leasehold tenure in New York state, where some years earlier the state government had intervened against tenants resisting collection of rent.[17] The only other surviving newspaper to comment on the affair, the *Semi-Weekly Advertiser*, edited by John Ross, a future Tenant League supporter, minimized the incident.[18] But Ross did emphasize the important role played by the issue of back-rents, and argued that since the land commission the tenants had looked upon arrears as remitted, or at least held in abeyance until a final settlement was reached. 'If the collection of the back-rents is persisted in,' he wrote, 'it is altogether probable that the terms "Tenant outbreak" and "Tenant League" will be comman [sic] in the vocabulary of P.E. Island.'[19] The stage was set for further confrontations.

Significantly, the next reported incident, at Peters Road, in southern Kings, also revolved around collection of arrears. In late January 1864, a letter in the *Examiner* from Lot 61 (adjacent to Lot 63), signed simply 'A Tenant,' reported that the land agent in the area had adopted a new and more stringent policy on back-rents. Instead of being content so long as arrears were not increasing, he demanded that all be paid at once, under threat of distraint.[20] The agent in question was probably George W.

DeBlois, who represented both Sulivan (Lot 61) and Cunard (Lots 63 and 64), the two largest non-resident proprietors, although it is possible that he was acting through a sub-agent.[21]

As recently as 1860 the 173 leaseholders in Lots 63 and 64 were, on average, 5.79 years in arrears, even if one accepted the generous assumption that all the rent due in the current year, 1860, would be paid in full.[22] Assuming that such occurred, and that the tenants continued to pay all of the rent coming due each year, and no more, then, all other things being equal, they would still be 5.79 years in arrears in early 1864. In fact, over the previous three years (1857–9) the average rent paid was £494 1s. 3d., or approximately 61.5 per cent of the annual rental of £803 3s. 7d. If the assumption is made that the tenants continued in 1860 to pay at that rate, they would have been, on average, 6.17 years in arrears after the autumn collections of that year. To make a further projection, assuming the same rate of payment through 1861–3, then by the winter of 1863–4 the Cunard tenants on Lots 63 and 64 would be, on average, 7.33 years in arrears.[23] After the report of the land commission, collection of 'old rent' on an estate where many people had arrears stretching back long before 1 May 1858 was widely regarded as provocative.

No statistical information on arrears is available for Sulivan's township, but it would be a fair surmise that the circumstances of his tenants were similar to those of the Cunard tenants, if not worse. The census-taker for 1861, Duncan Fraser, had declared that of the Lot 61 land occupied by tenants, 30 per cent was second class in quality, and 70 per cent third class.[24] In fact, although there were 142 farms listed in the census, he stated that 'I did not see more than four or five good farms on the whole Township.'[25] George Wightman, the researcher for the land commission, visited Lot 61, and clearly agreed with Fraser, whom he described as 'an experienced farmer.' Wightman concentrated his analysis on the ninety-two holders of written leases or agreements to lease, and among them he found only sixteen to be 'in fair circumstances'; the situations of the remainder were either worse or much worse.[26] This focus should not obscure the fact that, among the tenants, an unusually high proportion, 26.4 per cent, held their land by verbal agreement, a situation rife with insecurity for the occupier. Of all Island townships, only four contained more 'tenants at will' in absolute numbers and only one of those four had a higher per centage of tenants in this position.[27] Lot 61 probably presented as clear a case of combined poverty and insecurity as any township on Prince Edward Island.[28]

According to the *Examiner*'s informant from Lot 61, writing on Friday,

15 January 1864, 'Many were unable to make up more than a year's rent and were distrained on, and are now at the mercy of the Sheriff, while others of us had to sell off our cattle or the meal we had for the support of our families, and to sell them for half their value to raise the money.' But he reported that when serving writs on Peters Road 'this week,' the sheriff 'met with a warm reception from the women and children of that Orange locality. ... the Sheriff has been three times unsuccessful in his attempts to serve writs there.' In addition, the Peters Road tenants had 'formed a League with the Tenantry of the High Bank [on the neighbouring Lot 64] and are determined to resist the payment of rent.'[29]

The account from southern Kings is particularly significant for two reasons: the mixed character of the crowd in terms of sex and age, and the link with Orangeism. With respect to the role of women, it is important to emphasize that this letter is the sole evidence of direct involvement of women in collective resistance to authority in the history or immediate prehistory of the Tenant League and that there is no equivalent to the pattern documented in the Highlands of Scotland, where historian Eric Richards has stated unequivocally, 'Highland riots were women's riots.'[30] The only other evidence which bears on possible female involvement in the Tenant League is an episode on 9 October 1865. It resulted in a deputy sheriff having to face a charge of assault on Elizabeth Devine, a female complainant, which was later 'settled' in circumstances and for reasons that are not clear; although the information is fragmentary, it does bear upon one woman alone, and there is no reason to believe that it pertains to collective action by women.

Indeed, even the limited amount of testimony concerning participation of women in resistance to the sheriff in southern Kings is fraught with uncertainty and ambiguity. The fact that the author did not identify himself is not necessarily an indication of unreliability, since a tenant on the Sulivan estate, Lot 61, might have strong reasons not to risk annoying the land agent. But there is no independent confirmation of the details, and it is possible that, following precedents in Scotland and eastern New York, the 'women' in the January 1864 incident were men *dressed as women* in order to make the retreat of the sheriff all the more humiliating. Even if the demonstrators were in fact women, the incident stands out for its isolation.[31]

The reference to the presence of children is also noteworthy, and again ambiguous in its thrust. Several incidents during the Tenant League agitation generated reports of participation by boys; the January 1864 case is the only one involving children in which their sex is not specified as male.[32] The obvious intent of one narrator, assemblyman Francis Kelly, who had not been present personally on the occasion of which he spoke, was to mock the supposed faint-heartedness of a deputy sheriff who had reported being deterred from executing writs: 'He ... tells a yarn about what a few boys who were coming from school said to him, ... and about [a] man and two little boys who looked "ferocious" at him'; Kelly also implied that the deputy had not been sober, and that his own indiscreet statements had undermined his mission.[33] In the letter bearing on the January 1864 incident, the coupling of 'women' and 'children' – in juxtaposition to the adult male status of the sheriff – may indicate a similar desire to deride the authorities.

The association with the Orange organization is suggestive because in the late 1850s and early 1860s, despite the rarity of Irish Protestant immigrants in the colony, it had emerged as a militant Protestant social movement, closely allied to the Conservative party. In 1863 there had been an exceptionally acrimonious assembly debate on a bill providing for incorporation of the Grand Orange Lodge of the Island, followed by a mass petition for disallowance of the legislation, to which the British government responded favourably. A connection of an embryonic tenant movement with Orangeism would be a warning light to Roman Catholics at any time, but especially in the inflamed atmosphere of 1864.[34] Like Whelan in his initial comments on the Lot 29 incident, the anonymous correspondent took pains to emphasize that the tenants actually involved in resisting the claims of the proprietors in southern Kings were Protestants, not Irish Roman Catholics; he underlined the point with his reference to 'that Orange locality.'

Three days after the report of the disturbances in southern Kings appeared in the *Examiner*, Ross, who was reputed to be an Orangeman, published in *Ross's Weekly*, another newspaper of his, a report of a large meeting which Lot 63 inhabitants had held 'lately.' It passed resolutions demanding for local tenants the right to buy their farms on terms similar to those accorded tenants on the nearby Selkirk estate in September 1860. The sale of that property had been the second major transaction under the Land Purchase Act of 1853, and most of the former Selkirk tenants lived on townships in southern Queens immediately to the west of the lots Cunard and Sulivan owned in southern Kings.[35]

The persons present at the Lot 63 meeting promised to resist distraint proceedings, and expressed determination to prevent occupation of any farms seized for rent. The rationale of the last measure was to ensure that a landlord who evicted a tenant or squatter would be unable to

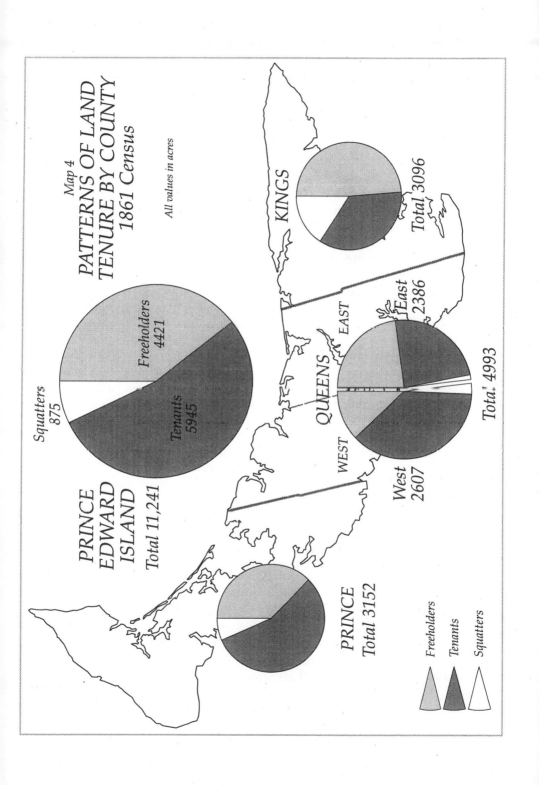

Map 4
PATTERNS OF LAND
TENURE BY COUNTY
1861 Census

All values in acres

KINGS
Total 3096

PRINCE
EDWARD
ISLAND
Total 11,241

Squatters
875

Freeholders
4421

Tenants
5945

EAST

QUEENS

East
2386

WEST

West
2607

Total 4993

PRINCE
Total 3152

Freeholders

Tenants

Squatters

profit from the action by installing a new tenant. In the meantime, the tenants at the meeting demanded a reduction in rent from 1s. 6d. sterling to 6d. currency and full remission of arrears. According to *Ross's Weekly*, at least seven hundred persons were believed to have pledged 'to resist the action of Sheriff, Bailiff, or Constable.' But the aspect of the proceedings Whelan found most remarkable was the declaration appended to the resolutions that 'the majority' of those present were supporters of the Tory government.[36] Making no attempt to disguise his amusement, he republished the report from *Ross's Weekly* under the heading 'Government Supporters in Arms Against the Government.' He also stated his belief that Orangemen were deeply involved, and although an outspoken opponent of Orangeism, he conceded that the people of southern Kings were 'plucky and resolute.'[37]

A few weeks later, under the heading 'Anti-Rent Association,' a letter from 'James' appeared in the *Protestant and Evangelical Witness*, reporting that the seven hundred who had taken the pledge had planned to march on 19 February into Georgetown, the county seat for Kings, on the occasion of a sheriff's sale of two farms. The demonstrators were to march with banners and pipers, 'not to molest the Sheriff or any body else; ... But to show their numbers, which would be sufficient, it was supposed, to deter any persons from bidding on their farms.' Had the demonstration occurred, it might have had some impact, for the population of Georgetown was probably less than one thousand. The undoing of this plan was 'bad roads,' which prevented mass attendance at the sale. The two farms, both on Lot 63, were sold to satisfy the demands of Cunard for £46 2s. 9d. and £29 7s. 8d., respectively.[38]

The news from southern Kings and the publication in late February of the correspondence associated with the Pope-Palmer delegation apparently had a catalytic effect on much of the Island tenantry.[39] The reverberations were particularly dramatic in Queens, the most populous of the three counties, which had also the highest proportion of tenancy; in fact, at the census of 1861 more than one-half of Prince Edward Island tenants lived in Queens. On 7 March, tenants and freeholders on Lot 34, Queens County, met at the Saw Mill Bridge, Covehead, and declared that it was time for tenants themselves to take over the task of negotiating with the proprietors. They proceeded to establish a committee which, entirely on its own authority, set a maximum price for the best land on the township: ten shillings per acre, with six years to pay, all arrears being forgiven. Until their terms were met, 'the tenantry will consider themselves justifiable [sic] in resisting to the utmost of their power, legally, all coercive measures that may be employed by the Pro-

TABLE 2.1
Tenurial status among land occupiers on the basis of counties in 1861, expressed in percentages and absolute numbers

	Freeholders		Tenants		Squatters		Total number of occupiers
	%	No.	%	No.	%	No.	
Prince	37.2	1,174	55.8	1,760	6.9	218	3,152
Queens	34.7	1,731	61.9	3,089	3.5	173	4,993
Kings	49.0	1,516	35.4	1,096	15.6	484	3,096
PEI as a whole	39.3	4,421	52.9	5,945	7.8	875	11,241

Note: Calculations based on totals derived from PEI, Assembly, *Journal*, 1862, app. A

prietor for the collection of rents.'[40] The proprietors for Lot 34 were the descendants of Sir James Montgomery, and in the same number of the *Islander* that reported the resolutions, an advertisement announced the determination of James Montgomery of Warwickshire to have all rents and arrears paid.

A week later, the tenants on the western half of Lot 29 met in the Crapaud East school room. In the course of a meeting that lasted four and a half hours they resolved to withhold rent until the proprietor, Viscount Melville's first cousin, Lady Georgiana Fane of 5 Upper Brook Street, London, consented to conversion of their tenures to freehold 'on just and moderate terms.' The matter of obtaining freehold must have been a particular source of frustration for tenants of both Melville and Fane on Lot 29, for the census of 1861 revealed that 98.7 per cent of occupiers of land were tenants. This was the highest per centage for any township on the Island, and a proportion so high as to indicate a policy of rigid resistance to sale on the part of the landlords. Moreover, 14.7 per cent of occupiers held their land only by verbal agreement. Like the tenants of southern Kings, Fane's tenants appeared to be supporters of the Tory government. Despite their dissatisfaction with the results of its initiatives on the land question, they went out of their way to commend its efforts.[41]

Eastern Queens County

Most striking of all was the activity in eastern Queens County. Tenants on Lots 35, 36, and 37 held a series of three meetings in taverns and hotels in Fort Augustus, Glenfinnan, and Ten Mile House. The purpose of the first, on 29 February, was, according to the report of the secretary, 'to discuss the propriety of forming a Tenant League.' Those attending decided that before proceeding further they would summon their local assemblymen, Opposition Leader George Coles and Kelly, also a Liberal, to explain their proposals for settling the land question. At the second meeting, on 10 March, with Coles and Kelly present, some three hundred tenants demanded the right to purchase their lands on the terms accorded the Worrell and Selkirk tenants. They explicitly rejected the terms proposed by the government delegates and by the proprietors led by Cunard. Finally, a committee was to be formed 'to correspond with the other tenants in Lots 48, 63, 64 &c., in carrying out the proposals which we have made to the proprietors.' The third meeting, held eleven days later at Ten Mile House, passed a set of identical resolutions.[42] Both

Coles and Kelly opposed resistance to rent collection, and at the Glenfinnan meeting, Coles, at least, 'strongly advised' against such action.[43]

The overwhelming majority of tenants in the area were Roman Catholics of Irish or Scottish extraction, with a history of militant action
against the local representatives of the leasehold system. It had been on
those townships, and also on the adjacent Lots 48 and 49, that an abortive Tenant League, with the declared intention of affiliating with the
'Irish Tenant League in Dublin,' had emerged in 1850-1. The landlords
on Lots 35 and 36 were the McDonalds of Tracadie, notorious for their
short leases, high rents, and ruthlessness in the use of distraint and eviction; freehold was highly uncommon on the two townships, accounting
for only 3.6 per cent of occupiers of land listed in the 1861 census.[44] So
unpopular was the McDonald family that one landholding member, a
priest, had been virtually forced in the 1830s and 1840s to leave first the
Tracadie area and then the Island. His elder brother Donald, a legislative councillor, was actually fired upon and wounded in 1851. With the
shooting incident, the ambushers had come close to crossing a line that
resisters of authority in the Highlands of Scotland observed, for Richards has noted that 'assassination was not employed in any Highland
protest'; in contrast, historian W.E. Vaughan has written that in Ireland
between 1857 and 1878 nine landlords were killed 'in connection with
disputes about the tenure of land.'[45] Donald's son, John Archibald, was
cast in the same mould as the rest of his family, with the result that tensions on the Tracadie estate were consistently high.

But the participation of the tenants on Lots 35, 36, and 37 was particularly significant for another reason. The Irish and Scottish Roman Catholics were unquestionably Liberal supporters, who had no hesitation in
condemning the Pope-Palmer delegation. Their organized presence in
the tenant movement, which had started in Protestant and Conservative
areas, meant that it was certain to transcend religious, ethnic, and partisan lines. Indeed, little more than a year after the most religiously divisive election in Island history, these tenants were forming a committee
to correspond and cooperate with the staunch Protestants of southern
Kings, who described themselves as government supporters. This fact
suggested that early in 1864 the emerging movement was already working consciously to overcome the quasi-tribal divisions that had governed Island politics since the late 1850s.

The Glenfinnan and Ten Mile House resolutions about corresponding
with other tenants specified Lot 48, which touched parts of Lots 35, 36,
and 37. On 7 March the tenants of that township had held a meeting in

the Mount Albion schoolhouse. The newspaper report of the meeting mentioned no one by name, an omission possibly related to the tenor of the proceedings. The preamble to the reported resolutions characterized the Island's leasehold tenure as a 'slave-holding system.' The purpose of the meeting was described as 'ominous,' and in making the standard demand for a right to purchase on terms like those accorded persons on the Selkirk estate, the Lot 48 tenants used some highly unusual language. They stated that 'the Tenantry of this Island, heretofore loyal, sympathetic and patriotic, have ultimately been driven by Proprietary oppression into extreme measures, if not desperation.' This declared volatility suggested that self-conscious radicals were playing a directing role, and contrasted dramatically with the wording of the report from the Glenfinnan meeting on Lot 35 a few days later, which cited the resolutions passed as evidence of 'the good sense and moderation' of the tenants involved. The Lot 48 meeting also proclaimed solidarity with the tenants of southern Kings, and the intention to resist collection of rent and arrears 'until a compromise be effected upon equitable principles.'[46]

Nine days later, on 16 March, tenants and freeholders on Lots 49 and 50, also in eastern Queens, met at the Mount Mellick schoolhouse. The published report, prepared by the secretary, Alexander McNeill, the long-time teacher in the district, was remarkably similar in rhetoric to the Mount Albion report. The mover of the first resolution was George F. Adams of Vernon River, a tavern keeper and postman. It denounced 'the present dissatisfactory slave-holding, rent-paying system,' declared solidarity with the tenants on Lots 35, 36, 37, 48, and the three townships in southern Kings, and advocated action 'in conformity therewith as a combined and united Tenant organization.' The other major resolution of the Mount Mellick meeting predicted 'ruin and misery, if not ... anarchy and confusion' should the Cunard proposals be adopted. The meeting closed with three cheers for the Queen.[47]

Thus, by 16 March, the very day a new session of the legislature opened, a militant and broadly based tenant organization was in the process of formation. The men who were to be its best-known leaders, Adams and McNeill, were already emerging. Island tenants were beginning to move from spontaneous resistance in the face of distraint proceedings to a self-conscious and calculated determination to refuse to pay any more rent. Some comments Kelly made in the House of Assembly indicated that this resolve was becoming firmer. He reported a conversation with tenant delegates from Lots 48, 49, and 50 in which he told them that, by such combinations they might go a little too far, and get some of their friends locked up in some of the jails of the Island. In reply, they assured him that there were already many hundreds in their League, and should any of them be so unfortunate as to get into jail, for anything they did in pursuance of the principles of their combinations, they would assemble in force and rescue the victims in what jail or jails soever they might be confined.[48]

The tenants on those townships were predominantly Protestant, and Kelly, an Irish-born Roman Catholic, stated that several with whom he had spoken were known to be government supporters. Hence the embryonic tenant movement was clearly cutting across the sectarian lines which had defined the support for political parties in the previous three general elections. Indeed, perhaps its most distinctive feature on the Island political landscape was its cultural and religious heterogeneity.

Between late March and 19 May 1864, when 'The Tenant Union of Prince Edward Island' was formally founded in Charlottetown, numerous tenant meetings were held throughout the rural areas of the colony. Although it is impossible to be precise about the total number, on the basis of newspaper reports it is certain that between 29 March and 16 May there were no fewer than thirteen of these meetings. Their geographical locations extended from Campbellton, Lot 4, twenty-nine miles from the north-west tip of the Island, to Rollo Bay Cross Roads, twenty miles from the eastern end of the colony. The reported districts and names indicate that the meetings and the incipient movement were drawing upon a fairly representative cross-section of the Island tenantry, including even the Acadian population.

Despite the broad representation, the overall pattern indicated a predominance of activity in Queens County, where seven of the thirteen meetings occurred. It is probable that because the meetings in Queens were held relatively close to Charlottetown, then the seat of all Island newspapers, they had the best chance of being reported in the press. But the concentration of organization in Queens becomes still greater when one takes into account the fact that each of the seven reported meetings in the previous month (29 February–28 March), from Crapaud in the west to Fort Augustus in the east, was also held in Queens. Thus, if one examines the entire period after the reports of the disturbances in southern Kings County and publication of the Pope-Palmer delegation correspondence, from 29 February to mid-May, fourteen of the twenty reported meetings were in Queens. Ten were in its eastern half.

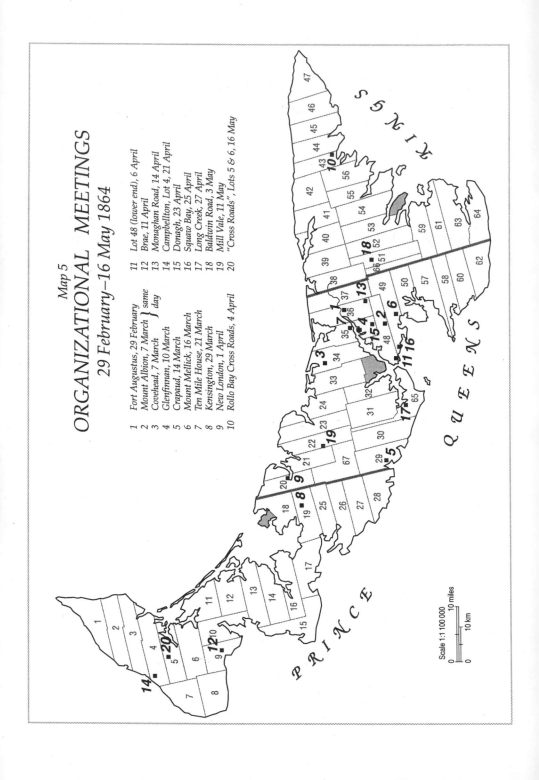

Map 5

ORGANIZATIONAL MEETINGS

29 February–16 May 1864

1   Fort Augustus, 29 February
2   Mount Albion, 7 March } same
3   Covehead, 7 March   day
4   Glenfinnan, 10 March
5   Crapaud, 14 March
6   Mount Mellick, 16 March
7   Ten Mile House, 21 March
8   Kensington, 29 March
9   New London, 1 April
10 Rollo Bay Cross Roads, 4 April

11 Lot 48 (lower end), 6 April
12 Brae, 11 April
13 Monaghan Road, 14 April
14 Campbellton, Lot 4, 21 April
15 Donagh, 23 April
16 Squaw Bay, 25 April
17 Long Creek, 27 April
18 Baldwin Road, 3 May
19 Mill Vale, 11 May
20 "Cross Roads", Lots 5 & 6, 16 May

Scale 1:1 100 000

0   10 miles

0   10 km

As already mentioned, Queens had the highest proportion of tenancy among the three counties. In Queens as a whole, at the time of the 1861 census the percentage of tenancy was 61.9, but for analytical purposes the county may be divided into western and eastern halves.[49] 'Western Queens' contained eleven townships and, at the 1861 census, 2,607 occupiers of land, while 'eastern Queens' had twelve townships and 2,386 occupiers. There were significant differences in patterns of land tenure between the two halves. For example, the proportion of freeholders in the western half was less than half that of the eastern. The western half had 72.2 per cent tenancy, and, in absolute numbers, more tenants than either Prince County or Kings County as a whole. Indeed, 31.7 per cent of all Island tenants lived on those eleven townships of western Queens. The pattern of leasehold predominance was remarkably uniform: in only one township, Lot 32, the closest – in fact, adjacent – to Charlottetown, did freeholders constitute a bare majority, 50.3 per cent.[50] Coincidentally or otherwise, Lot 32 was known for the poor quality of its land. In the 1861 census the enumerator for the township declared that 'Upon the whole, ... Lot 32 is very poor, and may be classed second or third quality.'[51] It was also the township in western Queens with the smallest number of land occupiers.

In contrast to the dominance of one form of land tenure in western Queens, the twelve townships of the eastern half of the county, with an overall tenancy percentage of 50.6, can be divided into three distinct areas. On the two northern townships constituting the Tracadie estate, the McDonald family held sway and resisted conversion to freehold tenure. But the purchase of the Selkirk estate in 1860 had reduced the incidence of leasehold on the southern four to 10.9 per cent. On each of the remaining six, freeholders were present in substantial numbers, although in each case a minority, ranging from 14.9 per cent on Lot 48 to 46.9 per cent on Lot 37.[52] There were a multitude of different estate owners on those six – Lots 33, 34, 37, 48, 49, and 50 – and indeed, according to testimony by Peter Robertson of Lot 48 before the land commission of 1860, 'there is scarcely a Township on the Island which has so many owners as Lot 48.'[53] Some of these can be characterized as minor proprietors, holding modest fractions of townships, and it was commonly held that small proprietors in Queens County were among the most obstinate for aspiring freeholders to deal with.[54] Historian Donald E. Jordan Jr has noted a somewhat analogous pattern in County Mayo, Ireland in the middle decades of the nineteenth century: 'owners of smaller estates charged higher rents. They operated on a much narrower margin of

TABLE 2.2
Tenurial status among land occupiers in eastern Queens and western Queens in 1861, expressed in percentages and absolute numbers

	Freeholders		Tenants		Squatters		Total number of occupiers
	%	No.	%	No.	%	No.	
Eastern Queens	47.2	1,126	50.6	1,207	2.2	53	2,386
Western Queens	23.2	605	72.2	1,882	4.6	120	2,607

Note: Calculations based on PEI, Assembly, *Journal*, 1862, app. A

profit than did the larger owners and could ill-afford generosity towards their tenants without seriously affecting their standard of living and putting their landownership in jeopardy.'[55] In the Prince Edward Island case, it is possible that for some small proprietors their Island properties represented a substantial proportion of their personal wealth – much more so than for Selkirk or Cunard, for example; and, like Jordan's small-scale Irish landlords, they may have believed that they were not in a position to make concessions.

The common feature of all the reported meetings after 29 February was an apparent determination of tenants to obtain the right to convert their land tenures from leasehold to freehold on terms they considered practical. A small minority of meetings passed motions referring to the proprietors as 'land claimants,' or to their 'claims.' Such phrasing implicitly raised the spectre of the Escheat movement, which had argued that landlords had forfeited title to their properties owing to non-fulfilment of the terms of the original grants in 1767, and therefore deserved no compensation. Yet, other resolutions that the same meetings passed expressed willingness to pay for freehold title – although the reluctance to accept existing titles as valid could not have been reassuring to the landlords whose properties were involved. Some resolutions explicitly stated a determination to resist the payment of rent and back-rent. One meeting, in Kings County, declared that 'every embarrassment should be caused to those Proprietors, their agents or bailiffs, who might have the hardihood to attempt seizing or selling the effects of any individual tenant in these localities.'[56] Almost all the meetings – seventeen of the twenty – expressed the desire that a tenant league, or union, be founded, or at least to establish linkages with tenants in other areas.

On 22 April an advertisement entitled 'Delegation Meeting' appeared in the Islander newspaper over the names of thirteen 'Delegates,' including McNeill as 'secretary.' All appear to have been from eastern Queens County, and indeed in the text of the advertisement they mentioned Lots 35, 36, 48, 49, and 50 specifically. Using language reminiscent of the resolutions passed at the Mount Albion and Mount Mellick meetings, the sponsors called upon all tenant organizations in the colony 'who ... intend withholding the further liquidation of rent or arrears of Rent' to appoint 'three discreet Delegates' per township to attend a meeting in Charlottetown on 19 May. The purpose of the meeting was described as 'the formation of one concentrated Federal Tenant Organization, the basis being unflinching fidelity, loyalty, union, sympathy and action.'[57]

The advertisement attracted some hostile comment, particularly from the editor of the *Islander*, William H. Pope, who was a central figure in Island public life. As well as an assemblyman and the colonial secretary in the Island government, he was a lawyer. In an editorial published one week after the appearance of the notice, he focused on the announced intention of some tenant meetings to resist the collection of rent 'legally.' He declared that 'There can be no *"legal"* resistance to the payment of rent, on the part of the tenants who hold lands under any proprietor.' After pointing out, step by step, the legal implications, he issued a warning to the tenants involved. 'Let no man be deceived in this matter. The law will be enforced even should it require the presence of a company of troops in every county in the Island.'[58]

Pope's intent was to deter tenant militancy, and no doubt he was especially concerned to have an impact in districts which had supported the Conservatives politically. It was their coalescence with traditionally Liberal tenants which was most ominous for the government of which he was chief strategist. Pope himself had been elected in eastern Queens in the previous year, and Lots 49 and 50, where the leadership resided and where the rhetoric was exceptionally militant, formed part of his constituency. As a representative in the assembly for the southern six of the twelve townships in eastern Queens, he might reasonably have expected that his words of caution would receive attention there. But matters had proceeded much too far for editorial advice of this nature to have significant deterrent effect.

3

A New Tenant Organization

We, the Tenantry ... pledge ... to withhold the further liquidation of rent and arrears of rent; and ... to resist the distraint, coercion, ejection, seizure, and sale for rent and arrears of rent.
From 'The Tenant's Pledge,' adopted at the founding convention of the Tenant Union of Prince Edward Island, 19 May 1864[1]

The Founding Convention

On 19 May 1864 in Charlottetown's largest hotel, the North American, seventy to eighty delegates and friends of the emerging tenant movement met.[2] Although they excluded the press, politicians, and general public, an authorized report of the proceedings appeared one week later in *Ross's Weekly*.[3] Subsequent correspondence in the newspapers provided further information. Those attending were of diverse political backgrounds, and there was at least some difference of opinion on basic questions, although it is clear that there was majority support for the decisions taken.[4] The meeting decided to petition the Queen against the Fifteen Years Purchase Bill, which had been passed by the legislature on 2 May, and to organize 'The Tenant Union of Prince Edward Island.' The union was to operate 'under the direction and guidance of a Central Committee,' which would wield considerable authority. Each township was to have a local committee, which would decide upon 'a fair and reasonable price' to be offered to the proprietors of their lands. Any difficulties in this process – for example, over the price to be offered or the rejection by a proprietor of a fair price – were to be referred to the Central Committee.

The convention made it clear that the new organization was not advocating expropriation without compensation by resolving 'That any tenant who shall refuse to make a fair offer ... shall forfeit the sympathy and all the advantages of this Union.' But delegates also left no doubt that they were serious about their determination to bring the leasehold system to an end, by publishing a 'Tenant's Pledge' to be taken by members. They were 'to withhold the further liquidation of rent and arrears of rent; and ... to resist the distraint, coercion, ejection, seizure, and sale for rent and arrears of rent.' Each who took the pledge was also to 'bear a proportionate share of expenses in connection with this organization.'[5] The membership fee was five shillings.[6]

It is a moot question why the organization chose the name 'Tenant Union' when many people in the colony, friends, foes, and uncommitted, already used the term 'Tenant League' in referring to the movement. The answer may lie in Irish history: by the end of the 1850s it had been clear that the Irish Tenant League, founded at mid-century, had failed badly, despite making an impressive start when more than forty Irish members of the House of Commons elected in 1852 professed support for its principles. In the words of Irish historian F.S.L. Lyons, it 'did not achieve a single item in its programme.' Its objectives had been framed in terms of tenurial reform, but its entire strategy had been predicated upon rejection of any notion of a national rent strike, and upon reliance on an extension of the franchise which had occurred in 1850. Indeed, the failure of the Irish league despite the election of its candidates would make it more difficult for other movements to argue in favour of parliamentary methods.[7] According to T.W. Moody, 'The collapse of the league brought deep despondency and deep distrust of constitutional methods.'[8]

The founders of the Tenant Union of Prince Edward Island were probably familiar with the history of the Irish Tenant League. They certainly knew that involvement in parliamentary politics was not what they wanted, and indeed that they did not even want parliamentary politicians to participate in their organization. Since they desired to distance themselves from the parliamentary approach, it would not be surprising if they deliberately avoided a name which connoted it. But habit triumphed: Islanders continued to refer to the new tenant organization as the Tenant League.

The convention elected a ten-member Central Committee which was to meet in Charlottetown on the first Tuesday in June. All ten were from Queens County; eight had been sponsors of the newspaper advertisement calling for the convention and the other two, Alexander Robertson and David Lawson, were from Lots 30 and 34, respectively. All but Robertson were from eastern Queens. It is clear from the surnames that a majority on the Central Committee were Protestants of English or Scottish extraction. There is no conclusive evidence as to whether the make-up of the Central Committee reflected accurately the general body present in terms of national, social, or religious origins. A petition against the Fifteen Years Purchase Bill, signed by fifty-two at the meeting, survives.[9] Yet those fifty-two do not include all the persons known to have been present. For example, Archibald McNeill, a Charlottetown journalist, attended (not in a professional capacity), and although not a delegate, spoke; he favoured the Fifteen Years Purchase Bill because of its provision for remission of arrears, and therefore opposed the petition, which he did not sign. He also objected to the talk of resisting rent collection.[10] The signers described themselves as 'Delegates chosen at various public meetings throughout the Island,' and hence the fifty-two may represent only those selected at district meetings, as opposed to friends of the movement, like McNeill, introduced by delegates – and only a portion of those selected, if some disagreed with the petition.

There were delegates from as far west as Brae, Lot 9, and as far east as Rollo Bay Cross Roads. But thirty-two of the forty identifiable through surviving newspaper reports of public meetings and other sources were from Queens County, and twenty-three from its eastern half. The three members of the Central Committee who appear to have been most prominent at the convention were all from Lot 49, a relatively prosperous area in eastern Queens, and one of their striking characteristics was occupational pluralism. None was primarily a tenant farmer. The general secretary of the Central Committee, Alexander McNeill, was a veteran schoolmaster. According to a genealogist, who is also his grandson, writing in 1982, he doubled as an auctioneer: 'I still have his auctioneer's mallet.' The son of a Scottish immigrant, McNeill had been born in 1821 at North River, Lot 32, a few miles to the west of Charlottetown. Before the land commission of 1860, he stated that he rented about fifty-two acres near his schoolhouse from proprietor Robert P. Haythorne, and that a few years earlier he had paid £150 for a 999-year lease to the land. McNeill wrote the petitions for the Tenant League, and would be presented with a gold ring engraved with its initials.[11] George Frederick Adams, the 'financial secretary,' as he later described himself, was primarily a tavern or inn keeper; but he was also a postman, and in addition, by his own account he had purchased leasehold rights from proprietors more than once.[12] The chairman of the convention, James Bulpit Gay of Pownal, owned a general store and had at least one ship engaged in the export trade.

The occupational identities of these key leaders of the league make it plain that the common basis of membership was not occupational, but programmatic. The founders took the view that a commitment to assist in ending the leasehold system through support of direct action was more important than any requirement that members be tenants. There does not seem to have been any sentiment to the effect that only tenants could lead or represent tenants. Furthermore, the information that survives about those who signed the petition is consistent with Lieutenant Governor George Dundas's comment that they 'are generally well to do': Robertson, for example, owned a store in Bonshaw, Lot 30, and in 1856 he had purchased a large farm for £1,278.[13] The question of the motivation behind the involvement of such persons, sometimes the possessors of significant property, in a controversial tenant movement is an obvious one. The governor, in elaborating on his theme that 'they have by no means the plea of poverty to shelter them,' drew the inference that 'It is downright lack of principle, therefore, that actuates these men.'[14] One could with equal reason conclude precisely the opposite: that the apparent absence of pressing personal financial need as a motive suggests instead a principled rejection of the leasehold system.

Yet the analysis should be taken at least one step further. Not only in Prince Edward Island, but in agricultural areas everywhere, the rural auctioneer, the country merchant, and the tavern keeper surrounded by farmers all had a common interest in the development of a farming population with ample purchasing power. In the Island colony, tenants and their political allies and supporters often argued that the leasehold system was draining the countryside of the money generated there, money which could otherwise be used to make purchases at auctions, country stores, or taverns. Such reasoning had made a reformer out of George Coles as a young brewer and distiller: an Island population with money in their pockets would mean effective demand for local products, and would stimulate domestic manufacture. Rightly or wrongly, the landlord and his demands were perceived as retarding the growth of business on the Island, and that assumption must have contributed to a certain lack of legitimacy which the landlords' demands faced among Islanders. It was commonly remarked that among a tenant's creditors, the one he was least likely to pay was the proprietor. Conversely, it would be easy for interested parties to believe that anything – such as a tenant movement – which further undermined compliance with the demands of the rent collector would increase the probability that other creditors would be paid.[15]

The most visible single member of the Tenant League to emerge from the 19 May 1864 meeting was Adams. His address on that occasion was ordered printed in a sympathetic newspaper, *Ross's Weekly*, 'and also in a separate form for general circulation.' It and several other addresses appearing in the press over his name make it possible to examine his position in some detail. When speaking at the North American Hotel on 19 May, he emphasized the active role played by freeholders in the early stages of the movement, and the necessity of unity between them and the tenantry; many freeholders were, of course, former tenants. Adams explicitly rejected the argument that the movement should wait for the next election (which would normally be held in 1867) and then make its weight felt at the polls. Rather, he argued that the issue could be settled before then through united action. He stated that the movement had no desire for political power, but aimed solely to resolve the land question. Another, more negative side of his message surfaced when he warned tenants who chose to stand apart from the struggle that 'their conduct will not be forgotten.' In arguing for 'the necessity of a united people' he cited examples from British history in which reforms were won not by the populace waiting patiently for the gift of legislators, but 'by the *people themselves* demanding them, and in many instances, shedding their *blood for them*, and that *too without a murmur*.' He appealed to all sympa thizers of the movement to lay aside political and religious differences 'until this all importance grievance is redressed.'[16]

Little information has survived about Adams's personal background. He has become an example of the maxim he cited when speaking on 19 May about the fate history reserves for leaders of popular causes which are not immediately victorious: 'then [they] are branded as mad men or fools.'[17] He has faded into obscurity, and almost all the surviving information on his life and actions has come from unflattering assessments by bitter political enemies. His date and place of birth are now unknown to historians and genealogists of Prince Edward Island. Contemporaries referred to him as an Englishman, and his father was a native of Gloucestershire; obituary notices indicate that they and George's mother arrived in 1850.[18] According to Dundas, Adams had come to the colony under an alias, although it is not clear from the context whether Dundas believed 'Adams' to be his original or his assumed name.[19] By the late 1850s his Half Way House in Vernon River was already well known and popular for providing meals and overnight accommodations on the road between Charlottetown and Georgetown. The diaries of surveyor Henry Jones Cundall for 1856–8 record numerous stops at Adams's establishment, which he described as 'a comfortable house & well kept for a Country inn.'[20]

In 1861 Adams had married Martha Ramsay of Rose Hill, Prince County, and in doing so he had united with a family which had been involved in lawsuits with the Sulivan estate over several decades. During the 1860s there were a number of legal actions between the two sides.[21] Those pertaining to Martha's brothers Hugh and John, in particular, received much publicity, especially because their defiance led to their temporary incarceration, and because Hugh published many letters criticizing the land agent, George W. DeBlois, in local newspapers. The tone of the correspondence, as well as its duration, suggests relentless contentiousness, although it is noteworthy that, according to testimony by John before land commissioners appointed to implement the provisions of the compulsory Land Purchase Act of 1875, he and Hugh claimed to hold their land 'adversely' (that is, contrary to the claims of another party), rather than as tenants, and that such a situation could easily breed quarrels and an appearance of belligerence.[22] Yet the appearance may have been an accurate reflection of reality: many years later Adams himself would be involved in litigation with Hugh and John, apparently over custody of his infant daughter.[23] In 1864 the Tenant League treated the Ramsays' case as a *cause célèbre*, and its Central Committee published a resolution supporting them at a time when they were in jail. It was preceded by a preamble declaring that they 'have been imprisoned in their noble defence of what appears to us to be the cause of the tenantry and the country generally.'[24] Not everyone agreed: in February, Edward Whelan had refused to publish in his newspaper a letter by Hugh concerning the dispute with DeBlois on the grounds that the points at issue were really a private matter.[25]

Adams's published addresses indicate that he had a reasonably good education for the time. He was definitely more coherent than McNeill, the general secretary and schoolmaster, who had attended the Central Academy in Charlottetown, in expressing his thoughts on paper. Again according to Dundas, Adams had been 'a wild Chartist' before emigrating to the Island.[26] Chartism is consistent with his views, which can be characterized as a radical variant of populism, with a definite willingness to contemplate militant direct action. His populism is apparent from the lack of class exclusivity in the movement for which he was principal spokesman: 'the people' would resolve the land question through militant direct action, without the intervention of governments or political parties. Although in an address at Mount Mellick on 29 June 1864 he recognized the existence of distinct economic strata within the tenantry, he did so for the purpose of concretely analysing the impact of particular legislation on tenants, not in order to create a basis for elevating some or denigrating others.[27]

The openness of the movement is apparent from the broad spectrum of persons whose sympathy and financial support it solicited. But this inclusiveness also provided the moral basis for stigmatizing those who stood aloof as the worst sort of traitors. The willingness to contemplate militant direct action, even in defiance of the law, is readily apparent from the 'tenant's pledge.' Although Adams's self-conscious radicalism is more elusive, it emerges clearly in one of his published addresses, in which he appealed to Islanders to recognize the connection between the land question and other issues.

These districts [in southern Kings County] have thus taught the people of the Island a lesson which I trust they will never forget, that is, whenever a people have a grievance, never to be afraid nor hesitate to at once take action and look for and have their rights, and always depending on a majority of the people afterwards in supporting them; this is a fact worth knowing, and one that will be acted upon hereafter on many questions besides the one we now have at hand.[28]

The most penetrating critique of the proceedings at the North American Hotel on 19 May came from Whelan, a fact acknowledged explicitly or implicitly by all shades of political opinion, from Dundas to the three pro-government newspapers to a Tenant League spokesman, John Bassett. The three newspapers immediately reprinted Whelan's most critical remarks, and Bassett conceded in a letter dated one month after the convention that his opposition was a significant stumbling-block for the organization in its early attempts to mobilize support among tenants.[29] As the veteran ideologue of Island reform, who had flirted with outright support for Irish revolutionaries in 1848 and had been involved in founding an abortive tenant movement as recently as the winter of 1860–61, Whelan was well equipped to understand the implications of the policies adopted on 19 May. His remarks in the next issue of the Examiner indicated that he was increasingly nervous about the course the Tenant Leaguers were adopting. He expressed concern over the exclusion of MPPs from the convention, and the lack of experienced leaders.[30] Clearly, Whelan's position reflected apprehension on the part of the Liberal leadership about the possibility of a 'third party' arising to undercut its traditional base of support. But it was also grounded in principle and in genuine concern for the Island tenantry. On 30 May, when he reprinted from *Ross's Weekly* some of the convention proceedings, he stated that although he had originally planned to reserve detailed commentary for a later date, upon further consideration and as an indication of his sense of urgency, he had decided to explain his position at once.

Whelan's masterful article entitled 'The Tenant Union – Its Advantages and Abuses' went far beyond his earlier focus on the political embarrassment caused to the Tory government. He began by endorsing in principle the formation of a tenant union, as embodying the natural unity of interests among Island tenants of Scottish and Irish extraction, a point he had been making since arriving in the colony in 1843. But he insisted that the legislature was the proper focal point for the tenants' struggle, and that the thrust of the tenant organization should be to ensure that their representatives were committed to the abolition of leasehold tenure. He warned that open defiance of rent collectors would initiate a struggle in which the sides were unevenly matched. The 'wealthy landlord, ... in addition to his social position, has the law of the land on his side.' Whelan went on to criticize Adams, whose views he declared to be 'in some points, very extraordinary and indefensible,' and to outline his own position:

Mr. Adams tells us ... that he is determined to resist the law of the land. This is very silly, and demonstrates most conclusively the incompetency of the men to manage the movement which they have inaugurated. We feel it our duty to tell them, without delay, and without circumlocution, that no man has a right to resist the laws for the attainment of any public object – (and certainly in any case for no private one), until all the resources under the Constitution have been exhausted. We fully agree with the Tenants' Union that it is the right of an oppressed people to resist the enforcement of arbitrary or tyrannical laws, – but a few conditions or drawbacks are annexed to the exercise of this right: Under representative institutions such as ours, the laws are presumed to be the emanation of the public mind – if they are bad and oppressive, the power which made them can repeal them – if the people will not unite to repeal, change, or modify the laws, it is not likely they will unite to resist them. Resistance, then, on the part of a fragment of the community becomes futile, criminal, and ultimately disastrous to those engaged in it. ... It may be no difficult matter to bag a Bailiff or an Agent now and then but to attempt to bag a Company or a Regiment of armed soldiers, would be another affair. Gongs, and tin kettles, and a few discharges of musketry, may frighten a worthless Bailiff out of a settlement; and, perhaps, it would be of no great harm if the dirty fellow was favoured with a bath in the nearest duck pond; but it is not the Bailiff alone that is outraged in this case. It is the majesty of the law – it is the authority of one of the greatest empires in the world that is thus impotently and foolishly defied. And no one will doubt that Mr. George Adams and the Tenant Union would cut a very precious figure in putting themselves in a hostile attitude against the British Empire.[31]

This 30 May 1864 article represented a turning-point in Whelan's public position on the movement. Over the years since the late 1840s he had become increasingly moderate and appreciative of the virtues of the British constitution. Yet this liberal constitutionalism was always mixed with a dash of pragmatic calculation or, perhaps, a concern, rooted in his knowledge of Ireland's tragic history, about the power equation. As a man who had never entirely shed his early Irish radicalism, he shared with Adams a keen appreciation of the fickleness of history. In 1860 he had written that what 'was criminal on the part of the few poor devils who were too weak to carry their point, becomes praiseworthy and heroical with the many who are strong enough to bear down all opposition. History shows that popular ebullitions have always been regarded in this way.'[32] When he condemned the 'Tenant's Pledge' as 'very obnoxious to every properly constituted mind,' he coupled that admonition, which was consistent with his refusal over the years to counsel non-payment of rent, with the warning that the pledge 'can only serve to retard rather than to advance the object of the Tenants' Union, namely, the settlement of the Land Question.'[33] But whatever Whelan's motives, his opposition to the policies of the Tenant League was a major problem and irritant for the leaders of that movement.

Politics makes strange bedfellows in times of great upheaval. On 6 June, Dundas, who had felt the sting of Whelan's barbed pen many times since arriving in the colony in 1859, summed up Whelan's role within the Island press on the Tenant League issue as follows: 'the Islander, Monitor & Protestant highly approve of [his] tone. They endeavour to imitate the Examiner, but having neither the talent nor the force of character of Mr. Whelan, their efforts are feeble.'[34] Whelan's stature would make him a target of the league. By July Bassett was threatening that if the Liberals returned to office he would not receive the queen's printership, a major patronage appointment.[35]

In the same issue of Ross's Weekly that included Bassett's letter, an editorial announced that henceforth the paper would be advocating a policy remarkably similar to that of the Tenant League. Its practice since its foundation in 1859 had been to abstain from all political and religious partisanship; filled with light reading, bad puns, and weak jokes, it often had no editorial. This neutrality did not save it from criticism, for its very commitment to neutrality could provoke the wrath of local controversialists.[36] The sole substantive issue on which it declared a position was the continuance of leasehold tenure, which it opposed with increasing vehemence. Having favoured the award of the land commission, and having seen it thwarted, John Ross had concluded that the time had come for 'the people to take the matters into their own hands and declare that they shall become free men.' He advocated a policy of compulsory sale and, like the league, insisted that the tenants must pay for the land; this would not be the old Escheat program of confiscation which had led to a dead end in the previous generation.[37]

On 7 June there was full attendance when the Central Committee of the Tenant League met from 2 p.m. to 7 p.m. in Charlottetown. It urged local committees to become active, and to meet before it convened again in four weeks. One of the activities Adams suggested in an address published by order of the Central Committee was preparation of lists of enemies. This category included 'every person who bids on, or purchases, any article that is sold for rent, or who acts in any way against the interest of any member of this union.'[38]

Adams apparently envisaged that these individuals be given a version of the shunning treatment later afforded to Charles C. Boycott, a land agent in County Mayo, Ireland who had undertaken ejectment proceedings against some tenants. First his servants and farm labourers deserted his premises after a visitation by a large crowd on 23 September 1880, demanding that they leave and never return. Subsequently the neighbouring population denied Boycott such everyday rights and conveniences as purchasing food and other necessities at local shops, having the local blacksmith shoe his horses, and having a laundress do the washing of the household. Postal service ceased for him – although he continued to receive threatening letters – and any member of the remaining five-person household who ventured off the grounds was harassed. The nightmare ceased only when Boycott and his family left Ireland on 1 December, after having departed from their house under military escort four days earlier. The extent to which the total withdrawal of services and cooperation was voluntary or the result of unsubtle pressure was a matter of controversy, but the effectiveness of the campaign was so starkly evident as to add a new word, serviceable as both noun and verb, to the English language. By 1882 the British government would pass a statute attempting, among other things, to define the crime of boycotting.[39]

The movement against Boycott would be in reprisal for specific objectionable acts by the land agent which had come after the emergence of a pattern of abrasive relations with the local tenantry. On 7 June 1864 in Charlottetown, Adams was proposing to go beyond retaliation and to use the weapon of social and economic ostracism more aggressively. 'We shall,' he stated, 'find out the men who are willing, and who can afford to live without us; for whosoever is not with us we take to be against us.' He even announced plans to canvass the towns of the Island, in order to determine who was on which side there. Referring to the enemies of the movement, and particularly to some he had seen in the capital 'within the last few days, ... who appeared quite in glee because some unfortunate tenants' horses were seized and taken from the plough, and brought to be sold for rent,' he had a prescription which rather eerily foreshadowed the fate of Boycott in Ireland. 'I hope for your own interest that you will always act to those despicable characters as they deserve; shun them in every way as you would a pestilence, and you will then see nearly every man a member of the union; or if not, we would advise them to leave the country.'[40] Later in 1864 Ross's Weekly published a letter from 'Observer' addressed 'To the Tenantry.' The writer argued that, following the award of the land commission of 1860, it was illegal to collect or pay rent or arrears, and advocated that any and all land agents, without exception, 'be shunned and detested by all right thinking people.'[41]

Such a strategy requires the capacity to control the behaviour of large numbers of people, and participation may be voluntary or involuntary. This procedure of deliberate polarization has parallels in other social movements, such as the Irish Land League, founded in 1879, a local branch of which organized the movement against Boycott. Sociologist Samuel Clark has noted that 'It was the policy of the Land League to divide communities into pro- and anti-League factions and to create as much ill-feeling as possible between the two.'[42] Clearly the leadership of the Prince Edward Island Tenant League was ready in 1864 to contemplate such a strategy.

But Adams was able to close his address on a more positive note. He reported that 'there are some members of this organization who, I have every reason to believe, will shortly become freeholders through our efforts.'[43]

Purchase of the Haythorne Estate

It is impossible to determine the degree of local activity, Island-wide, at this stage in the history of the Tenant League. But there was certainly evidence of activism in the part of eastern Queens County where Adams lived. Tenants on both Lots 49 and 50 formed three-member committees to approach the respective proprietors. On each of the two townships a substantial minority, in fact almost an identical proportion, of occupiers of land had been freeholders at the last census: 45.3 and 45.2 per cent, respectively.[44] In the case of Lot 50 the two landowners were sisters, Maria Matilda Fanning and Lady Louisa Wood, daughters of Edmund Fanning, lieutenant governor from 1787 to 1805. They resided in Bath, England, and therefore a formal written overture was necessary. When tenants on Lot 50 drafted a letter to Fanning and Wood, they referred to Haythorne, a resident landlord who was one of at least three proprietors on Lot 49, as having made an agreement with his tenantry already.[45]

At the Central Committee meeting on 2 August 1864, Adams announced that Haythorne had come to terms with his tenantry.[46] Haythorne did not have a large estate, and the purchase, by itself, represented only a tiny fraction of what was required in order to bring the leasehold system to an end. According to Dundas some years later, 5,502 acres were involved in the transaction, and this information is probably accurate, since at the time Haythorne was a member of the Executive Council, and it is almost certain that the lieutenant governor obtained the figure directly from him.[47] Assuming Dundas's report of the acreage to be correct, and the farms to be not much smaller or larger in size than the norm of one hundred acres, it is reasonable to infer that only fifty to sixty tenants were affected. Yet Haythorne's willingness to sell directly to his tenants, without the intervention of the government, greatly enhanced the prestige and credibility of the league; several weeks earlier, it had made publicly known the formation of a committee to approach him.[48] Although the terms, 12s. 6d. per acre plus arrears, payable over five or six years, were the subject of some controversy, it is clear from data Haythorne furnished later that they were more favourable to the tenantry than the provisions of the Fifteen Years Purchase Bill. The land was well situated and of good quality; the tenants were paying unusually high annual rents, and if they had bought under the terms of the bill, according to Haythorne, 'I should [have] become several thousand pounds richer thereby.'[49]

The news of the negotiated settlement evidently embarrassed the

Conservative government, whose members and supporters tried several times to prove that its terms were less advantageous to the tenantry than their bill. Although some critics continued to minimize the role of the Tenant League in the negotiations,[50] there can be no doubt that its part was crucial and that the mode of settlement was according to its prescription. Coles would eventually state in the House of Assembly that the leaders of the league had actually drawn up the agreement.[51] Furthermore, it seems to have been implemented to the satisfaction of both the landlord and his tenants; in 1868, Dundas stated that 'Mr Haythorne reports that the arrangement has been fairly executed by the Tenants, and that there is but one actual defaulter.'[52] The Haythorne purchase amounted to a major propaganda victory for the league, a concrete demonstration that its methods could work.

Later in August the Tenant League was able to present the lieutenant governor with two petitions against the Fifteen Years Purchase Bill, signed by 262 fishermen and 2,550 tenants and supporters, respectively. These documents did not impress the Colonial Office. Thomas Frederick Elliot, deputy under-secretary of state for colonies, referred condescendingly to 'these worthy fishermen.'[53] In any event, the petitions were too late. The decision to confirm the legislation had already been made, and any appearance of delay was strictly for cosmetic purposes; the Colonial Office did not want to appear to have acted hastily. Indeed, Arthur Blackwood, a senior clerk, revealed in a minute he attached to a despatch received on 20 June that Sir Samuel Cunard had already been told that the bill would be confirmed.[54]

Progress of the Tenant League

The phrase 'revolution of rising expectations' has become a part of common parlance in the contemporary English-speaking world, and scholars have used the phenomenon as a factor in explaining the emergence of movements of social protest, including the Irish Land League. In brief, the hypothesis is that unrest does not always spring from continuing deterioration in a group's circumstances, but rather arises when a problematic situation no longer seems inevitable, and particularly when a trend towards improvement is reversed, and reversed sharply. Thus, what was once tolerable becomes intolerable when expectations of betterment are frustrated or dashed.[55] Prince Edward Island tenants as a whole were part of a 'New World' in which 'rising expectations' were the norm. Furthermore, the testimony before the land commission of

1860 makes it amply clear that as Islanders moved past the mid-century mark, and particularly as the neighbouring republic drifted towards civil war over a pre-modern form of agrarian productive relations which most North Americans found morally and politically repugnant, they were increasingly conscious that their small society was an anomaly in this New World where every man should be his own master and 'independent.' The awareness of being part of something which transcended the Island increased as the American Civil War broke out, raged, and drew to an end, and as the serfs of Tsarist Russia gained their freedom.

References to the leasehold system on Prince Edward Island as 'serfdom' and 'slaveholding' were of course grossly overdrawn, and even wrong-headed. I know of no case, recorded in any form, oral or written, of an Island tenant being bound to the land. The comparisons also overlooked the formidable political power of the Island tenantry through the broad electoral franchise and responsible government, and their ability to punish, and thus deter, politicians who might be inclined to use the punitive powers of the state vigorously against them; slaves and serfs had no institutionalized political power modelled on Westminster. There had been slavery on the Island, and slaves and former slaves would have readily recognized the difference between slavery and leasehold tenure.[56]

But within the contemporary context against which the tenants of the Island in the 1860s measured their own situation, they had reasons to feel genuinely aggrieved. The prospects of the individual tenant on the land he rented from a proprietor like Cunard, Lady Georgiana Fane, John Archibald McDonald, Viscount Melville, Robert Bruce Stewart, or Laurence Sulivan were constricted by the leasehold system, which did not represent the ideal of independence as understood in English-speaking North America. Lack of autonomy and security made the Island farmer stand out from his fellows elsewhere. During hearings before the land commission of 1860, Islanders had expressed fear of their young people voting with their feet to avoid the dead end of leasehold. In the middle 1860s Island tenants reached the end of their patience. Thus arose the Tenant League.

The tenants who set in motion a new and militant movement in Prince Edward Island in 1863–4 were part of a society where hopes had been raised by the land commission. That is what they shared with other Island tenants, but every social mobilization has its initiators. Who initiated the Tenant League? It is worth examining the situation of those who were first in protesting at the local level and in reaching out beyond their own districts to form a colony-wide organization.

What distinguished the tenants of Lot 29, southern Kings County, and the McDonald estate particularly was that there seemed to be little realistic hope available for them in the midst of a world – and even an Island – which was changing rapidly. On the Worrell and Selkirk estates, the local government had purchased tenants' lands for resale to them, but on neither property had the owner pressed the tenantry particularly hard. Charles Worrell had not collected much rent, and the Selkirk estate, which had been founded for non-pecuniary motives, had had a reputation for unusual willingness to sell to individuals.[57] From a tenant's perspective, the really difficult landlords, the efficient ones, and the determined ones, had not budged greatly. Sales under the terms of the Fifteen Years Purchase Bill would be the exception, not the rule. Those who harboured such sceptical thoughts about that statute were right, as a mere 2,911 acres, less than one per cent of the eligible land, would be converted under it by 1868. This total represented just 52.9 per cent of the acreage transferred at a single stroke in the Haythorne transaction.

It was no accident that the initial rumblings of revolt occurred where they did. Lot 29 in western Queens County had, as noted, the highest per centage of tenancy in the entire Island. Lots 35 and 36, the base of the McDonald estate, had the highest proportions of tenancy in Queens outside of Lot 29.[58] The townships of southern Kings, owned by large non-resident landlords who had first suggested the land commission and then repudiated it, bordered lands of the former Selkirk estate which had been offered to the Island government for 2s. 4d. sterling a few years earlier.[59]

There was another group of founders, those on Lots 48, 49, and 50 in eastern Queens, who provided much of the organizational vigour and leadership for the league. An example would be Samuel Lane of Lot 49, a former postman and a farmer. The lieutenant governor described him as a Haythorne tenant who was 'particularly well off,' and information on the value of his crops in the Wightman report, prepared for the land commission, supports this assessment.[60] Lane chaired the meeting at Glenfinnan which was the first to make an explicit commitment to reach out to tenants in other areas, especially in southern Kings; he was one of the thirteen whose names appeared in the newspaper advertisement calling for the founding convention of the league; and his was the first of the fifty-two signatures on the petition for disallowance of the Fifteen Years Purchase Bill which emanated from the convention. Other excep-

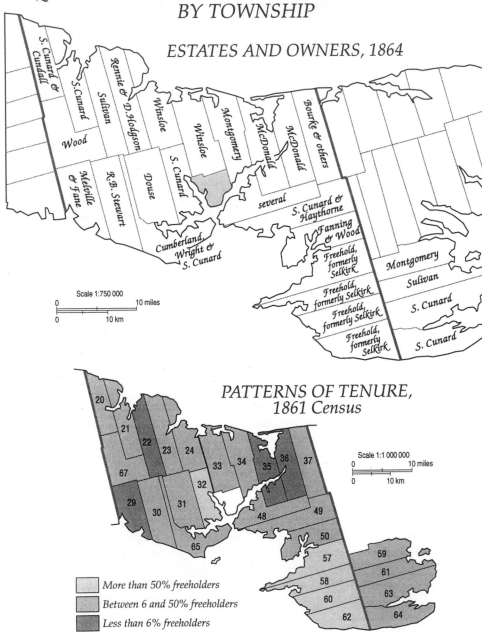

Map 6

QUEENS AND SOUTHERN KINGS COUNTIES
BY TOWNSHIP

ESTATES AND OWNERS, 1864

S. Cunard & Cundall

S. Cunard

Rennie & D. Hodgson

Sulivan

Winsloe

Wood

Winsloe

Montgomery

McDonald

McDonald

Bourke & others

Melville & Fane

R.B. Stewart

Douse

S. Cunard

several

S. Cunard & Haythorne

Cumberland, Wright & S. Cunard

Fanning & Wood

Freehold, formerly Selkirk

Montgomery

Freehold, formerly Selkirk

Sulivan

Freehold, formerly Selkirk

S. Cunard

Freehold, formerly Selkirk

S. Cunard

Scale 1:750 000

0   10 miles

0   10 km

PATTERNS OF TENURE,
1861 Census

20

21

22

23   24

67

33   34   35   36   37

32

29

30   31

48   49

65

50

57   59

58   61

60   63

62   64

Scale 1:1 000 000

0   10 miles

0   10 km

More than 50% freeholders

Between 6 and 50% freeholders

Less than 6% freeholders

tionally prominent members from the same part of eastern Queens included Adams, Gay, McNeill, and Robert Stewart of Lot 48.

Why did the farmers – and the general public, for Adams, Gay, and McNeill were not primarily farmers – in those townships react so eagerly to the promise of a new and effective tenant movement? There are at least two factors, high rents and geography. In the first instance, there were the proportions of leaseholders on those townships who were paying relatively high rents. The 1861 census recorded the numbers of leaseholders on each township who had agreements specifying rents of more than 1s. currency or sterling per acre. Within the colony as a whole, 21.5 per cent of leaseholders had done so. Among the forty-one townships with significant numbers of leaseholders, ten exceeded this per centage, most by a great margin. Of the ten, nine played an exceptionally important role in the history of the Tenant League. Five – Lots 23, 31, 36, 63, and 65 – accounted for eight of the fifteen incidents of resistance to authority or of 'agrarian outrage' noted in map 3; a sixth township, Lot 35, formed part of the Tracadie estate and therefore was linked to the three incidents attributed to Lot 36. If having to pay rent was unacceptable to Prince Edward Island tenants in the middle 1860s, having to pay unusually high rents was *a fortiori* a reason to support actively and vigorously the growing tenants' revolt. Lots 48, 49, and 50 were among the ten, and the percentages of their leaseholders who had high rents were 93.6, 100, and 78.5, respectively. The statistical evidence supporting an association between high rents and intense involvement in some form of Tenant League activity is clear and straightforward.[61]

The other part of the explanation is more complex. It is linked to the geographical position of Lots 48, 49, and 50, and the experiences of neighbouring townships in eastern Queens. Immediately to the north, the McDonald estate, proverbial for intransigence and arbitrariness, was a constant reminder of how unfair the system could be. Immediately to the south were the four townships of 57, 58, 60, and 62, where most of the land on the Selkirk estate, purchased on 17 September 1860, had been. On each of the four, acreages ranging from 11,691 to 14,600 had changed hands as a result of the Selkirk transaction, and between the censuses of 1855 and 1861 the respective percentages of tenancy on those townships had declined dramatically, as illustrated by table 3.1. The tenants of Lots 48, 49, and 50 had no Selkirk to deal with, but rather a considerable number of small-scale proprietors with a general reputation for inflexibility. In other words, for them, the Selkirk model of voluntary sale seemed to offer little hope, while the situation of the nearby

TABLE 3.1
Percentage of tenancy among land occupiers on four townships of southern Queens
County comprising 84 per cent of the land purchased from the Earl of Selkirk in 1860

	1855	1861	Change
Lot 57	60.2	10.7	−49.5
Lot 58	55.6	9.6	−46.0
Lot 60	49.4	22.6	−26.8
Lot 62	75.2	1.3	−73.9

Note: Calculations based on PEI, Assembly, *Journal*, 1875, app. E; 1856, app. D; 1862, app. A

McDonald tenantry, as well as their own high rents, reminded them of the need to settle the issue of leasehold tenure.

The Haythorne purchase in the summer of 1864 must have surprised many, both proprietors and tenants, because it was unprecedented on the Island: a proprietor selling through an arrangement with the collectivity of his tenantry. That was the formula the Tenant League advocated. Thus the procedure the new organization was recommending could work. The league gained support steadily, with little or no sign of setback, during the remainder of 1864 and into 1865. The movement had arisen as an extraordinarily spontaneous common response to a history of frustration with more conventional means, such as electoral politics, petitions, royal commissions, and delegations. Coming as it did in the wake of the bitter election campaign of 1863, in which the manipulation of denominational factors determined the outcome, the growth of such a body cutting across religious lines so soon afterwards was remarkable, and startled veteran local observers. That its program held genuine potential was even more of a challenge to old assumptions.

Many who took the initiative in protesting in 1864 and declaring that a new organization had to be formed were tenants to whom the prevailing system and public policies seemed to offer little hope. Their situation was anomalous and they knew it. They were frustrated, and seemed ready to resort to desperate measures as a result.

4

The Gathering Conflict

All the measures with regard to what is called 'The Land Question of Prince Edward Island,' have but served to engender and call into action the dishonest feelings & Illegal combinations set forth in the petition.
Lady Louisa Wood and Maria Matilda Fanning, writing from Bath, England to British Colonial Secretary Edward Cardwell, 31 October 1864[1]

I met another of Callaghan's sons on horse back with a Tin Trumpet in his hands, he turned his horse crossways on the road and endeavoured to prevent my getting along and ordered me to go back and dared me to go any further. I asked him the reason why I should go back, was I not at liberty to travel the road as well as any other person, he replied that I would find that out pretty quick that the devil much farther would I go until I would get my head in my hands.
Deputy Sheriff James Curtis to Sheriff John Morris, 14 March 1865[2]

The Hardening Attitude of Proprietors concerning Arrears

There were many signs that the Tenant League was making excellent progress in the months following its founding convention on 19 May 1864, and especially after publication in August of news of the successful negotiations for purchase of the Haythorne estate. The petitioning against confirmation of the Fifteen Years Purchase Bill was a lost cause, and the leaders of the league probably knew it.[3] But circulation of the petitions provided a wonderful means of stoking the fires of indignation throughout the countryside. That was the negative side of the coin, while the Haythorne purchase provided the positive message: that the Tenant League prescription could work – indeed that it had worked

already. The press covered the stages of the conversion of the Haythorne estate to freehold tenure, and the vigour and persistence of the
challenges against the terms of the purchase revealed how uneasy the
government and its supporters were.

A number of proprietors also seemed to be nervous, particularly
about the collection of arrears of rent. Under the terms of the Fifteen
Years Purchase Bill, a dozen landlords had forgiven arrears prior to
1 May 1858, but, as already noted, following the report of the land commission of 1860 many tenants had regarded that remission not as the
closure of the issue but as a minimum. This did not sit well with proprietors who had consented neither to the establishment of the commission
nor to the legislation. From their point of view, the land commissioners
had set a most unhelpful precedent with respect to arrears, one which
the Island government had explicitly built upon with the Fifteen Years
Purchase Bill.[4] In a letter written during the summer of 1864, one land
agent, whose employer had not been among the consenting landlords,
cited that statute as well as the growth of the Tenant League in explaining why he expected attempts to avoid the payment of arrears. He
stated: 'I am not very sanguine that I shall for the future be able to make
very large collections.'[5] Probably because of fear that their entitlements
were being eroded, several proprietors, including some who had agreed
to the pre–1 May 1858 remission, moved during the autumn of 1864 to
assert their rights to arrears.

The *Islander* of 28 October published a 'Notice to Tenants!' in which
George W. DeBlois, land agent for the Cunard, Sulivan, and Townshend
estates, covering, collectively, between 280,000 and 285,000 acres[6] – that
is, an area greater than one-fifth of Prince Edward Island – took a hard
line concerning rent and arrears. Tenants were to pay their annual rent
'together with a fair proportion of any back Rent that may be due'
before 21 November, or 'they will be sued without further notice.'[7] The
advertisement was surely intended to put tenants on the defensive, for it
provided no guidance as to the meaning of 'a fair proportion,' while
positively promising legal action in retaliation for non-compliance; presumably, it would be up to the individual tenant to make his case on an
individual basis to an agent who had been chosen, at least in part, for his
ability to extract payment in such situations.[8] A week later the same
newspaper included an advertisement from John Roach Bourke asking
'a portion of the arrears' due from tenants on the estates of Viscount
Melville, the Reverend John McDonald, and two other non-residents he
represented; the deadline was 25 November. One week after that, Lady

Georgiana Fane asked through her agent Henry Palmer (yet another son of James Bardin) for rent and 'settlement of Accounts.' Near the end of the month the agent of James Montgomery, Warwickshire notified tenants on Lots 34, 51, and 59 that he was demanding rent 'and other Amounts due ... forthwith.'[9]

Why did these proprietors act when they did? From one perspective, their assertion of their legal right to collect arrears at a time when much of the tenantry was in a refractory mood on the subject might seem like sheer provocation. Yet for anyone aware of the sentiments Island tenants had expressed repeatedly, and able to anticipate sequences and consequences of events, the rationale was absolutely clear.

Arrears fell into the category of what sociologist Samuel Clark has referred to, in an Irish context, as 'non-contractual privileges.' He has defined these as 'privileges that landlords were not legally required to grant, but did grant, at least partly in order to maintain tenant compliance and to reward good tenants.' One of the non-contractual privileges Clark has identified in Ireland was permission 'to fall into arrears during hard times.'[10] While landlords conceived of these extra-legal privileges as indulgences or favours, they also served as instruments of control, binding tenants to the leasehold system. The threat of their withdrawal was 'the principal means by which Irish landlords controlled the behaviour of tenants. ... It ... worked most effectively when it was left implicit, but sometimes it had to be made explicit.' As Clark has pointed out, 'a major problem with non-contractual privileges as a mode of domination is that, if granted repeatedly, they will often come to be regarded by subordinates as their due.'[11] From privileges they could become transformed, in the minds of tenants, into 'customary rights to which they had a traditional claim.' Non-contractual privileges would remain effective as part of a structure of domination only if they could be withdrawn.[12]

The analysis is readily transferable to the Island situation in 1864. Arrears existed, landlords who were not legally bound to tolerate them did so, and rather suddenly the legitimacy of their claims to them was being called into question. Over the years, tenants and landlords (or land agents) had often struck a deal to this effect: no legal proceedings would be taken to collect the arrears as long as they were not increasing, that is, as long as the annual rent was paid as it came due. Such an arrangement would be verbal rather than written, and the concession was certainly unenforceable if there was a falling-out, or if the landlord changed his land agent or his policies, or if there was a new landlord.

Any of these eventualities could lead to a demand to make progress in reducing the arrears.

Even assuming continuity in personnel or in general policy, the landlord or agent could still exercise discretion in individual cases, and hence a tenant's 'bad behaviour' could leave him open to retaliation if he happened to be in arrears. Although 'arrears' had become a dirty word in the lexicon of Island discussion on the land question, they were simply legal rents that had not been paid. To accept the waiver of all arrears would be a virtual invitation to tenants to let their current rent become defined as 'arrears.' From the perspective of an alert proprietor or land agent, relinquishment of arrears imperilled the entire leasehold system, for the logic entailed rewarding non-payment of rent. Consequently, it would be irrational for landlords to write off arrears entirely. In fact, the best guarantee of a rent strike would be wholesale remission of arrears.

The preceding reasoning is quasi-political. It centres on loss of popular legitimacy for an unenforced legal entitlement. There was also an explicitly political reason for landlords to be on their guard about the non-contractual privileges they granted in Prince Edward Island. One such privilege, namely, the right to use one-ninth rather than 50 per cent as the exchange rate on rental payments, had already gone through the stages of transformation into a customary right to which the tenant had a traditional claim and finally, for those landlords who had not enforced their rights over the years, into statute law. That three-stage transformation (non-contractual privilege – customary right – statute) was related to a legal doctrine which Island landlords had to take into account.

Laches is 'Negligence or unreasonable delay in asserting or enforcing a right.'[13] In essence, the principle at stake is analogous to that of squatters' rights or prescriptive rights: formal, legal rights to property, if not enforced, become unenforceable. The events surrounding the land commission of 1860 must have made any prudent proprietor reflect about the security of his title to occupied land on his estate in Prince Edward Island. One line of attack to pre-empt was the possible argument that, through neglect, the landlord had given the tenant to understand that arrears were forgiven. The history of the One-Ninth Act contained the lesson that if the Island legislature and the Colonial Office had a choice between supporting the rights of all Island landlords and supporting the rights of only those who had traditionally insisted upon their rights, they might choose the smaller group. The British government had already put the rationale for leaning this way on the public record of the colony.[14] Therefore, published reminders from land agents made a great deal of sense as a means to establish documentary evidence of proprietary vigilance which could prove crucial in legal, lobbying, or political struggles in the future.

In summary, the logic of the proprietors' situation demanded reassertion of their rights to rent, both current and old. To sleep upon their rights could be costly. Thus the remainder of 1864 represented a period of agitation and propaganda on the part of the league, and also a time when many of the proprietors, particulary non-residents, seemed to be taking a deep breath in anticipation of a coming struggle to exercise their rights. In fact, both sides – Tenant Leaguers and proprietors – thought of the issues between them in terms of rights. That is why each was so determined.

Resistance of Fanning, Wood, and Cunard to the Tenant League

In the autumn of 1864 Prince Edward Island witnessed a number of tenant meetings, particularly in eastern Queens County. One of the most important occurred on 26 September on Lot 50. The result was a petition, signed by 106 tenants and dated 29 September, making an offer of 10s. currency per acre to the proprietors, Maria Matilda Fanning and Lady Louisa Wood of Bath, England. In the text of the petition, the signers described themselves as members of the Tenant League, and stated that they would be paying no more rent. They asked the proprietors to reply to Alexander McNeill of the Tenant League; McNeill, who was not listed among the signers and who held land on the Haythorne estate, was probably not a tenant of Fanning and Wood.[15]

The response of Fanning and Wood is significant for what it reveals about the gap in sensibility between the owners and the tenants. On 31 October the two sisters forwarded the petition to the British colonial secretary, Edward Cardwell, as evidence of the unreasonableness of the tenantry. Their policy, they stated, was a purchase price of 20s. sterling, exactly triple the rate the tenants offered.[16] They had not been among the proprietors who consented to the land commission of 1860 or among those who agreed to be bound by the terms of the Fifteen Years Purchase Bill. In the opinion of the two landowners, 'all the measures with regard to what is called "The Land Question of Prince Edward Island," have but served to engender and call into action the dishonest feelings & Illegal combinations set forth in the petition.'[17]

At Cherry Valley, Lot 50, on 3 December, McNeill read the proprietors' reply to a meeting of tenants with Samuel Lane in the chair. In effect, Fanning and Wood stated that the tenants had the same right to purchase that they had had previously. On the preceding day the *Islander* had contained a notice to Lot 50 tenants from Charles Wright, their new land agent, requesting payment of arrears before 21 December. Faced with the blatant stonewalling of the proprietors, Edward Grant, one of the thirteen 'delegates' who had in April called for the founding convention of the Tenant League, described the reply as 'an insult to the tenantry and common sense.' The meeting unanimously passed a resolution which reaffirmed their refusal of pay further rent 'until a right of purchase is afforded ... on fair and just principles,' and which stated that those present were 'unflinchingly determined to stand or fall by the tenants' pledge – consequences notwithstanding.'[18] On 22 December Wright placed in the Island press a 'Further Notice' to the Fanning and Wood tenantry that they must pay their land tax by the due date in January 1865 or face the legal consequences. A standard provision of leases was that the landlord passed on to the tenant responsibility for any taxes due, and Wright seems to have been making a point of reminding the Lot 50 tenants that, in spite of everything, nothing had changed in their favour.[19]

Another resolution passed at the Cherry Valley meeting called for publication of the proceedings in *Ross's Weekly*, but the preamble to the two resolutions may have attracted more attention than anything else. It stated that the progress of the Tenant League had 'far exceeded our most sanguine expectations' and claimed that the membership numbered 'upwards of eleven thousand members.'[20] This appears to have been the first time the league made a public statement about the extent of its membership. *Ross's Weekly*, the one vigorously pro–Tenant League newspaper, repeated this figure of 11,000 in an editorial later in December. Since the total population of the Island at the census of 1861 had been 80,857, the number of 11,000 is extraordinarily large. If the Cherry Valley preamble and *Ross's Weekly* were correct, then an astounding proportion of Island adults had pledged to refuse to pay further rent and to support others in refusing to pay. A more precise idea of the significance of this figure may be gauged if one notes that at the census the number of occupiers of land was calculated as 11,439, and the number of males sixteen years and over as 22,247.

In reflecting on the significance of these numbers supporting a social movement in British North America in the second half of the nineteenth century, an obvious comparison would be with the level of support for the Knights of Labor, an organization which flourished in the 1880s and 1890s, and which historians Gregory S. Kealey and Bryan D. Palmer argue was the most important labour movement in Ontario until the late 1930s. They have calculated that 'the total peak memberships (at specific points in time with no account taken of volatility) across the province' added up to 'a minimum of 21,800 members ... [which] represented 18.4 percent of the hands employed in manufacturing in 1881 and 13.1 percent of those so employed in 1891.' They cautiously add that the actual numbers might be double what they have been able to document. Even doubling their figures and disregarding volatility (that is, assuming, for example, that the number of members in one local asembly in 1885 can be added to the number in another local assembly in 1888), this falls far short of the proportional level of support for the Tenant League, an organization which challeged the status quo at least as radically as, and certainly more militantly than, the Knights. The 'eleven thousand' figure for the Tenant League would be, in Kealey's and Palmer's terms, a static count, as opposed to a volatile count, and therefore all the more impressive.[21]

One remarkable circumstance regarding the numbers the Tenant League claimed is the fact that opponents were not inclined to quibble about them. This lack of controversy is noteworthy, for the Island press in the late colonial period was exceptionally lively, and no organization or movement was allowed to make empty claims of public support without vigorous challenges. The opponents of the league included the two pre-eminent journalists in the colony, Edward Whelan and William Henry Pope. But despite their unequivocal opposition to the league, they seemed to realize that the organization had spread so rapidly that there was no point in attempting to deny the extent of its support.

How would landlords other than Fanning and Wood respond to the rapidly changing situation? Of course Sir Samuel Cunard, who controlled the greatest quantity of land, would command much attention. He had largely dictated the approach of the Fifteen Years Purchase Bill and had proceeded to insist to the British government that its passage was essential to the maintenance of law and order in the countryside. The growth and declared plans of the Tenant League raised serious questions about the adequacy of Cunard's analysis and prescription. We have already noted that in October of 1864 he had reasserted his right to arrears of rent.

On 14 January 1865 the *Protestant and Evangelical Witness* announced a substantial reduction in rent – from 1s. 6d. to 1s. currency per acre – for the Cunard tenantry of Lots 63 and 64 in southern Kings County, where some of the first signs of unrest had occurred. The *Protestant* credited Lieutenant Governor George Dundas with having influenced Cunard's decision, but there was subsequent controversy over whether this was accurate.[22] Certainly Tenant Leaguers could be forgiven for interpreting the rent reduction as a victory for themselves, a result of their agitation in the same sense that the Haythorne purchase could be attributed to their work. Whoever deserved the credit, it was a matter of public record that the Cunard tenantry of southern Kings County were on average deeply in arrears, and had poor prospects of paying off the arrears; some also had unusually high rents. Therefore, they were prime candidates for a calculated reduction – an exercise in strategic leniency.[23]

The gesture towards the tenants of Lots 63 and 64 was soon balanced by the reply of DeBlois to a group of Cunard's tenants on Lot 48. They had written to him on 4 January proclaiming themselves to be Tenant League members and proposing terms of purchase: 10s. per acre, to be paid over six years.[24] DeBlois rejected their proposal and informed them that the only terms under which Cunard would sell were those of the Fifteen Years Purchase Bill (which were open to the Lot 48 tenants), but he went further than that: even if other terms were possible, the fact of their being Tenant Leaguers would preclude negotiations with them. That is, Cunard and DeBlois would not negotiate with tenants who openly professed membership in the Tenant League.

DeBlois condemned the 'object' of the organization as 'unlawful and seditious,' and stated that he would not forward their petition to Cunard. Furthermore, he declared his intention to 'enforce from you the payment of every shilling of your rent.' This, then, was the statement of Cunard's policy towards the Tenant League, for DeBlois can be fairly assumed to be his spokesman. DeBlois forwarded the correspondence between himself and the Lot 48 tenants to the *Protestant* for publication, with a covering note stating that he wished to disabuse Cunard's tenants 'in the different sections of the Island' of any misunderstanding concerning the terms open to them. Leniency might be possible in certain special cases – even for large numbers of tenants – but there were certain lines to be drawn.[25] The intent of DeBlois in rejecting the proposal from the Lot 48 tenants in the public manner he employed was explicitly *pour décourager les autres*.

Cunard's land agent had made it clear that he was open only to the terms of the Fifteen Years Purchase Bill, and that he would have nothing to do with tenants approaching him as Tenant Leaguers. Fanning and Wood had declared that their policy on the sale of land to tenants had not changed, and the frosty tenor of their subsequent proceedings was probably a result of their tenants' declaration of their links with the Tenant League. Thus, by the end of 1864 and the beginning of 1865 prominent proprietors had come forward to announce openly their resistance to the procedure of the league.

Events were moving closer to physical confrontation. In fact, on 29 December 1864 *Ross's Weekly* published correspondence involving a member of the Central Committee of the Tenant League, Robert Stewart of Lot 48, that concluded with an ultra-militant declaration. It seems that the proprietor, a Mrs Stewart, had rejected an offer to purchase at 15s. per acre. Robert Stewart promised resistance to the collection of rent, and stated that 'Should any sheriff or baliff [sic] covet the honor of martyrdom in maintaining landlord supremacy it is his business, not ours.'[26]

Deputy Sheriff Curtis Goes to Lot 36

On 10 March 1865 Queens County Deputy Sheriff James Curtis left Charlottetown with the intention of serving several writs in Fort Augustus and the Monaghan Settlement, Lot 36[27] at the suit of the Reverend John McDonald, an absentee proprietor residing in the United Kingdom. James Callaghan, a licensed tavern keeper, is the best known of the tenants Curtis was seeking. He was one of the early militants involved in formation of the Tenant League, and indeed the first recorded meeting for this purpose in 1864 had been held at his home on 29 February. Callaghan had been present at the founding convention of the league on 19 May 1864, and he signed the petition organized there against the Fifteen Years Purchase Bill. He appears to have been dissatisfied with his situation in Fort Augustus, for the *Herald* of 11 January 1865 had contained a notice advertising his desire to sell his chattels and his leasehold interest in 271 acres by 17 March. If there were not a private sale by that date, he would be holding a public auction.[28]

The proprietor was much better known than Callaghan to Islanders in general, although he had left the colony twenty years earlier. A native of the Island, McDonald, one of the famous landlord family of the Tracadie estate, had been educated in England and France, and had been engaged as a Roman Catholic clergyman in missionary work in Scotland before returning with colonists in 1830. He had been involved in conflict with tenants intermittently over the following fifteen years, first on his own estate and later, in another part of the Island, with tenants on the Cunard estate. Feelings against him had become so intense that his parishioners, led by a local assemblyman, had virtually driven him out. A man of considerable talent, he apparently shared in a family tendency towards high-handedness and confrontation. So bitterly was he angered by what he believed had been lack of support by his bishop that he never returned to the Island.[29]

Resident or non-resident, it was not in McDonald's nature to sleep upon his rights, and this desire to get what was legally owing to him led to Curtis's trip. The day was eventful, and Curtis did not complete his mission. His report to his superior, Queens County Sheriff John Morris, revealed the state of popular feeling in at least one part of eastern Queens. He encountered one obstruction after another. His vivid account, written in a vigorous prose style, gave a foretaste of the sort of welcome the forces of law and order could expect in the Island country- side by 1865.

When Curtis reached Callaghan's tavern, approximately sixteen miles from Charlottetown,

I enquired for him and was informed by his Wife that he was in Charlottetown. The Parties in the house did not know me at first but soon discovered who I was and what my business was. When [blank space][30] Callaghan son of the said James Callaghan ordered me out of the house, I replied that I would not go, ask- ing at the same time the Wife of the said James Callaghan if her husband was not a Licensed Tavern Keeper,[31] and receiving for answer that he was I ordered my dinner and food for my Horse. After dinner on making ready to leave three or four persons came around the Sleigh ... one was the before named son of James Callaghan ... [and] they told me not to dare to proceed any farther along the road and I replied that I had as good a right to travel the road as either of them and had my duty to enforce and would do it, that I was acting in the capacity of Sheriff and expected to be treated with civility.[32]

Curtis proceeded on his way, but related that shortly afterwards 'I met another of Callaghan's sons on horse back with a Tin Trumpet in his hands.' In normal circumstances, the tin trumpet – or the bugle, horn, or conch shell – was used at meal-time to summon those who were at work in the fields. But it was becoming a trade mark of the Tenant League and as such it had two functions. In the first place, it was a means by which supporters passed on the warning that suspicious persons were approaching. So important was its role in this respect that there is at least one documented instance of a local branch of the league resolving that each member should acquire 'a bugle to sound the note of alarm at the approach of *rent-leeches*.'[33] Secondly, once the sheriff or constables were in close proximity, the clamour created by dozens of horns could intimidate such persons and make their horses difficult to control.

Callaghan's son

turned his horse crossways on the road and endeavoured to prevent my getting along and ordered me to go back and dared me to go any further. I asked him the reason why I should go back, was I not at liberty to travel the road as well as any other person, he replied that I would find that out pretty quick that the devil much farther would I go until I would get my head in my hands.

Despite the threat, the deputy sheriff persisted, but soon

discovered three men in a field[,] two of them had a stick in their hand[,] the other had a Gun[,] they stood and looked at me and shortly proceeded to the road towards me, when they came to the gate they stood[.] I went on and never spoke to them and a little farther on I met a man on the road with a Pitchfork in his hands and two boys each of them with a stick, they said nothing but looked ferocious.

As Curtis continued, he 'found at every farm people with large sticks in their hands all apparently acting together as the trumpets were blowing in every direction.' At a cross roads, he made a right turn, but looking to his right he saw

a number of men armed cross the field on the angle caused by the two roads[,] most of them with sticks and apparently for the purpose of stopping me ... when I noticed what appeared to be their intentions I drove on faster and thus got clear of them and on looking back I discovered a crowd of Persons on the road behind me.

He concluded that it would be 'useless' to undertake such missions in that area in future 'without being backed up with a strong force.'[34]

There is no reason to doubt the accuracy of Curtis's narrative and analysis. Although his letter would be published in the *Islander* exactly five weeks after the event, the only recorded challenge came a year later, in

April of 1866. The challenger was Francis Kelly, who, along with George Coles, was assemblyman for the district. The circumstances may explain Kelly's desire to debunk the deputy's story, or at least to place a different slant on the events in the Fort Augustus area on 10 March 1865. In the House of Assembly, when members were attempting to apportion blame for what had happened over 1864–5, Kelly responded to the Conservative premier's statement that the trouble had originated in the constituency represented by the leader of the Opposition. Although Kelly did not claim to have been present, he professed to be 'fully cognisant of all that happened on that occasion.' He ridiculed Curtis, stating that Callaghan had been 'working on the road, a few yards distant from Curtis' when the deputy had demanded to see him and that Callaghan's wife had not been present at all. He would have fared better had he 'abstained from drinking and making a fool of himself at Taverns.' Those who had told him not to proceed were 'a few boys who were coming from school.' It was they who had deterred him from serving his writs.[35]

The Attitude of the Authorities in Charlottetown

Curtis later became a figure of great controversy, and the focus of much criticism and indeed derision. But Morris, the sheriff for Queens County, appears to have been a respected individual, and he treated his deputy's report on his mission to Fort Augustus and the Monaghan Settlement as entirely credible. On the day after Curtis wrote to him, he in turn wrote to James Colledge Pope, the leader of the government, and referred to men 'well armed' being summoned by horns. Morris used the phrase 'preconcerted no doubt ... for that purpose,' thus raising the spectre of conspiracy. He reported that Curtis's life had been threatened, and endorsed as 'quite correct' his deputy's judgment that a strong force would be necessary if he were to be able to serve writs in those districts.

The sheriff stated on his own authority that there were 'a large number' of Tenant Leaguers in Queens County 'whose real designs, I believe, are to resist the Laws, and to put the Sheriff and Constables at defiance.' Morris also reported that several magistrates and officers in the militia and the Volunteers were involved, at least to the extent of 'subscribing ... funds.'[36] He appears to have been correct. The modern historian of the militia of colonial Prince Edward Island believes that 'many ... members probably belonged to the Tenant League as well'; and in August, one member of the Central Committee would be dismissed as a Volunteer captain in eastern Queens.[37] Morris requested that Pope 'submit this matter [the defiance of the sheriff's authority] to His Excellency the Lieu[t]. Governor in Council' so as to 'put down such illegal proceedings ... and enable the Sheriff and other officers of the [Supreme] Court to enforce Her Majesty's Laws.'[38]

This was an extraordinary letter, and the recipient, James Pope, had been in the premier's office less than ten weeks. He had been a member of the Executive Council since 1859, and at the time of his appointment Whelan had written that he had 'a great deal of brass and perseverance, and no small share of talent.'[39] But he was a vigorous businessman involved in several branches of trade, and his primary concern was his many business interests. It is not at all clear that he had wanted to be premier in January of 1865, and indeed he would not even contest the next general election.

Pope had become leader of the government only because the Conservative party had split disastrously over the Island's response to the Confederation movement. Edward Palmer, who had been premier from 1859 to 1863, had taken the leadership of the anti-Confederates on the Island, and had driven his successor as first minister, John Hamilton Gray, a pro-Confederate, from office in December of 1864. In part, this was revenge for Gray's having pushed him out of the premiership shortly after the successful election campaign of 1863. But once Palmer forced Gray to resign as premier, the pro-Confederates, led by William Pope (James's older brother), had counterattacked so vigorously and to such devastating effect within the Executive Council that it was impossible for Palmer to resume the premiership. William, although widely acknowledged to be a man of great ability, was far too unpopular with large segments of the Island population to become leader himself. The Confederation movement, despite the adherence of some leading Tories like Gray, William Pope, and T. Heath Haviland Jr, had little support among Islanders in general, and consequently it was necessary that the next party leader be someone not publicly identified as a pro-Confederate. James Pope's position on Confederation was sufficiently ambiguous to make him acceptable, and in fact the Tories had few choices left. The tension within the Executive Council was such that in replacing Gray and Palmer, one pro-Confederate and one anti-Confederate were chosen.[40] Thus, when Morris wrote to the premier in March of 1865, the government was already divided, weakened by the resignation of key members, and plagued by the association of several prominent Conservatives with a deeply unpopular cause, Confederation.

The Island government would probably have preferred that the sheriff of Queens County handle the Tenant Leaguers without much visible involvement by themselves, for the last thing they needed was to be identified in the public mind with heavy-handed methods of rent collection. But the situation did seem to be spinning out of control. The Tenant League appeared to be growing by leaps and bounds, and now Morris had given the premier his written opinion that militia officers and the like were involved. The militia had never been a serious fighting body on the Island and it is difficult to discern its function; certainly, no official is known to have believed that it could be relied upon to assist the civil power in conflicts involving the land question. With respect to the Volunteers, in the words of Haviland, a member of the Executive Council at the time the legislation governing them was passed, it provided that 'they shall only be bound to serve in case of actual invasion. I would have liked to have inserted the word *rebellion*, as it is in the Imperial Act, but I wished to avoid even the shadow of a suspicion.'[41] Clearly there was no authority for using them to assist the sheriff in serving writs.

Yet some action seemed to be incumbent upon the local government, for Tenant League organizers could use the participation of such persons as magistrates in the league as evidence in asserting that its activities were legitimate. By January William Pope had begun to hint in the *Islander* at a coming purge of justices of the peace so as to quash that argument. According to a report in *Ross's Weekly*, at the Cherry Valley meeting of 3 December 1864 Thomas Beers, a justice of the peace and apparently also a tenant who was in arrears in his payments of rent to Wood, had seconded a resolution promising that the Tenant League, 'as a popular and irresistible body, will continue to withhold the further payment of rent' until the proprietors made a satisfactory offer of purchase. Pope proceeded to initiate a public controversy over Beers's role.[42]

Early in the new year additional reasons for concern surfaced. Traditionally, Charlottetown had been the stronghold of Island conservatism, and no one would have predicted that there would ever have been any reason to suggest that it could be regarded as league territory in any sense. But on 20 January 1865, with 'a crowded house' and after 'very animated' discussion, a meeting of the Charlottetown Literary and Debating Society, a non-sectarian organization formed at the end of 1864, unanimously passed a pro–Tenant League resolution.[43] While those sceptical of the league may have believed that members engaged in a certain amount of puffery when making their claims of popular backing, the resolution of the debating club suggested that public sentiment in the capital was surprisingly friendly to it. The last place in the colony where the league could have been expected to show strength was Charlottetown, and if spontaneous moral support was emerging there, the government had to be anxious.

Until January there had been little indication of support for the league in Charlottetown. The exception to this generalization was the response to an 'Appeal' to the inhabitants of the capital which the Central Committee had published in October of 1864. The aim was to solicit funds, avowedly for two purposes: 'first we want money to give us power to carry out our object, and secondly, we wish to know our friends from our foes.' The Central Committee had appointed two of its members, John Grant of Lot 50 and Robert Stewart of Lot 48, 'as a Committee to solicit subscriptions in Charlottetown for this Union.'[44] On 3 January 1865 the Central Committee voted their thanks to Charlottetown contributors, and declared that 'the canvass of the town has fully realized their expectations.'[45]

This sort of news must have been disquieting to many people in Tory Charlottetown, particularly because the solicitation of city residents does appear to have been at least moderately successful. In fact, both Conservative assemblymen for the capital would eventually give testimony on the subject. In 1866 Daniel Davies declared in the assembly that among the subscribers to the Tenant League in Charlottetown were 'doctors, ministers and others of most respectable standing in society.'[46] His colleague Frederick Brecken would testify that the leaguers had been exceptionally audacious: 'their agents came to public officers in this building [the Colonial Building, housing the legislature], and sought by means of intimidation, to extort subscriptions from them.'[47] Even if one discounts the lurid language of Brecken, the evidence of Tenant League confidence is striking.

Others would also report that inhabitants of Charlottetown donated to the league. Several merchants with Liberal connections would be named, although there were certainly prominent exceptions; for example, Coles, a leading brewer and distiller as well as a politician, refused when league delegates approached him for a donation.[48] But, at the very least, the reported financial contributions in the capital indicated that merchants, professionals, tradesmen, and others in Charlottetown recognized growing support for the movement in the countryside, and therefore sought to avoid alienating the Central Committee. In part, such donations would be interpreted as the actions of canny or opportunistic business people eager to curry favour with country customers; but they would nonetheless function as encouragement because they represented a degree of recognition and acceptance as legitimate.

The reference to 'ministers' is particularly intriguing because the documentary record yields little other indication of the attitude of clergy. In 1872 James Pope would protest against the appointment of the Reverend George Webber, a Bible Christian minister, as chaplain to the Legislative Council because he was 'a very worthy representative of the Tenant League and disloyalty – a regular Chartist.'[49] The comments by Davies and Pope are all that there is on the subject of involvement with the league on the part of clergy, with one exception: the role a Roman Catholic priest, the Reverend George-Antoine Bellecourt, played in agitation in 1866 for the early release of three Tenant League prisoners, two of them his parishioners, who had been convicted of offences growing out of resistance to the authority of the Queens County sheriff.

Soon after the publication of news about backing for the Tenant League in Charlottetown, the Conservatives had cause to worry about another bastion of support, namely, Orangeism. Despite occasional denials of partisanship, the hostility of the Orange leadership to the Liberal leadership was overt, and their endorsement of the Conservatives could be equally unequivocal.[50] At the most recent general election, in 1863, the Orange organization had been a bulwark of Tory strength as never before. But on 9 February 1865, at the annual assembly of the Grand Orange Lodge of Prince Edward Island, Grand Master David Kaye, a member of the colony's Executive Council, noted that some Orangemen were in the Tenant League. Although not denying the propriety of tenants associating, he stated that adherence to league obligations was contrary to Orangeism's emphasis on loyalty, which in his view included upholding the law. Therefore, membership in the Orange Association was 'incompatible' with participation in the league, and individuals would have to choose one or the other.[51]

The annual assembly of the Grand Lodge seemed preoccupied with this question of dual loyalties. The Committee on the Grand Master's Address endorsed Kaye's position with respect to the Tenant League; the Committee on Correspondence reported the presence of Orangemen in the league and advised the Grand Lodge to have them quit the league; Grand Secretary Thomas J. Leeming indicated his concern over the involvement of Orangemen in other societies and stated that Orangeism should come first; and the Committee on the Grand Secretary's Report asserted the primacy of Orangeism over everything except religion.[52]

William Pope, who had played 'the Orange card' with a vengeance in 1863, presented the Grand Lodge with a gift, a facsimile of a medal commemorating the Massacre of St Bartholomew in 1572.[53] But such gestures no longer seemed sufficient to keep Island Orangemen from supporting or pledging to support fellow-tenants in collisions with authority in the countryside. Kaye and Leeming noted a decline in commitment among the Orange brethren, and in the report of the next annual meeting both would concede that their admonitions against joining such organizations as the Tenant League, however coded, had not been well received. In 1867 Leeming would admit that in some districts adherence to the league had depleted Orange ranks – that is, faced with a choice between the league, with its belief in tenant solidarity, and Orangeism, with its appeal to Protestant solidarity, many Protestants chose the tenant movement.[54]

The contrast with the history of the Irish land question, and the later failure of the Land League to transcend Orange-Green divisions in Ulster, the stronghold of Irish Orangeism, is marked.[55] There are several factors in explaining the discrepancy. Most obvious is the relative recency of the Orange movement in Prince Edward Island, where the first primary lodge had been established in 1849 and the second in 1859; by 1862 there were fifteen, and the first large celebration of the Battle of the Boyne took place in 1860. Thus, in the middle 1860s the typical lodge was still a novel feature of the landscape in its district. Secondly, since few Orange adherents in the colony were of Irish origin, their attachment to Orangeism was relatively superficial – and, again, recent – compared with that of their Old World counterparts; it was also brief in comparison with their experience with leasehold tenure. Thirdly, Orangeism on the Island was grounded in a general sense of alarm among local Protestants over a number of issues, primarily with respect to church-state relations in education, and although these struggles were an enduring feature of Island history, they were non-violent. The very absence among Island Orangemen of Irish Protestants with deep roots in Orangeism and personal and family histories of bloody struggles against Irish nationalism must also be counted as a factor, for it would have made the local brethren more acceptable as allies for many Roman Catholics of Irish origin.[56] The political consequence of the compatibility between Tenant Leaguers and Orangemen on the Island was that another pillar of Conservatism appeared to be crumbling.

The rapidly evolving circumstances, together with Morris's letter to the premier following Curtis's trip to Fort Augustus and the Monaghan Settlement, virtually forced the government to take some action. If it did not act, and if the situation deteriorated further, the attitude of the London authorities, prompted by vigilant – or alarmist – proprietors, would become a factor. Indeed, the lieutenant governor and the justices of the Supreme Court were likely to ask awkward questions of the local government, if only to protect their own reputations and prospects for future honours and advancement. Dundas was an imperial civil servant with future appointments to be concerned about; Chief Justice Robert Hodgson was a future knight and a future lieutenant governor; and Justice James Horsfield Peters aspired to be chief justice.[57] Government members seemed to realize that somebody had to do something, but their reluctance, as politicians who would have to face the Island electorate, was evident. They would clearly have preferred that someone else take the decisive steps.

On 16 March the Executive Council met, and decided to order that the sheriff 'be directed' (the verb is significant, given later obfuscation) to exercise his powers by calling out the *posse comitatus*.[58] The term was drawn from medieval Latin, and meant, literally, 'the power of the county.' This device, which could be described as archaic by the 1860s, enabled the sheriff to call upon men above the age of fifteen to assist him in such duties as suppressing riots.[59] The very day that the Council met and instructed Sheriff Morris to call upon the *posse*, *Ross's Weekly* published an editorial article entitled 'The Tenant Union and the Courts of Law,' which unfortunately does not survive. That editorial would eventually result in a bill of indictment against John Ross, the publisher, for libel. Another newspaper later characterized it as 'reflecting on the Judge of the Supreme Court, and describing the Sheriff and Bailiff as *raiders* who went forth to execute the duties of their office without the Queen's authority.'[60]

More confrontations between the Tenant League and the forces of the sheriff were inevitable. The only remaining questions concerned time, place, form, outcome, and consequences. 'Whether' was no longer in doubt.

5

Popular Defiance and Samuel Fletcher

I have no doubt that the rescue of Fletcher from the Sheriff and his subsequent defiance of the authorities has tended more than any thing that has occurred to encourage that unlawful association.
Theophilus DesBrisay, JP, to Premier James C. Pope, 2 September 1865[1]

In the eye of the law the county was not ... an organisation of local self-government. ... but ... a unit of obligation. ... it was on the county itself, not on the individual officers, that rested the immemorial obligation of furnishing an armed force, ... as the *posse comitatus*, to put down any resistance to the keeping of the peace.
Sidney and Beatrice Webb, *The Parish and the County*[2]

Tenant League Demonstration in Charlottetown

The next display of Tenant League support occurred in Charlottetown on Friday, 17 March 1865, exactly one week after Deputy Sheriff James Curtis had undertaken his ill-fated trip to Fort Augustus and the Monaghan Settlement on Lot 36. It was also the day after the Executive Council had directed Queens County Sheriff John Morris to use the *posse comitatus* to deal with future problems in maintaining public order. This was the first time the league had made a public show of strength on the streets of the capital. It was St Patrick's day, but there was nothing particularly Irish about the demonstration and no surviving record suggests that anyone in the intensely nationality-conscious colony believed the date was a significant consideration in what happened.

The demonstrators, estimated by the sheriff to number five or six hundred, crossed the ice-covered Hillsborough River in sleighs and on horseback from Southport on Lot 48. This meant that they came from

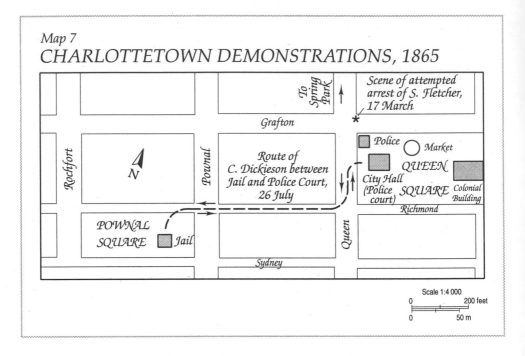

Map 7
CHARLOTTETOWN DEMONSTRATIONS, 1865

eastern Queens County; some may have travelled from parts of Kings County as well. They marched through the capital behind a band, waving flags with such inscriptions as 'Vox Populi,' 'Tenant Rights,' and 'Free Land for All.' The mood does not appear to have been hostile. The *Islander* of William Henry Pope, colonial secretary, stated that 'We heard no menaces, threats, or anything of the kind,'[3] and indeed all reports agreed that the procession was orderly. Yet the more conservative elements in Charlottetown could not have been pleased to witness a Tenant League demonstration in their streets. Some aspects of it must have been particularly galling. According to a Conservative assemblyman, the marchers stopped at the business establishment of Benjamin Davies, a well-known and outspoken radical who had been an assemblyman many years earlier, and applauded;[4] Davies had been the opening speaker at the 20 January meeting of the Charlottetown Literary and Debating Society which had unanimously passed a pro–Tenant League resolution.[5] But the sole breach of the peace during the event occurred when Curtis arrested Samuel Fletcher, one of the marchers, as the parade passed Apothecaries Hall, on the north-east corner of Queen and Grafton streets.[6]

Who was Samuel Fletcher, what is known about him prior to 17 March 1865, and why was he arrested? Born in 1821, he resided in Alberry Plains, Lot 50, and was a tenant of the absentee proprietors Maria Matilda Fanning and Lady Louisa Wood. Fletcher was clearly a supporter of the Tenant League. He had signed at least two of its petitions: the protest against confirmation of the Fifteen Years Purchase Bill and the 29 September 1864 demand that Fanning and Wood sell freehold title to their tenants for 10s. currency per acre. Moreover, he was participating in a public rally of the league, and was carrying the instrument of leaguer vigilance, a tin trumpet. The justification for his arrest was a writ for non-payment of rent; according to an affidavit of debt sworn on 4 March, he was exactly two years in arrears. Thus he was behaving as an orthodox member of the league in the most important respect of all, by withholding his rent.[7]

Curtis evidently recognized Fletcher and decided to seize the opportunity to arrest him. There is much reason to doubt his judgment in this matter. Under the specific circumstances, in the midst of a crowd of Tenant Leaguers gathered to demonstrate their strength and solidarity, the likelihood of failure was great. If Curtis failed, leaguers would interpret the incident as evidence of the authorities' impotence even in the capital, and the probable result would be an increase in the boldness of league supporters everywhere. Whether Curtis thought in such strategic terms and miscalculated his chances for success or whether he simply acted on impulse, without reflecting on the possible consequences, is unknown. In any event, when he made his move, Fletcher struck him in the face and attempted to escape. Others joined in striking the deputy sheriff, and knocked him to the ground. Fletcher fled, despite the intervention of apothecary and Justice of the Peace Theophilus DesBrisay on Curtis's behalf. DesBrisay had a grip on Fletcher briefly, but 'the prisoner' swung his trumpet at him, 'which blow I narrowly escaped.' DesBrisay lost his footing and Fletcher 'escaped in double quick time up the street.'[8]

This is the first occasion on which Fletcher emerged as an individual historical actor, and the last time he made a public appearance in the presence of Prince Edward Island authorities. There is no evidence to suggest that he was a leader of the Tenant League in a formal sense. Yet, because of the events on 17 March he became a celebrated figure, and clearly the object of great concern for the local authorities. He symbolized, in Tenant League eyes, the ability of the ordinary leaguer to evade the grip of the clumsy and ineffective forces of the sheriff. Capturing him became a significant objective in the minds of government leaders and law enforcement personnel precisely because of his symbolic importance. He personified successful defiance even in the centre of Charlottetown – in Sheriff Morris's words, 'when passing up the main Street nearly opposite the Police Station.'[9]

Fletcher could not have suspected that he would attain celebrity status when he went to the capital on 17 March, but he had done so, whether he liked it or not.

The Government Adopts a Posture

The dramatic march of the Tenant League through the streets of the capital and the sensational 'rescue' of an arrested person from the custody of the deputy sheriff represented a major escalation in the growing tension between the Island government and the league. The incident on the 10th, when the deputy failed to execute some writs, had happened in the countryside, and although there were threats and intimidation, there had not been actual violence. On 17 March in Charlottetown the challenge to civil authority had been taken a step further. Because of the location, the publicity was certain to be greater, and the 'demonstration effect' would be much more damaging unless something were done. The recapture of Fletcher seemed to be what was required.

There followed considerable manoeuvring within government and law enforcement circles in Charlottetown. On 18 March DesBrisay, who happened to be the father-in-law of W.H. Pope and himself a small-scale proprietor, provided Sheriff Morris with a letter giving his account of the Curtis-Fletcher incident on the previous day, and his own role in the attempted arrest. He also offered some advice: 'From what I witnessed on this occasion, and the apparent sympathy with these misguided men [the demonstrators],' he believed 'it would be perfectly useless for you [Morris] to look for assistance in the discharge of the duties of your office to the constitutional force of the County.'[10] The term 'constitutional force of the County' referred to the *posse comitatus*. As historians Sidney and Beatrice Webb explained in their book, *The Parish and the County*:

In the eye of the law the county was not ... an organisation of local self-government. ... but ... a unit of obligation. ... it was on the county itself, not on the individual officers, that rested the immemorial obligation of furnishing an armed force, ... as the *posse comitatus*, to put down any resistance to the keeping of the peace. ... From the standpoint of constitutional law the officers of the county were but instruments to secure the keeping of the King's peace, the due execution of the King's writs ... [etc.].[11]

In DesBrisay's view, rather than relying on the *posse*, 'you [Morris] must obtain a number of paid individuals from Charlottetown, as Special Constables, sworn to support and assist you.'[12]

This was an opinion Morris welcomed. In a covering letter he submitted on the same day to Premier James C. Pope, he commented on the events of the 17th and also on the instructions he had received to call out the *posse comitatus*. While acknowledging the propriety of such directions in a formal sense, he declared unequivocally that 'a party of Special Constables on whom confidence could be placed, and paid by the Government would answer a much better purpose.' He closed by stating that if his advice and that of DesBrisay were rejected, he would follow instructions and call out the *posse*. But no one who read his letter could doubt that if he did so he would be acting against his better judgment – virtually under protest, in fact.[13]

The concern of Morris and DesBrisay seems to have been effective execution of whatever attempt might be made to capture Fletcher. They were clearly sceptical about using the *posse comitatus*, and with reason. In the only recorded Prince Edward Island case that could be viewed as a precedent and guide, involving the shooting of a horse from under a sheriff in Kings County in 1859, there had been two unsuccessful attempts to call out the *posse*. In the end, a special force of constables was organized at a cost of £137.[14] The attitude of the Island government in the spring of 1865 was complicated, and finding the best means of apprehending Fletcher was not their primary consideration. Government members were acutely aware that they had a delicate balancing act to perform: on the one hand they had to avoid actions which would result in total alienation of the Island electorate, and on the other they had to prevent the situation with respect to law and order from deteriorating to the point that the British government became anxious. But there was more to their motivation than that necessary minimum of calculation. If they had simply wished to walk a tightrope between the Island tenants and Downing Street, then authorizing the Queens County sheriff to hire the constables he asked for would have been the answer.

The chief strategist of the Island government was W.H. Pope, colonial secretary, lawyer, assemblyman, editor of the *Islander*, and elder brother of the premier. William shared with his brother and his father, a former leader of the Conservative party, strong ambition and ruthlessness. In addition, he was a brilliant, enigmatic man identified with the Worrell Job, visceral religious controversialism, and Confederation. He had antagonized large numbers of Islanders, and was all but unelectable, with the sole exception of the general election of 1863. On that occasion, although known as an 'infidel' or free thinker in his personal beliefs, he campaigned as an ardent champion of 'Protestant rights,' which were supposedly threatened by Roman Catholic pretensions in educational matters. It was part of an eighteen-month battle he waged, commencing in 1861, once it became clear that the Conservatives would have difficulty making a political alliance with the Roman Catholic bishop. He indulged in extraordinary abuse of Catholic doctrines and institutions, permanently scarring his reputation with Island Catholics. By 1865, his public identification with the intensely unpopular Confederation movement had become, for many Islanders, just another example of his perversity. Yet no one of discernment underestimated his talents or his determination. When Pope had been appointed colonial secretary in 1859, Edward Whelan had written that he had 'more cunning, more perseverance in the pursuit of his object, (no matter what it is,) and more real talent than any other man in his party.'[15] By the middle years of the 1860s, among knowledgeable political observers, his name connoted win-at-all-costs politics.

In the affair of the *posse comitatus*, William Pope's cunning certainly emerges from the record, as do such qualities as vindictiveness. On 27 January 1865 he had published an editorial in which he hinted at use of a *posse* and military force. Although he thought the reports of between ten and fourteen thousand leaguers were exaggerated, he expressed alarm at the rapid progress of the organization in winning support among Islanders. He conceded that there were a 'large number pledged,' who could be counted in the 'thousands.' Among them, he was persuaded, there were 'men of nerve and resolution' who would attempt to carry out Tenant League principles, regardless of consequences, although he doubted that this hard core numbered one hundred. He attacked the league for '*terrorism*' in its recruitment methods, and in its solicitation of funds. '"If you do not join the league you are the enemy of the people, and its members will not deal with you," is the threat held out by the leaders of the organization to the miller, the blacksmith, the shopkeeper, and to every artisan in the settlement.' He was particularly scornful of those Charlottetown persons who, although not believers in Tenant League principles, had subscribed funds out of 'expediency.' Some others, he said, had turned leaguers away rather than give financial support to 'an illegal organization.'[16]

Ultimately, Pope argued, adherence to the Tenant League would prove futile for tenants. He stated that on 25 January a Cunard tenant belonging to the league was served with a process and arrested, and he proceeded to outline that tenant's legal position. Even if systematic resistance were mounted to the sheriff, the final result would not be in doubt. 'One officer, or a dozen, may be driven from a settlement, but, depend upon it, they will return with a force sufficient to enable them to execute the law; and those who oppose them will in the end be worsted.' He cautioned people in the countryside that 'acts of violence against those attempting to collect rent, ... would probably lead to very serious consequences.' The repercussions could include the 'heavy costs' of having to pay the expenses of 'a Military force sufficient to secure the execution of the process of the Law Courts in every part of the Island.'

It was a major article, drawing fully upon Pope's powers of logic and persuasion. In making his argument against the Tenant League he did not misrepresent the line it took, which surely signified the earnestness of his desire to appeal to those it attracted. To put it another way, he realized that the league had such broad support that his task was not simply to mobilize opponents and the uncommitted; he had to argue with committed leaguers and attempt to convince them that the way they had actually chosen – not some caricature of it – was wrong and disastrous.

In the course of the article, Pope warned that, 'If the Sheriff should be interfered with in any district, he may call upon the inhabitants to assist him; and if they refuse, they will render themselves liable to fine and imprisonment.'[17] This was the first suggestion that the authorities might resort to the *posse comitatus* to deal with the league. The *posse* was, as already noted, an ancient device. It would have been considered somewhat outmoded even a century earlier. In a study entitled *Public Order and Popular Disturbances 1660–1714*, historian Max Beloff noted that use of the *posse* rested on 'the legal doctrine of the responsibility of all citizens,' and that where those disobeying the law had popular sympathy, 'the posse could not be relied upon for more than nominal obedience.' As evidence of the unreliability of such civil forces as the *posse* even as early as the period of the Stuart Restoration, Beloff cited 'the large number of instances in which military help was called for almost immediately on the outbreak of disorder.'[18] In England, 'military help' usually meant the militia, an option effectively closed on Prince Edward Island because of the overlap in membership between the militia and the Tenant League.

The *posse comitatus*, virtually a dead letter in England, was unlikely to be an effective force in Prince Edward Island in 1865, especially if recruited in the countryside. What was William Pope's motive in suggesting such an expedient? This is where an understanding of his nature is useful, for his record indicates that he was cynical with respect to ends and means. He was a man who, while not changing his private religious convictions, could encourage the most alarmist Orange thoughts concerning Roman Catholic fellow-citizens, and a few years later encourage the same Catholics to seek public funding for their denominational schools – to offset the infidelity of the era, he said. His inconsistencies could usually be traced to shifting political alignments and opportunities; long-term consistency and the negative consequences for others of his actions and the positions he advocated mattered less to him than the continued dominance of his social class and his political party.

Apparently the fact that some Charlottetown residents gave financial support to the Tenant League angered Pope, and after the thwarting of Curtis on Lot 36 he again referred in the *Islander* to the *posse comitatus*. He stated that those called upon by the sheriff must go, 'it matters not who they are,' and emphasized the responsibility of persons who had encouraged the league 'by their money and their countenance.'[19] Just exactly what he had in mind came into clearer focus on 24 March:

It is ... proper that those who have encouraged the Tenant League in their unlawful designs, should be called upon to assist the Sheriff to overcome the evils which they have assisted to create. If some fifty or sixty persons shall be *weekly* called upon by the Sheriff, and required to proceed with him, to the neglect of their business, and at their own cost, and very great inconvenience, the country will very soon discover that the Tenant League is a much greater evil than they at first imagined.[20]

Involuntary participation in the *posse*, therefore, was to be an instrument of punishment for those who had gone along with the league. William Pope was known to be resolute and ruthless, and few doubted that the man who had been behind the audacious Worrell Job and the virulent attacks on Roman Catholicism in the early 1860s was capable of extreme measures.

When the Executive Council considered Sheriff Morris's letter of

18 March, along with DesBrisay's account of what had happened in Charlottetown on the previous day, it rejected Morris's advice, and ordered that he be directed to carry out the instructions he had already received – that is, to call out the *posse comitatus*. The Council did authorize him to hire up to five special constables *in addition to* calling out the *posse*, but clearly the constables were not to be a substitute. The Council also advised that, given the assault on Curtis, and 'the Evil results which must necessarily ensue from the unlawful combination' to resist payment of rent, the lieutenant governor should issue a proclamation commanding the general population, but especially those in positions of legal responsibility, to desist from involvement.[21]

The proclamation Lieutenant Governor George Dundas issued on 22 March referred to the violent obstruction of an officer of the law, and, without using the terms 'Tenant League' or 'Tenant Union,' warned that resistance to rents was resistance to the law. It commanded 'Magistrates, Sheriffs and other Ministers of the Law, and all Constables and Peace Officers, *and all other loyal subjects* of Her Majesty, ... to give their prompt aid and assistance, when lawfully required, ... in arresting and bringing to justice' offenders.[22] The sole precedent in the history of the colony for such a proclamation was one which Lieutenant Governor Sir Alexander Bannerman had directed against Orangemen and Ribbonmen, naming both, in 1852; his action been prompted by the publication of the Orange oath in a newspaper.[23] The proclamation of 22 March 1865 would be reprinted regularly in the Island press, and subsequently Dundas would tell the British colonial secretary that it was 'drawn in language as forcible as the Law Officers could use.'[24] On the day after the lieutenant governor issued his proclamation, the premier stated in the House of Assembly that 'The laws, ... must be obeyed, and if there is not force enough in this Island to cause them to be respected, we will have to ... procure assistance from abroad. The last shilling in the Treasury will be expended to maintain their supremacy.'[25]

On the same day that James Pope was publicly warning of his willingness to take strong action, the clerk of the Executive Council was writing to Morris to inform him that he must call out the *posse comitatus*.[26] The *Islander* reported on the next day that 'At this moment the Sheriff is calling out the *posse comitatus*.'[27] As the local poet John Le Page, who often wrote satiric verses on public occasions, expressed it,

Said *Johnny* ——[28] we must be quick –
He holds the Sheriff's *status* –

I'll call, said he, to go with me,
The Posse Comitatus! ...

And there and then, he took his pen,
And wrote, for the occasion,
His summonses to chosen men,
To meet him at his station;
Ready to fight, for law and right,
Laying aside pretences,
To go with him, in marching trim,
And bear their own expenses![29]

As the general public in Charlottetown became aware of their liability to serve in the *posse* and also realized that action was imminent, some approached the lieutenant governor to have him prevent resort to this extraordinary measure. But he declined to interfere in the decision to summon the *posse* during this 'emergency.'[30] One newspaper reported that public meetings were held, and lawyers consulted, all to no avail.[31] The government was determined.

There was a saying in the Pope family concerning the brothers James and William that in political matters 'William made the snowballs for James to throw.'[32] In the case of the *posse comitatus* decision, the government, nominally controlled by the premier, was directed by his brother, the colonial secretary. It was absolutely committed to doing something, but although the defiance by Fletcher had provoked the action it was about to take, that measure was as futile as could be imagined if the objective was to catch Fletcher. The real intent was to punish those summoned to serve in the *posse* with a forced march of fifteen or twenty miles 'through mud and snow.'[33]

The *Posse Comitatus*

On Friday, 7 April, approximately 150 residents of Charlottetown accompanied Sheriff Morris to the community of Vernon River, Lot 50, and beyond.[34] Their ostensible purpose was to execute an arrest warrant against Fletcher at his home in Alberry Plains, some eighteen miles from the capital, near the Kings County line. When the *posse* assembled it was of course composed of conscripts, not volunteers. Indeed, on the basis of available evidence, it appears that those who knew or expected that they would be called upon were indignant. Several reports suggest that some

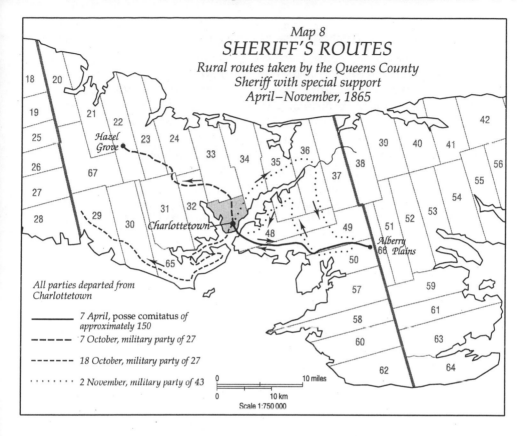

Map 8
SHERIFF'S ROUTES
Rural routes taken by the Queens County
Sheriff with special support
April–November, 1865

All parties departed from
Charlottetown

——————— 7 April, posse comitatus of
approximately 150

– – – – – 7 October, military party of 27

- - - - - - 18 October, military party of 27

· · · · · · · · 2 November, military party of 43

0   10 miles
0   10 km
Scale 1:750 000

simply did not appear, and there is no record of action being taken
against them subsequently. According to the *Herald*, 'Protests and
expressions of sympathy for the tenantry were embodied in resolutions
by a majority of the *posse*, ... and forwarded ... to the District wherein
Fletcher resided, so that no show of resistance should be offered.'[35] In
effect, they were raising the white flag lest they be misunderstood.

Sheriff Morris selected members of the *posse*, and his choice of participants was controversial. An anonymous letter in the *Examiner* from
'One of the Posse' alleged that he displayed bias in his selection, that he
chose primarily those suspected of supporting the Tenant League financially, and that he appeared to take some pleasure in the execution of his
task. The writer also accused him of behaving 'inhumanly' in not officially furnishing details of the itinerary until the *posse* had assembled in

Southport, across the river from Charlottetown, since that oversight meant that those called upon were not able to prepare adequately for the expedition.[36] Some were caught entirely unprepared, having expected the exercise to conclude in Southport. This lack of prepared-ness on the part of individuals was an important consideration, because the spring roads were in poor condition, 'almost impassable,' in the words of one newspaper editor.[37] The bad roads and the way Morris executed his task lend support to the theory that the expedition was to serve a punitive function, as suggested by William Pope two weeks ear-lier. Those called included the veteran radical Davies and John Ross, the only newspaper publisher in Charlottetown who supported the league. It seems that, under the leadership of Davies, members of the *posse* sang a pro–'tenant rights' song at one point, although the number who sang along would be disputed later.[38]

The fact that the call-out was undertaken against Morris's better judgment and advice may account for the inclusion of John Ings, the queen's printer, who was son-in-law of executive councillor James Yeo Sr and, as publisher of the *Islander*, the employer of William Pope in his editorial capacity. Morris may have been taking some revenge upon those who had forced him to go through with an operation that was virtually certain to make him unpopular with many people and that was also likely to make him an enduring butt of ridicule within the colony as a whole. An additional function of calling out people like Ings would be, as the *Examiner*'s anonymous correspondent pointed out, to give a veneer of impartiality to the proceedings. This, plus possibly a desire to deal out some 'poetic justice,' may also help to account for the inclusion of Charles Wright, land agent for Fanning and Wood, the proprietors to whom Fletcher owed rent. Wright had the misfortune of being thrown from his wagon into the mud at the beginning of the day. According to Ross, some 'boys' had tampered with his con-veyance. Afterwards Wright would apply in vain to the government for compensation; they 'declined to entertain' his claim for £2 19s. 6d. expenses.[39]

The expeditionary force was motley group. Some proceeded on horses, some on foot, and some in wagons. Ross, writing many years after the event, asserted that 'The cavalry ... was largely composed of proprietors, their agents and friends, who were anxious to see the rebel (?) [Fletcher] captured,' and suggested, by inference, that the infantry were largely sympathetic to the Tenant League.[40] Henry Jones Cundall, surveyor and land agent, and a member of the *posse* who was on horseback, noted in his diary that 'Banners floated with Tenant League mottoes from various places, ... & every fellow had a trumpet in his hands which they blew as we passed along.'[41] There was no possibility that Fletcher would be surprised by their arrival. At a blacksmith's shop near Vernon River, according to another participant, 'a grotesque battery had been erected, which was composed of several pieces of old stove pipe pushed through as many pieces of old board, behind which stood the gunners, with wooden legs, straw bodies, paper faces, and night caps, etc., etc.'[42]

The expedition failed. Fletcher was not apprehended, and in fact he was never captured, although as late as January 1866 the government was eager to arrest him. Members of the *posse comitatus* which had left town at 8 a.m. straggled back into Charlottetown either late at night, or on Saturday morning, depending upon whether they had travelled on foot, on horseback, or by wagon, and how their means of conveyance had held up. It had been a gruelling experience, which few relished. Almost no one suggested a repetition.

Cundall and others recorded that several accidents occurred, involving horses, wagons, and human fallibility. It was not a pleasant day, no doubt in part because of political differences among the conscripts. On the basis of press reports it is clear that 1865 featured a 'slow' spring in Prince Edward Island, and it is not surprising that the mud played a major role in people's recollections of the day in later years.[43] None the less there were lighter moments. According to an anonymous poet in the *Herald*:

Six of the Leaguers we did meet,
Who, blowing horns before us,
Vow'd that before we went they'd treat
The Posse Comitatus.

A quart of gin they did bring in:
If that's the way you 'sail [assail] us,
'Twill be our aim to come again
The Posse Comitatus.[44]

The futility of the expedition in terms of the professed objective is obvious, and the explanation for the failure is equally evident. But no surviving document explicitly sets out why Fletcher was chosen as the quarry on 7 April 1865, rather than some relatives and neighbours of

James Callaghan in Fort Augustus and the Monaghan Settlement.[45] The matter is worthy of consideration because the Island government had decided to resort to the *posse comitatus* in dealing with the Tenant League *the day before* the 'rescue' of Fletcher in Charlottetown. Presumably, if the Curtis-Fletcher incident had not occurred on 17 March, the *posse* would have been called out to pursue one or more others identified with the league. Quite possibly this would have been persons around Callaghan, for it had been the unsuccessful attempt to serve him with a writ which had prompted the sheriff to write on 15 March to the Executive Council appealing for help.

Several reasons qualify as possible explanations for the decision to proceed against Fletcher rather than Callaghan's associates. In the first place, as already noted, the violence Fletcher committed was actual rather than merely threatened. Secondly, the authorities knew exactly who had escaped custody on 17 March, but they did not know with absolute precision the identities of those threatening Curtis on his trip to Fort Augustus a week earlier. Two sons of Callaghan had been involved, yet Curtis had not been able to discover their given names. Thirdly, there was more than one crown witness to the events on 17 March, whereas a week earlier Curtis had been alone in a hostile environment which was unlikely to yield helpful testimony. All these reasons are important, although, on the other hand, Callaghan had been a more prominent activist than Fletcher. Balancing the last-mentioned consideration was the fact that the very location of the scuffle between Curtis and Fletcher on 17 March had created a certain notoriety for Fletcher, and given the government an additional incentive to bring him to account before the law. In the words of the chief justice, the failed attempt to arrest Fletcher had taken place 'within a few yards of the police station'[46] in Charlottetown. That counted. The event was more difficult to ignore.

But there was another, broader reason to pursue – or at least seem to pursue – Fletcher rather than the relatives and neighbours of Callaghan. The route of the march on 7 April brought the forces of law and order into the very heart of the home territory of prominent and visible Tenant League leaders like George F. Adams, James B. Gay, Samuel Lane, and Alexander McNeill. The sheriff appears to have stopped twice at Adams's tavern, 'Half Way House,' in Vernon River, on the way to and from Fletcher's farm. At Adams's, on the way back, he called the roll of those on horseback. In spite of the mockery and taunts the *posse* encountered, the act of stopping at Adams's and calling the roll may be seen as making an important symbolic point through a show of authority 'in that snug parlor where so much Tenant League plotting has been done,' as Whelan put it.[47] When returning, the horsemen met those marching on foot under Curtis's command. According to Cundall, those on foot were ordered to continue on to Adams's. There appears to have been little compassion shown on that day.

Post-Mortems on the *Posse*

The *posse comitatus* incident led to no arrests, no trials, and no convictions, and was not repeated. But it has occupied a minor place in Island history, which was established the following Monday when Whelan published an account in his weekly newspaper, the *Examiner*. Perhaps it was inevitable that the march of the *posse* would be treated as a comic episode, but if not, Whelan sealed the matter. With brilliant wit and biting satire, he lampooned Curtis mercilessly for acting impetuously on 17 March and choosing the wrong place and the wrong time to lay hands on Fletcher. Whelan summarized the results as follows: 'Sam Fletcher was an insignificant individual. He is now the hero of the hour; and the ease with which he has turned his back upon the concentrated force of the county, ... [has] thereby covered it with ridicule.' The event was 'nothing but a farce.'[48] The *Herald*, edited by Edward Reilly, a bitter rival and former employee of Whelan, was the other newspaper to publish a fairly detailed account. In Reilly's concluding words, it had all been 'a most gigantic farce';[49] he and Whelan agreed on that, and agreement between them was unusual. A third weekly newspaper, the *Protestant and Evangelical Witness*, edited by David Laird, who was frequently at odds with both Whelan and Reilly, referred to the *posse* expedition as 'burlesque.'[50] In the press, derision seemed to be the order of the day.

Aside from the newspaper accounts, particularly Whelan's, the major traditional source in constructing an image of the *posse* incident has been a poem by Le Page. He had been born in Pownal, Lot 49, a hotbed of sympathy for the Tenant League and a district through which the *posse* had passed on its route to Vernon River and Alberry Plains. His poem of sixteen octaves, entitled *The Calling Out of the Posse Comitatus*, appeared in print sometime in 1865 and reinforced Whelan's interpretation of the incident as farcical.[51] But there was a double edge to Le Page's satire: *inter alia*, he censured Adams for allegedly being absent from the 17 March demonstration in Charlottetown.

They, through the town, march'd up and down,
Their horns defiance sounding;
While over head their banners spread,
With *loyal words* abounding!
But, at their van, we missed the man,
Erst foremost in the cause, Sir,
Who taught them well how to rebel –
Obedient to the laws, Sir!

George Adams who, strange things could do,
And say, and write, – turn'd tail, Sir, –
And went away, St Patrick's day,
To drive the Georgetown mail, Sir![52]

When Ross wrote about the day in his memoirs, published in 1892, he had 'all the prominent members of the Central Committee' present, and in fact seated on 'the official conveyance.' He knew, according to his narrative, because the demonstration marched to his office door, and 'the Committee of Management' (presumably an *ad hoc* group constructed for the occasion of the demonstration) went inside to see him and conducted him to the vehicle carrying the official party. Adams, of course, would be included in the Central Committee group; whether he was actually present is unknown, but Le Page's verse touched upon a sensitive point, namely, the degree of personal commitment displayed or not displayed by the leadership. Ross reprinted Le Page's poem in a series of articles on the Tenant League which he published in *The Prince Edward Island Magazine* in 1900, but the octave commencing 'George Adams' was absent, and in writing of the 17 March demonstration this time, he did not mention the Central Committee.[53]

The *posse comitatus* inspired other satiric poetry as well, for on 26 April the *Herald* published an anonymous poem, *March of the Posse Comitatus*, fourteen quatrains in length, to be recited to the air of *The Girl I Left Behind Me*. It appears that the students at St Dunstan's College also found the incident amusing. The *Herald* reported that on 12 July the usual closing exercises ended 'with a side-splitting comedy, called, "The Posse Comitatus."'[54] Furthermore, Fletcher may have shared directly in the fun, since it seems that on 7 April a straw effigy of a man dressed in old clothes awaited the *posse* at the main gate to his farm.[55] It is not known with certainty where Fletcher was when the *posse* came to get him. Ross wrote that 'Fletcher lay concealed nearby, laughing in his sleeve at the discomfiture of the famous *Posse Comitatus*.' But Ross published that statement in 1900, and it is not clear how or whether he knew that Fletcher had personally witnessed the arrival of the *posse* at his farm – or whether this was an embellishment which simply served as another means for Ross, as an old Tenant League supporter, to ridicule the authorities.[56]

Almost immediately after the conclusion of the expedition, blame began to be apportioned. Whelan faulted Attorney General Edward Palmer, who, he claimed, favoured the idea of calling out the *posse*, and Curtis. The *Herald* and the *Protestant* argued that the Executive Council, not Palmer, should be held responsible. Neither disputed that the resort to the *posse* had been a mistake. To a certain extent this newspaper squabbling over responsibility can be attributed to the political cross-currents of the time. Whelan was an ardent supporter of the Confederation movement, and was inclined to blame Palmer for whatever he could, since Palmer, more than any other politician, had ensured that the Island would not be entering Confederation in the near future. The editors of the *Herald* and the *Protestant* opposed Confederation vigorously, and tended to lionize Palmer. Therefore they were predisposed to absolve him from all culpability.

When all the surviving documentation is taken into account, there is no persuasive reason to believe that responsibility for the decision to use the *posse* lay anywhere but with the Executive Council. Palmer, although a member of the elective Legislative Council, no longer belonged to the Executive Council, and cannot be held accountable for the decisions of its members; and the determination was made at a level higher than Curtis. The mastermind behind it all appears to have been William Pope, whose younger brother, the premier, listened carefully to his counsel in matters of political tactics and strategy. William had outlined the rationale for using the device of the *posse* in his *Islander* editorials, and the vindictive strain which runs through the episode bears his distinctive mark.

But on Friday the 14th, one week after the incident, William Pope's *Islander* more or less squirmed in embarrassment, and said that the sheriff, not the Executive Council, was responsible for calling out the *posse*. Pope did not criticize the decision, but he did censure the sheriff for the manner in which he had executed his duty. He suggested that Morris should have called out more of Fletcher's neighbours and fewer Charlottetonians, and 'taken from Charlottetown only those gentlemen who had subscribed money to aid the Tenant Leaguers to resist the Law.'[57]

His focus on the Charlottetown residents who had given financial support to the Tenant League represented a reiteration of his earlier position on the punitive nature of participation in the *posse comitatus*. As the poet Le Page interpreted this line of reasoning,

> Instead of ten, – two hundred men
> Shall snap rebellion's fibres;
> And, pleas'd or not, those rogues shall trot,
> The Tenant League subscribers.[58]

Pope went still further in his attempt to point the finger of responsibility elsewhere. In the same issue of the *Islander* he published five letters in order to document the supposed necessity of using the *posse*. One, that of DesBrisay, does not support the argument, but the others are consistent with it. There is one crucial letter which is missing from the *Islander*: Morris's of 18 March, in which he argued against use of the *posse*. Inclusion of that letter would have undermined Pope's claim that the sheriff had been responsible for choosing it as the means of pursuing Fletcher.[59] Clearly, Pope and the government had decided to make him the scapegoat for their insistence on the use of the *posse*.[60] Under the circumstances, it was disingenuous and ignoble, but such conduct was in character with William Pope's record. He even had the audacity to contrast unfavourably the efficacy of a *posse* of 150 with a body of five constables for the purpose of apprehending Fletcher – which had been precisely Morris's point in his now-suppressed letter of 18 March. Such brazenness was also in character for Pope; he did not readily recognize limits once he set out to make a point. Morris must have been pleased when his one-year term as sheriff expired a few weeks after the calling out of the *posse*.

Just as the march of the *posse comitatus* did not result in the capture of Fletcher, it did not stop the spread of the Tenant League. Some Charlottetonians may have become more discreet in displaying sympathy for the league, but, as a movement in the countryside, where the core of its support lay, it showed no signs of weakening in the months after the incident. In fact, Reilly believed that the tenants' cause had 'gained ground by this raid,' and he congratulated them in the *Herald* on their wisdom in not offering resistance which, he warned, would have led to 'martial law' and an end to the separate political status of the Island.[61] The government continued to be mocked in one way or another. Late in May, James B. Gay published an advertisement in the *Royal Gazette* indicating that he had in his possession a revolver someone had lost on 7 April.[62]

The twin images of official impotence and successful popular defiance of authority became imbedded in the consciousness of Islanders regarding the *posse comitatus* incident. In the following year in the assembly the respected lawyer Joseph Hensley described the *posse* as 'floundering along the road, some on foot, some on horseback, some in carriages, but all alike liable to have their progress arrested by the depth and tenacity of the mud ... nothing but a perfect burlesque upon the civil and administrative authority of the land.'[63] Within the popular tradition of Prince Edward Island history, most writers when dealing with the land question tend to focus on the two incidents surrounding Fletcher: the attempt to arrest him in Charlottetown and the use of the *posse comitatus*. The two events seem to have fired the interest and imagination of this branch of historical writing to an extent unmatched by anything else in the long history of the land question, and some authors have tended to embellish. For example, two, Errol Sharpe and Reg Phelan, have militant demonstrators wrecking a sheriff's office; in the primary sources there is no evidence for such a demolition.[64]

The march of the *posse* has even inspired art. More than a century afterwards, Island folk artist and retired farmer A.L. Morrison published a volume of paintings based on scenes from Island history, including one entitled *Sam Fletcher's Fort*. It portrayed members of the *posse* advancing cautiously on a mock cannon and a straw effigy at Fletcher's farm. The accompanying note by the artist describes the event as 'a farce in Island history.'[65]

The Continuing Search for Samuel Fletcher

It is probable that Fletcher had set in motion more than he realized. Months later, when defiance of the sheriff in the countryside had reached a crisis, William Pope's *Islander* would trace breakdown of law enforcement to Fletcher's successful evasion of legal authority.[66] An anonymous correspondent to Whelan's *Examiner* would unite the themes of absurdity and the deterioration of law and order: 'the sublime farce of the *posse comitatus*, ... set the whole country in a roar of laughter from East Point to West Cape. This ridiculous parade of empty authority on the part of the Government did more than anything else to increase the baneful influence of the Tenant Union.'[67] It was more than a joke.

Exactly one month after the last of the *posse comitatus* inductees returned to Charlottetown, the lawyer for Wood and Fanning moved in the Supreme Court for 'an attachment against ... Samuel Fletcher, for contempt of Court.'[68] The Court assented, and subsequently the prothonotary ordered Thomas W. Dodd, the new sheriff of Queens County, to 'attach' Fletcher 'to answer us for certain Trespasses and contempts,' namely, his escape from Curtis in Charlottetown on 17 March. The only further available information on this initiative is an undated statement by the sheriff that Fletcher had not been found.[69] It is not known how diligent or how passive Dodd had been in his search.

On 2 September the Executive Council read a letter from DesBrisay, who, as a justice of the peace, asked that he be authorized to hire constables to capture Fletcher. DesBrisay stated that it was important to arrest him because of his open and public defiance of the law on 17 March. Council agreed, and authorized DesBrisay to proceed 'with the least possible delay.' At the same meeting, Council read a letter from Charles Wright, the agent for Wood and Fanning, asking why Fletcher had not been proceeded against 'although a Warrant issued by Theophilus DesBrisay Esquire J.P. has been out against him for some months'; knowledge of Wright's letter may have prompted DesBrisay's show of interest.[70] In any event, these two letters suggest that Fletcher was becoming a symbol of popular and successful resistance to authority. If such was the case, then the government was at least partially to blame, since it had insisted upon sending the ineffectual *posse comitatus* after him on 7 April. DesBrisay's undertaking in September seems to have come to naught, although expenses were claimed in December.[71]

The Executive Council considered the case of the elusive Fletcher once again on 16 January 1866. It had been told that he was still on the Island, and apparently at Dundas's urging, a justice of the peace in Fletcher's vicinity – land agent and proprietor John Roach Bourke – was to be directed to arrest him. The result was another claim for expenses in April of 1866, but there was no arrest.[72] If one includes the original confrontation with Curtis in Charlottetown, and the directive to Dodd, about which little is known, this was the fifth attempt within a year.

In a document dated 31 December 1866 Fletcher assigned the 999-year lease for his one hundred-acre farm on Lot 50, 'together with all the Buildings and Improvements thereon,' to Patrick Cairns of Lot 49, in exchange for £270 local currency. Leaving his home on Lot 50 could not have been a decision he made lightly, for he had been there at least since 1853 and his farm must have been well developed for the leasehold interest to fetch £270. To convey some idea of what that sum meant, Fletcher's annual rent for the farm had been the equivalent of less than £6 local currency. The witness was McNeill. Then Fletcher slipped from the public records of Prince Edward Island.[73]

One prominent Tenant Leaguer, Manoah Rowe, would assert in a public letter dated 23 December 1867 that Fletcher had left the Island to avoid arrest. In his words, 'Every school boy knows that Samuel Fletcher was hunted day and night by the constables to take him until he left the Island.'[74] Rowe's letter has a certain credibility, for its wording suggests that it was the *renewed* searching for Fletcher which prompted him to leave. This is consistent with the report in January of 1866 that he was still on the Island. If Rowe's letter was correct, and if the inference drawn from it is valid, then it is probable that Fletcher decided to leave in the early months of 1866, when it became clear that the periodic harassment was likely to continue, and electoral defeat of the Conservatives was still many months in the future. Furthermore, he seemed to be taking a considerable risk by remaining on Lot 50, where, if the authorities tried enough times, they might succeed in capturing him. On 24 January 1866 the sentences of three Tenant Leaguers convicted in the Supreme Court of involvement in acts of physical resistance to peace officers had been harsh, in the view of many members of the public: between one and two years' imprisonment and between £20 and 50 in fines. The sentencing had clearly been calculated to make examples. This news would give Fletcher additional incentive to leave, for if arrested, tried, and convicted, he could expect no leniency from the judiciary. He had certainly disposed of his farm by the end of 1866; the arrangement with Cairns may have been made, and Fletcher may have departed, considerably before 31 December.

Where did Fletcher go? It is known that he died in Taunton, Massachusetts, approximately forty miles south of Boston, near the Rhode Island border, on 8 November 1870. He had been enumerated in the Massachusetts census of 1870 as living in Taunton with his wife and eight surviving children. He and two of his teenage sons worked at Britannia Mill there; Britannia is a white alloy of several metals, used principally in tableware. His two oldest sons were listed as carpenters, his oldest child, a daughter, age twenty-two, was at home with 'No Occupation' listed. The three youngest children, ages eight to thirteen, were in school.

Fletcher had sought refuge and a better life in the United States of America. His 'estate value' was listed as $5,000.[75] It may be said that he died in obscurity in a foreign land. That would be true, but it is quite possible that obscurity – getting on with his life – was all that he desired after the incident in Charlottetown on 17 March 1865. The Conservatives, led by the Pope brothers, had made it clear that he would not be allowed to remain in peace on Prince Edward Island. William Pope, in his editorial a week after the *posse* undertook its march, had written, somewhat prophetically, that 'five constables would have had no difficulty in securing their man, or compelling him [Fletcher] to leave the Country.'[76] Five attempts to take Fletcher were enough to persuade him to depart.

It is clear from the evidence on record in Prince Edward Island why Fletcher would want to leave. Enough was enough. The tin trumpets and continuing vigilance of neighbours provided a fair degree of security, and indeed the experience of one Tenant Leaguer, Joseph Doucette, in the late summer of 1865 would indicate that, for a while at least, Tenant Leaguers could supply those whom the authorities were exceptionally keen to capture with considerably more protection than that. But by the time Fletcher left, the league was disintegrating as an organization, and the forces seeking him had indicated by their actions that their search was relentless.

Why did Fletcher go to Taunton, Massachusetts in particular? Tenant Leaguers collaborated in warning each other of approaching sheriffs and constables, in protecting fugitives, and, as would become apparent by the autumn of 1865, in spiriting 'wanted' persons off Prince Edward Island. James B. Gay, a member of the Central Committee of the Tenant League, had lived in Taunton for an undetermined period of time in the 1850s,[77] and, given the pattern of cooperation, it is possible that Gay had used connections in Taunton to ensure that there was a place for Fletcher and his numerous family there.

Samuel Fletcher was one of the 'ordinary' Prince Edward Islanders who personified the strength and commitment of Tenant Leaguers. While it is true that he did not ask to be singled out and made into an icon of the league, he did follow its prescription, and not pay his rent, as well as signing league petitions. That is, he demonstrated more personal commitment to the league cause, in terms of putting his home, his personal 'property,' and the security of his family at risk, than William Cooper ever did to the Escheat movement.[78] He also went to Charlottetown on 17 March when he could have stayed home. Then, after the events of that day, thanks to Curtis's reckless action and his own refusal to be taken without a struggle, the ordinary leaguer became a symbol. Had Curtis bided his time and not acted on 17 March, almost certainly the *posse* would have marched to Fort Augustus and the Monaghan Settlement in search of some of Callaghan's relatives and neighbours, or somewhere else in response to some other incident – and Fletcher could have continued his existence as a farmer in Alberry Plains, Lot 50. Such is the way chance may alter the lot of individuals caught up in major social movements.

6

Landlords on the Retreat and under Attack

This lawlessness does not involve a contest simply between Mr. McDonald, of Tracadie, and his tenants, or between any other proprietor and his tenants. It is a question of whether the majesty of British law shall be upheld, or whether it must give way to the violent menaces of a rough, not over-intelligent, and unconstitutional organization like the Tenant League.
Edward Whelan's *Examiner*, 12 June 1865

The Decision to Grant No More Leases on the Cunard and Sulivan Estates

The march of the *posse comitatus* on 7 April 1865 had not put an end to unrest in rural Prince Edward Island. In fact, the twin images of official incompetence and successful popular defiance added fuel to a combustible situation. The tenantry was emboldened, and at least some of the proprietors showed signs of losing heart. On 2 May in Queens County an incident took place that resulted in charges being laid against three men for assault on Constable Bernard McKenna of Lot 48 'in the execution of his office.'[1] Little evidence survives concerning this specific occurrence, but the probability is great that it was related to McKenna's work in serving legal processes growing out of landlord-tenant relations.

On 13 May the *Protestant and Evangelical Witness* included a 'Notice' over the name of George W. DeBlois, acting as land agent for Sir Samuel Cunard, Edward Cunard, and Laurence Sulivan. He announced that the Cunard and Sulivan estates would not be giving out any new leases. Rather, they would be selling land at twenty or more shillings per acre.[2] These properties constituted 20 per cent of the Island. The change in policy was a significant turning-point in the history of the land question, and although Sir Samuel Cunard had died on 28 April in London, the relevant decisions appear to have been taken before his death.[3]

It is necessary to ask 'why?' and 'why then?' The change in policy could be traced in the first instance to the dilemma in which the Fifteen Years Purchase Bill had placed the twelve 'consenting' proprietors. If they wished to lease land to prospective tenant farmers, they would enter a situation in which they might be compelled by law to sell the same land to the same farmers at the terms specified in the bill, namely, fifteen years' rent plus a composition of arrears. 'Might' is an appropriate qualifying word, because section 6 stated explicitly that 'nothing in this Act shall extend to any lease made after the passing of this Act'[4] – that is, the formula for a purchase price applied only to land already leased. But when advocating passage of the legislation in 1864, Attorney General Edward Palmer, then also a member of the Island cabinet, had presented it to the Legislative Council as 'the first invasion of the rights of the proprietors ... the thin end of the wedge.'[5] Landlords whose freedom to sell part of their property had been limited by the Fifteen Years Purchase Bill could easily visualize an extension of the encroachment on the rights of property, through future legislative action, to cover their other lands.

Thus, by virtue of the rigid terms of sale that the statute of 1864 imposed on the consenting proprietors, and the precedent created, even Cunard and Sulivan were resigned to selling rather than leasing in future. At first glance this may seem ironic, for Cunard had helped to push the legislation through, using the argument that the alternative was dangerous agitation in the countryside. Whether he had foreseen in 1863–4 where the restrictions of the Fifteen Years Purchase Bill would lead, in terms of estate policy, is difficult to determine. He was certainly an experienced magnate, but at the time he was responding to a danger of disorder which he perceived as immediate and as requiring that a concession to the tenantry be made without delay. Perhaps, under the pressure of circumstances, he did not think through the logic of the situation as thoroughly as he might have done on other occasions. In any event, whether he was witting or unwitting as to the consequences of the bill, the change in policy to which it led marked recognition that the leasehold system had reached the limits of its expansion as far as the Cunard and Sulivan estates were concerned.

But only a superficial analysis of the causes behind the decision to stop granting leases would conclude that the chain of causation ended with the Fifteen Years Purchase Bill. That law was, as already noted, a direct result of the fear of rural disorder. In the words of William Henry Pope when he was a delegate in London early in 1864, 'reasonable concessions ... [were] essential, in order to secure the Colony, generally, from those much-to-be-dreaded evils which necessarily result from widespread agrarian agitation.'[6] The act had been intended as a preventive measure; but, as the product of a proposal Cunard made in December of 1863, it did not meet the standard of reasonableness Pope stated was necessary to avoid breaches of the peace. The growth of the Tenant League and the defiance of the forces of the sheriff indicated that the legislation had not worked as Cunard had hoped, and that Pope's prediction had been accurate.

There would be more agitation, more pressure, and probably more reform measures. The proprietors would have to protect themselves as best they could. Ceasing to grant leases was one means of self-protection. It defended their interests as property owners in two senses. If leasing might lead to compulsory sale according to the formula of the Fifteen Years Purchase Bill, it would be preferable to avoid that trap and preserve the freedom to sell on the best terms that could be negotiated. Secondly, farmers who did not become tenants would never become discontented, rent-resisting tenants, and therefore would not be a direct problem to land agents and proprietors.[7]

Such is a valid interpretation of the underlying logic in the change in policy from the perspective of the proprietors and others interested in protecting property rights and limiting agrarian agitation: a wise strategic move. But although this explanation of why the change occurred and why it occurred when it did is satisfactory as far as it goes, it is incomplete. In its own round-about way, the change in policy was also a step towards the end of leasehold tenure on Prince Edward Island, and a decision driven, ultimately, by fear of the tenant movement. Because of the size of the estates affected, the significance of the decision DeBlois announced in the *Protestant* on 13 May was immense. It was a sign that the old order was crumbling. More than managerial wisdom lay behind the disintegration, and indeed the best that strategic wile could suggest in this case was damage control: no more leases, and (1) therefore no further sales under the terms of the Fifteen Years Purchase Bill, and (2) therefore no additional disgruntled tenants. Popular agitation was the root cause of the change in course by the Cunards and Sulivan, since it was the source of the potential injury to their interests as property owners that they wished to prevent.

This virtual surrender by the Cunard and the Sulivan estates *as lease-*

hold estates with a potential for growth can be chalked up as another significant victory for the forces to which the Tenant League gave shape. Even proprietors with the largest estates and the best connections in London seemed to be conceding that the leasehold system did not have a future on Prince Edward Island.

Sale of the Estate of the Reverend James F. Montgomery

There were no better representatives of the proprietary system in its absentee form than the Montgomery family, descended from Sir James Montgomery (1721–1803), an original grantee. The evidence of the 1861 census indicates that the various members of the family who held land on Lots 34, 51, and 59 pursued conservative policies: 999-year leases and sterling rents at one shilling per acre. On the same day that news of the decision of the Cunard and the Sulivan estates to grant no more leases appeared in the *Protestant*, tenants on the estate owned by the Reverend James F. Montgomery of Edinburgh met on Lot 34. At the 1861 census only 21.9 per cent of occupiers of land on the township were freeholders, an unusually low proportion for eastern Queens County; 74.3 per cent were leaseholders, 3.8 per cent held their farms 'by verbal agreement,' and there were no squatters. Three members of the Montgomery family had title to the proprietary land.

The purpose of the meeting on 13 May 1865 was to consider a letter from the Reverend James F. Montgomery, formerly a practising lawyer, to a committee of residents.[8] He offered his part of Lot 34 for sale to his tenantry in exchange for £4,000 currency, with remission of all arrears, 'provided always, that the present year's rent, and of all subsequent years, until the transaction is completed, are regularly paid.' The offer was conditional upon his entire property on the township, amounting to 6,014 acres, being sold in a single transaction for which the tenants would bear the expenses; he would also have to be paid in a lump sum. This price for his portion of the township was considerably less than the formula of the Fifteen Years Purchase Bill would yield if applied to his rent roll – £1,146 less, by his calculation, taking into account only the fifteen years rental and ignoring arrears and expenses. Unless the tenants adhered to his conditions, he would insist on the terms of the bill as a minimum price.[9] In setting his conditions Montgomery was conforming to the conventional belief that one should not sell off parts of an estate piecemeal. As he explained, 'to sell separate farms is very disadvantageous for me; as in the event of many being sold, there would remain

TABLE 6.1

Tenurial status among land occupiers on three townships identified with the Montgomery proprietary family in 1861, expressed in percentages

	Freeholders	Tenants	Squatters	Total	Number of Occupiers
Lot 34 (Queens)	21.9	78.1	0	100	237
Lot 51 (Kings)	28.2	71.3	0.6	100.1	174
Lot 59 (Kings)	39.8	59.7	0.5	100	206

Note: Calculations based on PEI, Assembly, *Journal*, 1862, app. A

only the skeleton of a property which would be troublesome and expensive to manage.'[10] Already, some months earlier, according a petition presented to the House of Assembly, he had declined an offer to buy farms individually, but in so doing had encouraged his tenants to combine to purchase his entire property.[11]

After a discussion in which 'doubts,' 'difficulties,' and 'fears' were raised and apparently put to rest, the meeting accepted Montgomery's conditional offer. In a series of five resolutions which passed unanimously, those present declared, *inter alia*, that the proposal was 'exceedingly liberal and generous,' and that a letter thanking the proprietor should be written. The tenants at the Lot 34 meeting clearly saw themselves as part of a larger process, for in one resolution they expressed the hope that other proprietors would emulate Montgomery's conduct.[12]

What prompted Montgomery to sell when he did? What did the sale represent? In sworn testimony before the land commission of 1875–6, a first cousin, also named James F. Montgomery, stated that 'His [the Reverend James's] principal reason for selling was that, having entered the church, he did not wish to hold land where there was so much disturbance, and so much ill-feeling.' He went on to assert that his relative 'was ignorant of the value of the land, never having been here. I do not consider that he received half the value of his land.' Later in the hearings he offered this explanation for what he believed to be an unreasonably low purchase price: 'He is pretty wealthy and it was not a matter of moment to him. He is otherwise independent.'[13] The fact that the land commissioners of 1875–6 had the task of setting compensation rates for lands which were about to be expropriated, including the witness's own part of the same township, may have coloured his testimony concerning the value of the land his cousin had sold in 1865. But there is certainly other support for the view that the terms of sale were liberal.[14]

Did the Tenant League play a role in negotiating the terms? The Reverend James Montgomery wrote a letter that was published in the Island press in October of 1865, while the arrangements for the sale were in process, denouncing the league and stating that 'I will make no terms with any person who belongs to that association.' He also indicated that he had been 'assured that the tenants of Lot 34 have entirely held aloof' from the activities of the league.[15] The letter was addressed to his agent, T. Heath Haviland Jr, who was, in addition, solicitor general of the Island, an assemblyman, and a member of the Executive Council. Haviland declared that his intent in publishing it was 'to disabuse the public mind relative to the report that he [Montgomery] is about to sell his Property, through the influence of the Tenant League.'[16]

If the testimony by his first cousin a decade later is to be believed, the Reverend James Montgomery was aware that there was strife on the Island, wished to avoid conflict with his tenants because of his change in career, and was sufficiently affluent that he did not concern himself with maximizing the financial return he received. Neither the letter read at the meeting of 13 May 1865 nor his former occupation as a lawyer indicates the sort of other-worldliness that would render him oblivious to material affairs. But if the clergyman-lawyer's published letter to Haviland is to be taken at face value, then he appears to have been imperfectly informed about proceedings on Lot 34. Tenants on that township had put their position on record in March of 1864. At a public meeting a committee of seven had been 'appointed' to set in motion a process of negotiation with the proprietors for the purchase of their lands. If the process failed, the committee resolved that tenants on Lot 34 would be justified in 'resisting to the utmost of their power, legally, all coercive measures that may be employed by the Proprietor for the collection of rents.'[17]

This would become the standard doctrine of the Tenant League, and the committee of seven included David Lawson and Alexander Robertson, who would soon become founding members of the league. Lawson would also be a founding member of the Central Committee.[18] At the meeting on 13 May 1865 where Montgomery's letter to residents was read, another committee of seven was struck, including Robertson, although not Lawson. Possibly Lawson was too prominent a member of the league for it to be prudent, in view of proprietary sensitivities, to have him publicly associated with the negotiations; on 30 August he would be dismissed as a captain in the Volunteers.[19] But whatever the considerations surrounding composition of the committee, the purchase fitted the league prescription.

While it may have suited Montgomery and his Tory land agent to make a public show of distancing themselves from the Tenant League, there is no evidence that anything came of his threat not to sell to members. Making the sale and eliminating the threat to the tranquillity with respect to temporal matters which he as a clergyman apparently desired seem to have been more important than vetting the credentials of individual buyers. The letter to Haviland, though, would be a convenient device for opponents of the league, like Haviland, to use in countering its propaganda; and providing the letter would be fitting for a respectable lawyer and clergyman like Montgomery, since it is unlikely that he would have any reason to wish to assist the league or be associated in anybody's mind with it. Certainly the letter appears calculated to establish distance between himself and the league, for he wrote that 'I have learned more than I had hitherto known of the improper and illegal proceedings of ... the Tenant League. ... [which] I think it is the duty of every lover of order, to discourage and check to the utmost of his power.'[20] It is conceivable that he had come under some pressure to provide such a disclaimer, for Lieutenant Governor George Dundas, on leave in Great Britain, was in Scotland late in the summer and visited him, quite possibly around the time he would have penned the letter. The wording and tone coincided exactly with Dundas's sentiments.[21]

The transaction was not finalized until a meeting of tenants on 14 June 1866. According to a report in the *Examiner*, the price per acre amounted to 13s. 9d. currency, which meant that the cost of a 100-acre farm was £68 15s. Edward Whelan wrote that 'When it is remembered that in times past a sovereign [£1 sterling, which meant 30s. currency] an acre has been refused for certain farms on this estate, it is to be regarded as [a] matter of congratulation that such eligible terms have now been secured.'[22] A letter from Robert Robertson, the chairman of the meeting, a schoolteacher around the age of retirement, referred to 'the long delays and many obstacles encountered,' without being more specific.[23] In the previous month Kenneth Henderson, a cabinet member, had stated in the Legislative Council that 'considerable impediments [which] were thrown in the way ... arose from the general spirit of landlordism.'[24] Possibly Henderson meant that other proprietors pressed Montgomery not to sell.

No doubt, some landlords were as unhappy as the tenants were pleased over Montgomery's announced plan to sell prime land in a prime location for considerably less than the terms of the Fifteen Years Purchase Bill. Their displeasure would have increased upon reading a letter in the *Islander* of 22 December 1865 from Robert P. Haythorne, the former proprietor of part of Lot 49 and, by an odd juxtaposition, also a tenant of Montgomery on Lot 34. Haythorne expressed the hope that in the event of another land commission, the price paid to Montgomery 'may serve as a maximum standard of value.'[25] Given that the estate was widely held to be among the best and most valuable parcels of land in the colony, the other proprietors would have interpreted such views and the impending precedent as a threat to their pecuniary interests. In any event, Henderson credited Dundas with persuading Montgomery to carry through with the transaction. But many years later, in testimony under oath before the land commission of 1875–6, Haythorne stated regarding the Montgomery purchase that 'I concluded it myself in Edinburgh.'[26] At the very least, it seems that Montgomery received more than one visitor from the Island between May 1865 and the conclusion of the purchase.

Although certitude concerning all details is impossible, given the nature of the evidence, enough is known to ascertain the significance of the sale in the general history of the land question. Certain points are beyond debate. In the first place, the purchase conformed to Tenant League doctrine. Secondly, the transaction was highly unusual. As Haythorne stated before the land commission of 1875–6, 'Only Mr. Montgomery's and my estates were sold direct to the tenants. ... without the intervention of the Government.'[27]

Haythorne's generalization appears to have been correct, although there had been one other incident in which a landlord apparently tried to duplicate his transaction. In a letter dated 6 May 1865 and read two days later to 'a local meeting of the members of the Tenant Union of Lots 49 and 37,' John Roach Bourke offered to his tenants on Lot 37 terms similar to those Haythorne had given on Lot 49. According to a published report, negotiations had been conducted between Bourke and George F. Adams; the meeting passed unanimously a resolution leaving further arrangements in the hands of Adams. At least some of the press put a highly optimistic interpretation on this information. Edward Reilly treated the purchase as a *fait accompli* and cited it as evidence that 'the Tenant Union is working admirably and doing good work.'[28] Yet in the end no bargain of this sort for land on Lot 37 was concluded.

It is always difficult to explain the non-occurrence of an event (in this case, the non-purchase of Bourke's land by his tenants on the Haythorne terms) with assurance. But there were two obvious reasons that the Bourke scheme came to nothing. The value of the Lot 37 land was much less than that of the Haythorne estate, and Bourke's tenants were considerably poorer – that is, less able to pay – than Haythorne's. Whelan had been sceptical about Bourke's proposal from the beginning. Under the sarcastic heading 'A Landlord's Generous Offer,' he argued that many occupiers of land in the area, known as Little Hell, were squatters who had never acknowledged Bourke's title; by his reasoning, any 'purchase' would therefore be a major *coup* for Bourke, who would be collecting money where he had been unable to do so before.[29]

Later in 1865 the ubiquitous Bourke, wearing his hat as land agent for Viscount Melville, spent some time with Adams and tenants on Lot 29 in preliminary discussions of another possible direct purchase. But matters never reached the stage of negotiations, and certainly nothing concrete resulted. Perhaps the most noteworthy aspect of both stories regarding Bourke is that civil relations were possible between Adams, a leader of the Tenant League, and a proprietor and land agent, even one who was also a justice of the peace, and even after the arrival of soldiers on 6 August 1865 to bolster the civil authority. On 21 August Bourke and Adams, who lived in adjacent districts in eastern Queens County, had travelled together from Charlottetown to a meeting twenty miles west of the capital on Lot 29, in a carriage hired by Adams. The Tenant League was clearly a credible body as a possible instrument of negotiation between landlord and tenant, and when Bourke wrote to the *Islander* a few days later stating that a purchase was unlikely, given the price the tenants were willing to offer, there was no hint of acrimony in his references to Adams.[30]

Despite much newspaper attention to Bourke's activities with regard to Lots 37 and 29, only two direct purchases conformable to the prescription of the Tenant League were ever – from 1767 forward – made. Both the Haythorne and the Montgomery purchases were arranged less than a year after the formation of the league. Was the Tenant League involved? Even a *prima facie* assessment suggests that it is improbable that Tenant Leaguers would be uninvolved in the only two estate sales which adhered to their prescription, especially when the timing and Tenant League activism in eastern Queens, where both occurred, are taken into account. It is worth considering the testimony of both their friends and their enemies.

On 21 June 1865 tenants from the three parts of Lot 34 ('central,' 'northern,' and 'southern') met, and passed a series of resolutions. The first referred to the agreement between the Reverend James Montgomery and his tenants on the southern portion as having been made 'not *directly* under the auspices of the Tenant Union.'[31] The inference is plain: that the league was involved *sub rosa*. This was the view of persons friendly to it, and indeed those present raised £5 9s. 7d. for its support, although passing a resolution declaring themselves 'not members of the Tenant Union.' Curiously, the mover of this resolution was Lawson, a member of the Central Committee. The secretary of the meeting was John Bassett, a league activist from Lot 22, and he and Lawson made the first two speeches which were reported. Thus, regardless of formal disavowals, which in fact had a perfunctory appearance, the Tenant League could definitely make its weight felt on the township, and there were residents who were willing to support it financially.[32]

As a general rule, Whelan was not inclined to give much credit to Tenant Leaguers. But in September of 1865 he wrote, in the midst of an article severely critical of the league and especially its leadership, that 'They have been in treaty for it [the estate of the Reverend James Montgomery], we know.'[33] Dundas, who was equally unfriendly to the league, if not more so, would acknowledge in a despatch to the Colonial Office in 1868 that the league 'claimed' that it had facilitated both the Haythorne and the Montgomery purchases. He did not express an opinion, positive or negative, about the accuracy of the claim, a significant fact, given the way he usually reviled the league and its leadership.[34] The failure of Whelan and Dundas to dismiss outright reports of Tenant League involvement suggests that leaguers played an important role, for neither had a record of passing up opportunities – Whelan in his newspaper, Dundas in communications with Downing Street – to denigrate the pretensions or accomplishments of the league.

How does this fit with Montgomery's letter of 25 September 1865 pledging not to deal with league members? What about the Ninth Commandment? A lawyer and clergyman, if pressed, would probably fix upon the fact that the actual mechanism for the purchase involved a transfer of funds in cash from the Union Bank to him, and payment by the tenants of the price for each farm to the bank in six annual instalments, with provision for payment of interest on unpaid balances to the lending institution. Technically, Montgomery did not deal directly with his tenants; the bank did. Such an explanation of how it was possible to transfer title of his land, effectively, to members of the league and receive cash in return without actually selling to or 'making terms with' them would probably not be beyond his ingenuity.[35]

One suspects, from Montgomery's indifference to the real value of the land, that he was taking a strictly results-oriented approach to the process. He wanted to be rid of the entire property at one stroke, to make a clean break, and to do so expeditiously. This attitude was also visible in his failure to press tenants for arrears. At the time of sale the total was apparently £1,200 (the equivalent of three and one-half years' rent), and even Haythorne, who emerged as a spokesman for the tenantry on the estate, stated more than once that the tenants were able to pay.[36] The shortfall resulted either from inattention to collection or disinclination to pay or some combination of both. Between 1861 and 1864 the arrears had increased by £292 4s. 7½d.,[37] and in the spring of 1865, when Montgomery decided to sell, future prospects for collection were worsening and the likelihood of embarrassing strife was increasing. Getting completely clear of a potentially troublesome property and doing so in a way that did not leave him in a continuing and possibly conflictual relationship with his erstwhile tenants were his priorities. If a visit from the Island's lieutenant governor and support for the general principle of property rights suggested the expediency of a letter disavowing support for the Tenant League, he could comply at no cost to himself.

The Montgomery purchase was yet another step towards the eradication of leasehold tenure on Prince Edward Island, and it should be grouped with the Haythorne purchase. Regardless of any mechanism the coy could use as camouflage, each transaction followed the prescription of the Tenant League: a group of tenants purchased directly from the landlord without waiting for the intervention of the Island government. Montgomery's decision to sell, like the change in policy by the Cunards and Sulivan, is attributable to tenant militancy and what it could produce. Even on the assumption that no Tenant Leaguers played any direct role in the proceedings (which runs contrary to the thrust of circumstantial evidence), the Montgomery purchase, one of a unique pair of such agreements, cannot be understood without reference to the league and the climate it created.

Arson on the McDonald Estate

By the time reports of a possible purchase of Melville's half of Lot 29 appeared in print, the pace of events had quickened greatly, and discussion of issues surrounding the Tenant League had become considerably more acrimonious. Much of the deterioration in civility could be traced to a fire on the property of landlord John Archibald McDonald in Glenaladale, Lot 36 on the night of 27 May. A barn and a stable, valued at £150 and uninsured, burned. The fire was apparently the work of a person or

TABLE 6.2
Tenurial status among land occupiers on the two townships identified with the McDonald proprietary family in 1861, expressed in percentages

	Freeholders	Tenants	Squatters	Total	Number of Occupiers
Lot 35 (Queens)	4.6	94.8	0.6	100	173
Lot 36 (Queens)	2.7	93.1	4.3	100.1	188

Note: Calculations based on PEI, Assembly, *Journal*, 1862, app. A

persons angered by the action of the sheriff's bailiff, accompanied by the proprietor, in serving writs on known Tenant Leaguers for arrears of rent. According to Whelan, 'The tin trumpets [of leaguers] were blown, but too late.'[38]

Some months afterwards Chief Justice Robert Hodgson, acting as administrator in the absence of Dundas, stated that the affidavits which reached the Executive Council concerning the fire 'ascribed the destruction of the barns to the serving of the Writs.'[39] Dundas, who had been on the Island when the fire occurred, reported to British Colonial Secretary Edward Cardwell that 'it was suspected that the act was that of an incendiary connected with the League.'[40] George Coles, as one of the assemblymen for the area, took the view that the fire was accidental. Somewhat incredibly, despite the opinion of virtually everyone else who spoke out on the matter, he was adhering to that line even in 1868.[41]

An earlier fire in Glenaladale had already destroyed McDonald's residence. It had occurred on 14 April in the middle of the afternoon, and a crowd of two hundred or more neighbours – summoned, ironically, by the trumpets of Tenant Leaguers – had worked to save much of the furniture from the flames. Newspaper reports gave no hint of arson on that occasion,[42] although there had been a history of fire-setting on the estate. In 1851 persons hostile to John Archibald's father, Donald, had set a series of fires which destroyed several buildings. The son was similar to the father in his tendency to make enemies. Both were known as unyielding, hot-tempered men who did not hesitate to use harsh means against tenants.[43] In testimony before the land commission of 1875–6 John Archibald stated 'Where they [tenants] did not pay I either sued or distrained. ... I would take a pound of money even where it would have to come from a pound of flesh.'[44] An anonymous letter in the press from a resident of an adjacent district revealed the depth of bad feeling in the area. Although accepting implicitly the theory that the 27 May blaze was the result of a deliberate act, 'Philos' suggested the possibility that the owner had set both it and the earlier fire. Regarding the house, the writer reported that 'Dame Rumor says it was insured for twice its value.'[45]

In a proclamation dated 6 June, Dundas, with the advice and consent of the Executive Council, offered a reward of £500 for information leading to conviction of the person or persons responsible for the 27 May fire. But charges were never laid and the reward was never paid. From the beginning, Whelan was sceptical about the motives of the Island government, and concluded that the magnitude of the reward indicated they did not really expect to apprehend the Glenaladale arsonists and have to pay out. For most Islanders, £500 was an enormous amount, and he believed that those who could be tempted to inform would be lured for much less. In other words, if the local government had anticipated a successful investigation and subsequent payment of the reward, they would have offered a considerably smaller sum – a plausible inference, since in the following year, after an immeasurably more costly fire in Charlottetown, involving the destruction of four city blocks, the same government would offer £300, that is, 40 per cent less, for information.[46]

The *Islander* of W.H. Pope published a lead editorial article entitled 'Agrarian Outrage,' a term associated particularly with Irish experience. Historian Michael J. Winstanley has noted that in Ireland 'From 1844 a separate record of "agrarian outrages" was compiled, although the criteria for distinguishing these from other rural crimes always remained obscure.' Incendiarism, the maiming of animals, and the sending of threatening letters were the most frequent occurrences which the authorities classified as 'agrarian outrages.'[47] The very fact that separate monthly statistics were kept on a county-by-county basis indicates that 'outrages' were a significant feature of life in Ireland, whereas in Scotland, historian Eric Richards has written, concerning the Highland Clearances, that 'Through eighty years of discontent, I can find only three instances of "outrage."'[48]

Pope attributed the Glenaladale fire to the serving of the writs, resulting in turn from compliance with Tenant League principles on the withholding of rent. He concluded that 'The conflict between the Tenant League and the Law is fast approaching a crisis,' and made an explicit comparison with Ireland. 'Houses and Barns may be burned, Landlords, Agents, Bailiffs and Constables may be beaten or murdered here, as they have been in Ireland, but by and by all these violences will result in the discomfiture of those who commit them.'[49] Although acknowledging the relatively hard lot of tenants on 'the McDonald Estates,' he asked rhetorically whether 'by shooting at the Proprietor, or burning his Barns, they will obtain indulgences or benefits?'[50]

Moving on to the more general situation on the Island, Pope warned that the government intended to uphold the laws, and that not even the election of a crypto–Tenant League government would lead to suspension of law enforcement. If the local government would not support the law, the imperial government would do so at the financial cost of the inhabitants.[51] This line of argument suggested that Pope was beginning to perceive a potential electoral threat in the widespread backing for the movement. By the end of June 1865 both he and Whelan, bitter opponents of many years' standing, were hinting at a reconstruction of political parties. Pope had commenced the dialogue with an editorial in which he praised Whelan for having spoken to the tenantry 'in language at once vigorous, honest, and kind,' and reprinted a lengthy passage from an *Examiner* editorial, with italics and capitalization added in several places for emphasis.[52] Whelan, for his part, noting that two of sixteen Liberal MPPs (twelve assemblymen and four legislative councillors) had contributed money to the league, defended his own consistency, and alluded to the possibility of 'moderate Liberals and Conservatives' forming a coalition.[53] Pope responded by reprinting in full the musings by Whelan on the coalition idea, and by commending his conservatism in a conciliatory and even flattering tone.[54]

Shortly after the Glenaladale fire, Whelan had mounted a concentrated attack on the Tenant League. Although he stated unequivocally his belief that the blaze had been set deliberately by persons antagonistic to the landlord, and reported also that the league was being blamed, he did not accuse it of direct involvement.[55] He declared that the Tracadie tenants

are, taken as a whole, the poorest class of their kind in the Colony. Their rents are high, their leases shamefully short, and their arrears of rent have risen to such an amount as to render liquidation impossible. Of personal property, stock and farm implements, they have little or none; and improvements about their farms are almost imperceptible. They never had any heart to make improvements – how could they on a forty year's lease, and a rental of two shillings an acre, beside the arrears that have been accumulating for years? It was a cruel thing to entice unfortunate people to settle upon any property on such conditions. Poverty, squalor, and discontent, might surely be expected to be their companions through life. There was no inducement to thrift and industry, by which property might be accumulated, because that property would inevitably go to meet the demands of the landlord, which, in many cases, had been created by previous holders of the farms.[56]

Although acknowledging the especially difficult situation of tenants on the Tracadie estate, Whelan emphatically refused to excuse 'the abominable crime of arson.' Unleashing his pen and using language calculated to have an impact in rural Prince Edward Island, he declared that 'The house-burner and the murderer are twin ruffians, who carry on a partnership concern for the delectation of the Arch-Fiend, and against whom all society proclaims unrelenting warfare.' He also attacked the 'tenant's pledge,' and used a phrase, namely, 'the deluded tenantry,' which had been, over the years, the property of conservatives who wished to emphasize the gullibility of simple tenants misled by demagogues. He warned that resistance to the law would 'be put down with disastrous consequences to themselves.' The association between the crime on the Tracadie estate and the doctrine of the Tenant League was, in his view, clear. In response, Ross's Weekly censured him in two successive issues. He would not have been surprised, for he had stated that he expected to be 'reviled' and 'execrated' for his condemnation of the league as 'based upon a most dishonest and unsound principle,' namely, resistance to the law.[57]

The Glenaladale fire and the attacks by Ross's Weekly marked a turning-point in Whelan's position on the Tenant League. His article on 19 June, two and one-half columns in length, is particularly important. As well as a critique, it is the most penetrating commentary on the changing geography of league strength ever to appear in the Island press. This was a performance characteristic of Whelan, who often produced his best political analyses when under the most intense political pressure. He claimed that the league had become essentially a Queens County organization; that it was collapsing in southern Kings, its area of initial support; and that its only backing in Prince was on Lot 16, part of the Sulivan estate. In terms of the major contours of league support, Whelan's analysis probably contained much truth, and certainly fits with other evidence. Prince County seems to have played a relatively minor role throughout the history of the league, and there can be no doubt about the formidable strength of the organization in Queens. All recorded breaches of the peace during 1865 associated with the league occurred in Queens. This was a dramatic contrast with the history of the old Escheat movement, for much of the confrontation connected with it had occurred in the extreme eastern and western parts of the Island.

Whelan's comments regarding a decline in support for the Tenant League in southern Kings County are most suggestive. After the convention of 19 May 1864, when the formal organization of the league was established, the people of southern Kings had had little control over the movement. Eastern Queens delegates had taken over the leadership, a fact which might weaken support elsewhere. Between 19 May 1864 and publication of Whelan's editorial exactly thirteen months later, there were also developments in southern Kings which might, for an entirely different reason, reduce the attractiveness of the league for tenants of the area. The league was an organization advocating drastic measures to deal with severe problems, and as the severity of tenants' grievances in southern Kings diminished, and as their hopes for relief through more conventional means rose, so might support for the organization decrease. On 2 August 1864 the Island government had announced that under the terms of the Land Purchase Act of 1853 it was acquiring the land Robert Montgomery owned on Lot 59; according to a report of the commissioner of public lands, 4,243 acres were involved. At the 1861 census only 39.8 per cent of occupiers of land on that township, immediately north of Sulivan's Lot 61, had been freeholders, and such an acquisition, which meant resale to tenants, would have been welcome news.[58] In May of 1865, under the same statute, the government purchased the property George and William Montgomery owned on Lot 51, which amounted to 4,151$\frac{1}{2}$ acres; and under the agreement with the proprietors, the tenants affected by the purchase were to have £1,200 in arrears remitted. On Lot 51, whose southern boundary touched Lot 59, only 28.2 per cent had been freeholders in 1861.[59]

These two purchases by the government had occurred on townships where, at the 1861 census, the incidence of freeholding had been far below the general Kings County level of 49 per cent. They may have combined with the rent reductions for tenants of Sir Samuel Cunard in southern Kings, announced by DeBlois in January of 1865, to undercut support for the league in the area. Moreover, by late May of 1865, eight thousand additional acres on Lots 51 and 59 were being advertised for sale, apparently on behalf of James Montgomery of Warwickshire, uncle of the Reverend James and one of the proprietors who had consented to both the land commission of 1860 and the Fifteen Years Purchase Bill.[60] With some measure of redress for their grievances, and without influence in the organizational apparatus of the Tenant League, it is understandable that the tenants of southern Kings would become less militant, and reduce their support for the league, just as Whelan asserted.

Whelan replied in strong language to the attacks by *Ross's Weekly*, while emphasizing his belief that John Ross had not written the articles and no longer controlled the newspaper.[61] It appears that *Ross's Weekly* was misrepresenting Whelan's published views on the origins of the Glenaladale fire. Unfortunately, the relevant issues are missing, but there is no mistaking Whelan's sense of outrage. He termed one purported direct quotation 'a gross forgery' and denounced a statement characterizing his editorial article of 12 June as 'another deliberate falsehood.' He defended his record in promoting the interests of Island tenants, and tried to deflate leaguers' claims respecting their contributions to eradication of leasehold tenure on Lots 37 and 49. He even stated that the Tenant League 'has done nothing for the tenantry.' Belittling the leadership, he declared that 'The League ... has no men of mark or influence in its ranks.'[62] The next week he went further: 'no man of character or influence cares to see his name used in connection with it,' a statement which disparaged the rank and file as well as the leaders. Emphasizing his distance from the movement, he stated that 'never – never have we countenanced the abominable socialist and seditious doctrines of the League.' The organization itself 'is shunned as we would shun a leprous thing, by the lovers of law and order generally.'[63]

Three days later a meeting of Tenant Leaguers at Tracadie Cross Roads, near McDonald's home, adopted a resolution declaring Whelan 'a bitter enemy of the tenantry of P.E. Island.' The mover was James Callaghan, the tavern keeper Deputy Sheriff James Curtis had been seeking on 10 March.[64] Whelan was soon attacking more and more Tenant League leaders, over more issues, and in stronger terms. Thus, one significant consequence of arson on the McDonald estate was a drastic worsening of relations between Whelan and the Tenant League; already, over the issue of the planned purchase of the Bourke property on Lot 37, Whelan had had words of scorn for the claims of the league. From the time of the Glenaladale fire onwards, the relationship became one of incomprehension at best, but more often visceral enmity.

Whelan had been the leading reform journalist on Prince Edward Island for more than two decades, and the major social movement of the time was passing him by. It had, as he noted, excluded Coles and himself from the start, on the basis of their being legislators. The rationale for the exclusion is important, because it underlined the difference in perspectives, the league committed to direct action and lacking faith in the legislative process, and the traditional Liberal leadership dedicated to working through the parliamentary system and recoiling from extra-parliamentary means. Whelan emphasized where the logic of league tactics led. Resistance to the civil power, if successful, 'must be put down by some means, and we know of no other but military force. ... and it is not difficult to guess where the blowers of the tin trumpets would be in a contest with the military.'[65] There seemed to be no basis for accommodation between Whelan and the league.

The rhetorical heat had increased in the wake of the fire on the McDonald estate. Divisions within the Liberal party were becoming visible, as Coles and Whelan, partners since the struggle for responsible government in the 1840s, put radically different interpretations on the fire. Although they shared the same general viewpoint on the Tenant League and a commitment to parliamentary institutions, the Glenaladale conflagration drove a wedge between them. To Coles, the party leader, it was a case of some of his constituents – loyal Liberal voters and Roman Catholics who had stood by him, an Anglican, during the controversies over religion, politics, and education in the elections of 1858, 1859, and 1863 – being blamed for an event respecting which nothing had been proven conclusively. But Whelan, the leading Liberal journalist, rejected this theory categorically: 'is there any reason to suppose that it [the fire] was accidental? Not the shadow of a reason.'[66] To him, it was a clear case of arson and a sign of dangerous deterioration in standards of behaviour, for which the league leadership bore at least partial responsibility.

Ironically, the controversial fire took place just two weeks after the Island press published news of major advances for the tenant movement, namely, the change in policy by the Cunards and Sulivan, and the decision of the Reverend James Montgomery to sell. It is impossible to determine with precision whether the fire and the debate surrounding it accounted in any way for the aborting of the two initiatives associated with Bourke. But the explanations Bourke and Whelan offered concerning Lots 29 and 37, respectively, seem likelier.

The retreats of the Cunards, Sulivan, and the Montgomerys heartened the tenantry and raised concerns for landlords who planned to continue to give out leases. The Glenaladale fire alarmed some proprietors. Lady Georgiana Fane wrote to Cardwell that 'incendiarism has begun.'[67] The fire did not deter McDonald from claiming his rights; on 19 June in an advertisement dated from Charlottetown, he called upon his tenants to pay up at once 'as no other application will be made.'[68] On 10 July DeBlois announced that the collection of arrears of rent due to Edward and William Cunard, sons of Sir Samuel, under the terms of the Fifteen Years Purchase Bill, that is, for arrears accrued since 1 May 1858, 'will be strictly enforced.' Those who failed to make immediate arrangements with him at his office in Charlottetown would be sued 'without further warning.'[69]

Tenants felt new strength, proprietors felt new fear. Both sides were ready for increased confrontation. Tenants had reason to believe the leasehold régime was on the retreat, but landlords were prepared to insist upon their legal entitlements and to set in motion the machinery of law enforcement.

7

The Arrest and Bail Hearing of Charles Dickieson

It has been reported to me that immediately Sheriff's officers are observed to enter a settlement, tin trumpets are blown in all directions, many of the inhabitants assemble, surround the officers, blow trumpets in their faces and insult and defy them.
Administrator Robert Hodgson to British Colonial Secretary Edward Cardwell, 2 August 1865[1]

A man came up on a grey horse ... in a gallop. The man was Charles Dickieson. He said as he galloped past 'You buggers will see that you shall take bail.'
The examination and evidence of James Curtis, Deputy Sheriff of Queens County, taken this 26th day of July 1865[2]

Rescue him [Dickieson]! rescue him!
The reported urging of a Tenant League leader from the Lot 49 area on 26 July 1865 in Charlottetown[3]

The Serving of Legal Processes and the Case of George Clow

The procedures that came into play when a proprietor or land agent took legal action against a tenant for non-payment of rent were central to the next significant sequence of events in the history of the Tenant League. The landlord could call upon the sheriff or his deputy to deliver to the tenant a writ of *fieri facias* or *fi. fa.* It directed the sheriff 'to levy from the goods and chattels of the debtor a sum equal to the amount'[4] specified in the writ; once the sheriff had made the seizure, he was to sell the goods by auction. If the debtor – the tenant in this case – evaded the distraint, the next step was the more drastic action of a writ of *capias ad satisfaciendum* or *ca. sa.*, which directed the arrest of the defendant who failed to satisfy a judgment against him for a sum of money.

These were the major legal weapons, short of eviction, used against tenants. If it was likely that assistance would be necessary, the sheriff or his deputy might enlist bailiffs or constables hired on a *per diem* basis. Given their position on the front line of law enforcement, sheriffs, deputy sheriffs, bailiffs, and constables were certain to become the foci of strife and bad feeling as tension escalated. The *Examiner* reported on 3 July that 'a few days ago' on Lot 31 in western Queens County Deputy Sheriff James Curtis had been 'prevented by a mob from executing the command of the Supreme Court.'

Other evidence of Tenant League activism abounded in the summer of 1865. The Central Committee met on 4 July, and enrolled several new members, apparently enlarging its membership, for there was no mention of anyone being replaced. The account which was published afterwards listed ten branch meetings scheduled between 11 and 19 July. The geographical range was impressive: from Lot 1, at the north-western end of Prince County, to southern Kings. The Central Committee reported that organizers James Laird and Alexander McNeill were engaged in a 'western tour,' and stated that the league was making good progress in fund-raising. It clearly had some money on hand, for it voted £30 in support of the Ramsay family, George F. Adams's in-laws, who had been engaged in a long-standing dispute with the Sulivan estate.[5] In addition, the *Examiner* reported, on the authority of private letters received by Edward Whelan, that Laird and McNeill were collecting 30s. a day in league funds during their tour. The editorial included the question, 'why is McNeill allowed to desert his school for several weeks during a season when it ought to be opened?', and suggested that the affected parents approach the Board of Education.[6]

The launching of a new weekly newspaper in Charlottetown in early July gave further indications of changes in politics and in the state of public opinion on the Tenant League. On 8 July David Laird, a younger brother of James, the league organizer, founded the *Patriot*. Laird was president of the Charlottetown Literary and Debating Society, which had unanimously passed a pro-Tenant League resolution on 20 January.[7] The *Patriot* was to replace the *Protestant and Evangelical Witness*, which had commenced in 1859 at a time of acrimonious debate over the place of the Bible in the schools and which had specialized in unmasking 'popery.' Laird had been editor of the *Protestant* from the beginning, and his paper had normally been aligned with the

Conservatives in politics. His new venture was to be non-sectarian, and although no copies of the first issues survive to provide direct evidence, the *Islander* of William Henry Pope strongly criticized him, accusing him of opportunism and lack of moral courage for failing to take a stand against the league.[8] Laird's hesitation about expressing unequivocal opposition to the league suggests both ambivalence on his own part and concern about offending potential readers. Neither the successful establishment of such a paper, eschewing the sectarian emphasis which had won two consecutive general elections for the Tories, nor its unwillingness to condemn the league boded well for the Island government.

In the midst of this shifting in political and newspaper alliances, on 15 July Curtis proceeded to Lot 31 with an assistant to serve processes on William Large and George Clow, tenants who occupied adjacent farms in the district of North Wiltshire. The owners in 1865 were the heirs of William Douse, the former land agent of the Selkirk estate, who had purchased 'all the unsold portions' of the township from his employer in 1855;[9] as historian H.T. Holman speculates, this acquisition 'may have been prompted by inquiries about possible purchase by the government of the whole estate.'[10] In other words, had Douse not acted, the Lot 31 tenants would have become freeholders following the sale of the Selkirk estate to the Island government in 1860. The claims against Large and Clow were for debts of £54 0s. 9d. and £46 6s. 9d., respectively, plus costs; in each case, the costs exceeded the debt.[11] Both men were known as Tenant League activists, for it had been reported in the press that on 7 June they had moved resolutions at a meeting held to establish a branch of the league on their township.

The potential appeal of the Tenant League message on Lot 31 was great. At the most recent census 56 per cent of land occupiers were tenants or squatters, and an air of grievance and frustrated hopes hung over the non-freehold parts of the township. Rents were unusually high, as 79.7 per cent of leases specified more than 1s. sterling per acre. The leases were also exceptionally short, a factor which produced in a sense of urgency about resolving the land question. Among townships with twenty or more leaseholders, Lot 31 had the highest proportion of leases with terms of under thirty years; in fact it had 30.3 per cent of all such leases in the colony. Fewer than one-third of leaseholders had the usual 999 years or 'perpetual' tenure.[12] Before the land commission of 1860, Patrick Wynne, a Lot 31 tenant who would become a delegate to the founding convention of the Tenant League, testified he was paying

£3 12s. currency annually for forty acres on a twenty-one-year lease due to expire in three years.[13]

Large and his friends apparently learned of the planned visit by Curtis, and removed his goods and chattels, making it impossible for the deputy sheriff to carry out an effective distraint. The community solidarity involved in warning Large and assisting him in his evasion of the effect of the *fi. fa.* is not surprising. At the 7 June meeting one resolution had directed 'That every member provide himself with a bugle to sound the note of alarm on the approach of *rent-leeches*';[14] and as the *Examiner* had reported, the deputy had recently been encircled by 'a mob' on the same township.

Curtis also failed to make a seizure from Clow. In fact, Clow and a group of Tenant Leaguers intercepted him before he could reach the Clow farm, and escorted him back to Charlottetown, with Clow carrying the Tenant League flag. It must have been a humbling experience. According to the *Islander*,

On reaching the Town the procession paraded through the streets and proceeded to one of the Taverns, where, we understand, speech making commenced. While the Tenant Leaguers were celebrating their triumph over the Deputy Sheriff, the Proprietor and the Law, the *Fi. Fa.* which the Deputy Sheriff had in his possession for Mr. George Clow – and which he had kept secret to himself until he returned to town – was returned, and a *Ca. Sa.* handed to the Sheriff [Thomas W. Dodd], who, forthwith, arrested Mr. Clow and committed him to the Jail of the County. No attempt at rescue was made by the Leaguers, who appear to have returned peaceably to their homes, doubtless, less jubilant than they were when they entered the City.[15]

Whelan's account of the arrest squared with that of the *Islander*:

Clow remained for some hours at a tavern enjoying himself, and in the meantime a *Ca. Sa.* (a writ against the body,) was placed in the hands of the High Sheriff [Dodd], who promptly and effectually executed it on the person of Clow, although the latter made a cowardly attempt to elude the officer by a race through some back yards.[16]

Even the pro-league *Ross's Weekly* criticized Clow for his reckless 'piece of bravado' and declared that the purpose of the Tenant League was to purchase land from proprietors, 'not to take advantage of every

occasion to insult the Sheriff.' The serving of writs had been transformed into another demonstration of league power, although Curtis
had ultimately turned the tables on Clow. The potential for official
alarm seemed to make *Ross's Weekly* draw back. 'We earnestly hope that
we may not have a repetition of another settlement coming to town, and
braving the power of the law.'[17] Too many exhibitions of league
strength in the capital, the centre of formal political authority, could be
counter-productive, by raising the frightening prospect of 'dual power,'
that is, a situation in which some body such as a strike committee (during the Winnipeg General Strike, for example), or a group devoted to
civil disobedience as a tactic, such as the league, comes to exert such
power that it becomes a 'dual authority,' rivalling the established political organs of the state. In such a circumstance, involving a fundamental
challenge to its prestige and prerogatives, the state is almost certain to
feel impelled to destroy its rival, lest it be displaced or rendered entirely
impotent. For the league, there was a delicate balance between, on the
one hand, exerting enough physical force in the rural districts to neu-
tralize effectively the civil authority and, on the other, flaunting such
intimidating might under the noses of the officials in Charlottetown that
the lieutenant governor would call upon the British to provide military
support for the sheriff.

Clow may have been the first Tenant Leaguer arrested and taken into
custody for actions related to the league. He appears to have been one of
those 'ordinary' Prince Edward Island farmers who, by the summer of
1865, was willing to defy the law openly over landlord-tenant issues. His
march, with his friends, back to Charlottetown in the company of the
deputy sheriff may be seen both as an act of defiance in the spirit of the
demonstrators who marched through the capital on St Patrick's Day, and
also as a taunting gesture somewhat after the example of those persons
in eastern Queens County who erected mock forts, straw effigies, and the
like to greet the *posse comitatus* on 7 April. But Clow was not so fortunate
as Samuel Fletcher; late in July, when Whelan visited the jail in Charlottetown, he was still incarcerated.[18] Although there is no certitude on the
point, it is quite possible that he was one of those leaguers who could
pay their rent, but had reached the point that they simply would not do
so. In any event, he does not appear to have been evicted. The letterbook
of land agent Henry Jones Cundall for 1868 and 1869 indicates that Clow
was still a tenant on the Douse estate. On 14 April 1869 Cundall was
warning him to pay up promptly, or face 'legal measures.'[19]

The Battle of Curtisdale and the Capture of Charles Dickieson

News of the trip by Deputy Sheriff Curtis to Lot 31 to execute writs on Large and Clow, and of the arrest of Clow, appeared in the Island press by 21 July. But more dramatic events, arising out of an expedition by Curtis and three assistants to the district of New Glasgow on Lot 23, were already overshadowing the Lot 31 story. At the census of 1861, Lot 23, immediately north of Lot 31, had recorded one of the largest populations of any township on the Island, and an unusually high tenancy rate of 79.4 per cent; in fact, there were more tenants on Lot 23 than on any other township in the colony. The presence of only one squatter and not a single tenant 'by verbal agreement' among 306 occupiers of land suggested that the owners had exercised tight and formalized control over access to the land for many years.[20] This interpretation fits with evidence that the Rennie family, who had owned most of the township since 1810, had been 'developing' (as opposed to inactive or speculative) proprietors. In 1819 David Rennie's land agent and stepson, William Eppes Cormack, later to become famous as an explorer of the Newfoundland interior, had taken Scottish immigrants from Glasgow, Scotland to found *New* Glasgow in Prince Edward Island.[21] Before the land commission of 1875–6, John Ball, a surveyor intimately familiar with the estate, related under oath that one of the Rennies loaned passage money to the immigrants, and paid out of his own pocket for roads and bridges, without reimbursement. He knew the latter fact 'of my own knowledge.'[22] In 1854 Daniel Hodgson, a prominent member of Charlottetown officialdom, had acquired 4,732 acres of the township from David Rennie's son Robert.[23]

The tenants on Lot 23 were commonly considered to be in comparatively comfortable circumstances, and if Ball's report was accurate, they were unusually fortunate among Island tenants. In his opinion, expressed in 1875, they 'are all well off. ... it is a very valuable estate – I would say one of the finest in the Island.' But with respect to lawsuits for rent by the Rennie family, he reported that 'There were actions upon actions at the time of the Tenant League.'[24] During the land commission hearings of 1860, a delegation from the township had presented a statement of grievances which included the claim that accumulation of rent arrears was inducing out-migration among the younger generation.[25] Complaints about arrears leading to an exodus of the young were usually linked in this period to reports of inability to procure freehold tenure, and such may have been the case on Lot 23. According to *Ross's*

Weekly, publishing after the trek by the deputy sheriff to New Glasgow, all the Rennie estate tenants in 'the settlement' (New Glasgow? all of the Rennie estate?) had offered to purchase their farms at 15s. per acre, but the proposal had been refused.[26] Whether or not this offer and this rejection actually occurred, the story is consistent with testimony by Ball that David Stuart Rennie (a son of David Rennie) had once, 'many years' before his death in London on 6 February 1865, sold to a tenant at fifteen years' purchase, but had ordered that there be no more sales at that rate.[27]

In addition to the difficulty in obtaining freehold, there were two other serious problems for tenants on Lot 23. One grievance that the 1860 delegation emphasized was 'short leases'; in Island idiom, a short lease was anything less than perpetual tenure, and the census of the following year did reveal that almost one-third of tenants on the township had leases of less than 999 years' duration. In such circumstances, the delegates argued, 'every year as it passes renders the tenure of less value to the tenant and more to the proprietor.'[28] Furthermore, rents for many tenants on the Rennie estate were higher than usual, since 23.9 per cent of tenants on Lot 23 had leases specifying currency rents of more than 1s. per acre. These were probably set at 1s. 6d. or 2s.[29] Following D.S. Rennie's death, one-third of the estate was in the hands of his widow Susan, and two-thirds in the hands of his brother Robert, who resided in Charlottetown.[30]

The objective of the deputy sheriff and his bailiffs on 18 July was to serve James Proctor and Charles Dickieson, two tenants on the Rennie estate, with writs of *fi. fa.* for debts of £16 11s. 7d. and £37 6s. 9d.[31] The solicitor for the estate, Edward J. Hodgson (son of Daniel Hodgson, the other proprietor on the township), who had sought the two writs, was mindful of 'the extensive organization formed to resist the officers of the law in the discharge of their duties,' and had urged the sheriff to ensure that enough men were sent to do the job, that they proceed with caution, and that all provocation of the tenantry be avoided.[32] The first clear indication the party had that they would encounter resistance was the behaviour of John McLean, a man on horseback whom they met on the road. In the words of Jonathan Collings, one of the bailiffs, McLean 'closely scrutinized them.'[33] He recognized Curtis, who reported in an affidavit sworn the next day that when he

had passed them for a short distance he turned and followed them and commenced blowing a tin trumpet. ... Deponent [Curtis] knowing that an association had been formed by the people in and about New Glasgow to oppose and resist any attempt to enforce the Law, at once suspected that McLean was giving some signal, and this suspicion was confirmed by hearing the signal answered from farm to farm. ... deponent and his Bailiffs pressed on quickly to Proctor's house.[34]

Although Proctor was absent, his wife was at home, and since she promised to send for him, Curtis decided to wait. When Proctor did not appear after 'a length of time,' the deputy distrained a horse, wagon, harness, and saddle, and departed.[35]

From the route Curtis took, it appears that he intended to visit Dickieson's farm next. But he did not reach it. At New Glasgow bridge, on the way to Dickieson's farm, he found that

about twenty men were collected on horseback, and in waggons, blowing tin trumpets – fresh arrivals were coming in and joining them every moment. ... the most violent of all and the ring-leader, who seemed to be issuing instructions to them, was one Charles Dickieson, against whom this Deponent [Curtis] held a *Ca. Sa.* ... as soon as Dickieson recognized ... Jonathan Collings, he began to swear at and abuse ... Collings; but ... Deponent at once charged ... Collings to make no answer or reply as he was most anxious to avoid any altercation. ... Dickieson and his companions followed this Deponent and his Bailiffs as they proceeded [several miles] on their way to Wheatley River Bridge, ... Dickieson cursing and swearing at Deponent, and encouraging the people who followed him to rescue the property of Proctor.[36]

According to Collings, who also reported that Dickieson recognized him and subjected him to verbal abuse, 'the people ... hooted and yelled after them, the numbers increasing as they proceeded.'[37] Curtis and his party evidently wished to be rid of this noisy throng, for at his examination before two justices of the peace on 26 July he recalled that after passing New Glasgow bridge 'we ... gave the crowd an opportunity to pass – they would not pass – proceeded on again. Some of the crowd of persons drove ahead, and some stopped behind.'[38]

In the midst of the commotion, Proctor overtook them, and asked the deputy whether he would take security to release his property. Curtis stated that he was willing to do so, if sufficient were offered. Proctor then offered Dickieson as security, but Curtis refused to accept him. When Proctor suggested a second candidate, a merchant named Alexander McMillan who was not present, Curtis assented, and they made arrangements to meet with McMillan the following morning at the deputy's home. Upon Proctor's request, Curtis returned his saddle on the spot. Proctor then departed, apparently satisfied. Yet that was not the end of the confrontation. According to Curtis, 'as soon as ... Dickieson saw that Deponent [Curtis] appeared unwilling to act harshly, and had promised to give up Proctor's property, ... he became more abusive and excited, and repeatedly urged upon the people to rescue the horse and waggon.'[39] At one stage, Dickieson 'came up on a grey horse ... in a gallop, ... [and] said as he galloped past "You buggers will see that you shall take bail."'[40] But at the old Rustico Road, the Tenant Leaguers ceased following the deputy's party.

Curtis may have decided to take a circuitous route back to Charlottetown in the hope of shaking off the Tenant Leaguers, and he seemed confident that he had done so after they took a different road. The group of four 'proceeded slowly until we reached Curtisdale [a hotel in Milton on Lot 32] – stopped there for 10 minutes for some refreshment.'[41] But when Curtis emerged from the hotel

he found that they [the Tenant Leaguers] had gone round another way and had stationed themselves on the bridge [over Curtis Creek, running off North River] a short distance below Milton Church. ... they placed a horse and cart across the ... bridge, and another in the water at the side of the bridge, and ranged their horses in such a manner as made it utterly impossible for even a foot passenger to pass.[42]

In his examination before the justices of the peace, Curtis would recount that he 'saw several of the crowd go to the fence adjoining the road and arm themselves with sticks. ... The crowd was very abusive, made violent threats.'[43] This roadblock, with even an obstruction *in* the water (presumably to prevent any attempt to ford the creek), was more than a token or ritualistic gesture of defiance. The leaguers did not intend to allow the deputy sheriff's party to pass with the distrained goods.

Nonetheless, after some hesitation, Curtis and the bailiffs advanced towards the bridge. On the following day, Collings quoted the deputy sheriff as saying 'Gentlemen, please allow me to pass.'[44] The leaguers replied that they would not do so unless he surrendered the remainder of the distrained property. Curtis, who was driving Proctor's horse and wagon, then approached the blockade. One of the Tenant Leaguers grabbed the bridle of the horse, causing it to stop. Curtis alighted from the wagon, but a rush was immediately made upon Deponent [Curtis], and Deponent was struck several times upon the head and body with large sticks or longers,[45] which had been pulled from the fences by ... Dickieson and his companions. ... Deponent was half stunned by the blows he received; ... [and] owing to the confusion, is unable to state positively who struck Deponent; but ... most positively swears that he clearly and distinctly saw ... Dickieson throw a stone at Deponent, which struck Deponent on the head, and immediately Deponent's face was covered with blood from the wound caused by the blow from the stone. ... Deponent was partially stunned and blinded by the blood, and unable to offer much resistance. ... a rush was made, and the horse and waggon were carried off by the people.[46]

When examined before the two justices of the peace, Curtis stated also that 'Immediately after receiving the blow [he] thinks that he struck the Prisoner Dickieson on the chin with a stick. Much about the same time [he, Curtis] ... was struck several blows on the left arm, ... The first blow struck shattered the bones in deponent's arm.'[47]

As this was happening, Collings, who had served as a bailiff and constable for several years, made a crucial intervention. He was standing up in the wagon preparing to drive through the blockade if Curtis succeeded in clearing a path. From his perspective, the attack was 'murderous.' He saw the deputy

struck on the head and body three or four times with large sticks or longers which they had pulled from the fence by the road side; ... deponent [Collings], owing to the confusion, is unable positively to swear who struck the blows; but deponent most positively swears that he distinctly saw ... Dickieson throw a large stone, which he held in his hand, with all his force at ... Curtis; that the stone struck Curtis in the head, and his face was immediately covered with blood; ... deponent ran at Dickieson and seized him by the hair of his head and held him; at the same time deponent drew a pocket pistol and said he would fire; ... some one of them (but whom deponent does not know)[48] came forward, evidently with the intention of rescuing Dickieson; ... deponent told him to keep back; ... he paid no attention and prepared to attack deponent, whereupon, as soon as he came within striking distance, deponent struck him in the face with the pistol, and laid open one side of his cheek; and deponent further saith, that after ... Curtis had been struck on the head by the stone thrown by Dickieson, a rush was made, and Proctor's horse and waggon, which had been levied upon, was carried off in the direction of New Glasgow, the people yelling and hooting; ... this deponent still held hold of Dickieson, and with the assistance of [Henry]

Chowan [another bailiff], placed him in Curtis' waggon; ... as soon as the people perceived that Dickieson had been arrested, they drew back a short distance, and prepared to make a rush for the purpose of rescuing him. But ... deponent stood on the bridge, and cocking the pistol, told them that he would shoot the first man who came near him; ... this seemed to frighten them, and in a short time they withdrew; ... deponent then bound up Curtis' arm which hung helpless at his side, and they proceeded to town, and deposited Dickieson in the County Jail.[49]

The events of 18 July are noteworthy for several reasons. They were intrinsically important because they constituted a significant escalation in the confrontation between the forces of the law and the Tenant League. The deputy sheriff had sustained serious injuries. The editorial in Whelan's *Examiner* stated that 'his head was cut in several places, and one of his arms, we are informed, was pounded almost to a jelly. He is incapacitated for service, for a considerable time.'[50] Collings reported hearing at least one gun discharged, and, by his own account, inflicted a bloody wound on a Tenant Leaguer. The leaguers, after rescuing goods from Curtis's party, had only ceased attempting to free Dickieson from custody when Collings pointed a cocked pistol – which, he later stated under cross-examination, had been empty – at them.[51] There had been a major test of nerve between the two sides, with violence that would eventually result in grave legal consequences for Dickieson and others: trials, convictions, and sentences.

The only two persons whose first-hand contemporary accounts of the Curtisdale affray are known to exist are Curtis and Collings. Their versions agree in all essential points. The newspaper accounts are consistent with those of the law officers. The only contrary interpretation, more or less contemporary, of Dickieson's behaviour on 18 July is the argument his lawyer, Charles Palmer, made at his trial, namely, that his actions really amounted to an over-zealous attempt to give bail for the seized goods.[52] At best, that can be described as an extraordinary understatement. If in fact Dickieson intended to give bail, his actions went far beyond convention and the law in such matters, and at the time even the Tenant League apparently attempted to distance itself from his behaviour. On 26 July the *Herald* reported that 'we learn the Union repudiates the conduct of Dickieson in resisting the officers of the law,' and on the following day the pro–Tenant League *Ross's Weekly*, although emphasizing that it had heard only one side of the story, censured his 'open opposition to law.'[53] Some years later, in the Legislative Council, John

Balderston, a founding member of the league, stated that it 'immediately passed a resolution condemning the conduct of those who were concerned in [the incident at Curtisdale].'[54]

In the initial report of the incident, W.H. Pope's *Islander* expressed 'regret that Dickieson should have so far forgotten himself as to behave in so ruffianly a manner; as hitherto we have always understood him to be a quiet, respectable man.'[55] If this was a sincere expression of sentiment, then the situation must have seemed ominous to people like Pope: quiet and respectable farmers, normally on the right side of the law, were attacking the sheriff's officers as they attempted to do their duty. Indeed, the repercussions were weighty. The encounter of 18 July set in motion a train of events that led to the summoning of British troops to assist the sheriff and the civil authorities in Queens County.

Thomas W. Dodd, the high sheriff of Queens, wrote to W.H. Pope, the Island's colonial secretary, on 19 July, reporting that 'within a few miles of town' his deputy had been 'attacked and badly beaten by overpowering numbers, and the goods levied upon by him rescued.' His conclusion was that 'I am ... completely powerless to execute any Writs placed in my hands.'[56] Two days after that, Theophilus DesBrisay, JP informed Premier James C. Pope in writing that, following a complaint by Collings, he had issued warrants for the arrest of persons who had participated in 'the riotous assemblage' at Curtisdale, but that he had insufficient force at his command to have the warrants executed.[57] It is evident that after the incident at Curtisdale the civil authorities of Queens County believed they no longer had the power to enforce the law.

In the midst of this, Lieutenant Governor George Dundas departed for several months' leave in the United Kingdom. After more than six years on the Island, he was evidently not going to allow the events unfolding in the countryside to postpone or prevent a visit home. On 21 July, the day after he left, Chief Justice Robert Hodgson became administrator and met the Executive Council. Since the lieutenant governor was absent for almost five months, many of the most dramatic moments in the history of the Tenant League occurred when Hodgson was the official with, in the eyes of London, ultimate responsibility for the government. Born in 1798, he had been a fixture of the old, pre-responsible government régime on Prince Edward Island. A brother of proprietor and prothonotary Daniel Hodgson, he had been active for a generation as a lawyer, land agent, and official, and possessed sufficient bipartisan prestige that in 1850 when it had looked as though the

Reformers might assume office, they offered to allow him to remain as attorney general and executive councillor if he would join them. He had declined and had gone on to become chief justice two years later.

Despite whatever possibilities the Reformers of 1850 may have seen in a relationship with Hodgson, there is no doubt that he was conservative in orientation. By 6 September 1865, in a letter to a former colleague on the Executive Council then residing in the United Kingdom, Joseph Pope, the father of James Colledge and William Henry, Hodgson was confiding that he doubted the 'applicability' of responsible government to the Island under conditions of universal (male) suffrage.[58] He had been an executive councillor or judge almost continuously since 1829, and as he moved into his late sixties he also had an appetite for honour and office that had not been sated. The first native-born judge and the first native-born chief justice of Prince Edward Island, he would yet become the first native-born knight and the first native-born lieutenant governor. Although he could not have relished the position in which he found himself in the summer of 1865, he was also determined to get out of it with the minimum of damage to the peace of the colony, and to his own standing. There would be no apparent conflict in perspective between him and the local government. In fact, Hodgson seemed to share or mirror their sensibility and their sense of strategy with remarkable precision. He showed vigour in areas where the executive arm of government could act, the councillors hoped, with relative political impunity, and he hesitated at precisely the same measures that made them hesitate. Perhaps this was natural, since Hodgson was, as much as any of them, a charter member of the local élite.[59]

The affray at Curtisdale was the first matter commanding the attention of Hodgson and the executive councillors. It had occurred close enough to the centres of power in Charlottetown to cause grave concern among officialdom. When the Executive Council met on 21 July, it authorized Dodd and DesBrisay to employ as many special constables as necessary, an authorization that remained in force until 19 December, after Dundas returned. Acting with the advice of the Council, Hodgson removed James Laird, the Tenant League organizer, from the list of justices of the peace; the announcement in the 'Extra' of the *Royal Gazette* on the next day mentioned 'the Tenant Union' by name, and linked the dismissal to the 22 March 1865 proclamation by Dundas.[60] On 25 July the Board of Education, a nine-member body appointed by the Island government and controlling all public schools, directed school visitors to report all cases of teachers neglecting their duties and participating in

Tenant League meetings. One day later, probably not by coincidence, a meeting at the Mount Mellick school apparently passed a unanimous resolution supporting Alexander McNeill, their teacher and also the secretary of the league, who had accompanied Laird on the 'western tour.' One day after that, the board gave notice that all teachers associating with the league in future would be removed from the roll of licensed teachers, and denied payment of their salaries.[61] This was a significant sanction: firing people from full-time jobs.

The evidence suggests alarm on the part of the Island authorities. Dundas, writing some weeks later to the British colonial secretary from Scotland where he was on leave, and commenting on the case of Laird, stated that 'I know that it was the intention of the local Government to visit with its utmost displeasure any man, who in any way countenanced, or abetted the League, and whose position gave the authorities any hold over him.'[62] Vigorous as these measures may have seemed, proof of the intimidating power of the league continued to accumulate. On 22 July in the heavily Irish Roman Catholic district of Fort Augustus in eastern Queens an incident occurred involving Constable Bernard McKenna. On his way to serve a summons, he was coerced by 'a number of persons ... blowing tin trumpets' into signing a document stating that he would not act as constable or serve any legal processes 'for twelve months or until the Tenant League was settled.' Two weeks afterwards he swore that he did what he was told 'under fear of my life and under fear of my property being destroyed'; he was a resident of Lot 48, adjacent to the township where Fort Augustus was located.[63] At approximately the same time as McKenna was pressured into signing his bizarre pledge, another incident of harassment of a constabulary force apparently occurred. It had set out in search of persons for whom arrest warrants had been issued, and in this instance the harassment appears to have been non-violent. But the pattern was now familiar. In the words of Administrator Hodgson: 'On its becoming known that the constables were in the settlement, numbers of persons, carrying with them tin trumpets, assembled and surrounded them, blew their tin trumpets and followed the constables within a few miles of the city.'[64]

These additional encounters between the Tenant League and the forces of the law must have deepened the disquiet in Charlottetown. Certainly official consternation increased greatly with events in the capital on 26 July, which indicated that league defiance of the civil power was neither an isolated event nor confined to rural districts. Suddenly even the efficacy of such measures as hiring dozens of special constables was moot – especially in light of such an incident as that involving McKenna.

Demonstration and Disorder in Charlottetown

On the afternoon of Wednesday, 26 July, Dickieson was brought before two justices of the peace, DesBrisay and Charlottetown mayor T.H. Haviland Sr, both proprietors, for examination.[65] The physical process involved taking the prisoner from the jail on Pownal Square to the city hall, which also housed the police court, on the western portion of Queen Square, at the centre of the town. The scheduling became known in the countryside, and on Wednesday morning Tenant Leaguers converged on Charlottetown. They formed a crowd numbering in the hundreds, possibly totalling a thousand, and gathered near the jail and in the streets between it and the city hall, popularly referred to as 'the old court house.' The fact that Wednesday was a market day added to the number of country people in the capital. The location of the market building, known as the Round Market, on the northwest part of Queen Square increased the potential for turmoil, for it 'was the meeting place where people of all classes from town and country met to buy and sell.'[66] The *Islander* estimated that nine-tenths of those in the growing crowd were sympathizers of Dickieson.[67]

In the view of Whelan's *Examiner*, the Tenant Leaguers gathered in the streets of Charlottetown with the intention of making a demonstration in favour of Dickieson.[68] But the *Islander* would report that their plans were more audacious. On the day before these proceedings 'the authorities had notice that a rescue would be attempted.'[69] Regardless of the accuracy of this information, the *Islander*'s version is a good clue as to the state of mind within the Island government, since the editor, W.H. Pope, was also the colonial secretary and an executive councillor, as well as the brother of the premier. The source of the intelligence appears to have been an informant of Sheriff Dodd, and on the day of the hearing the sheriff became apprehensive as the crowd grew in numbers. He requested in writing that pistols and ammunition be issued to his constables. Pope, as colonial secretary, brought the request to Hodgson. The administrator assented after some deliberation, and according to the *Islander*, in the course of making his decision he consulted the executive councillors who were in Charlottetown. In Hodgson's view, there was no alternative. He accepted Dodd's analysis of the situation in the streets of the capital, and as a veteran member of the bar and bench he was convinced that, given the lapse of time since the arrest, cancellation of the court appearance would result in Dickieson's going free. He later reported to British Colonial Secretary Edward Cardwell that 'It was absolutely necessary, in order to prevent Dickieson from being brought up by Writ of Habeas Corpus and immediately discharged, that he should be taken before the Magistrate on the 26th July.'[70] There could be no retreat from the plan for an examination on that date without surrendering to the force of numbers in the streets – and that was not an acceptable option.

While acceding to the sheriff's plea for the issue of arms and ammunition, Hodgson also urged, through Pope, 'the utmost forbearance, in the case of their [the constables] being attacked.'[71] Dodd and about twenty-five special constables, each armed with a loaded cavalry pistol as well as a baton, accompanied the prisoner to the city hall around 2 p.m., and met no resistance, although, in the words of the *Examiner*, the leaguers 'gathered round the escort.'[72] At the examination, Curtis and Collings gave evidence; the charge was investigated; the witnesses were cross-examined; and bail of £200 was given for Dickieson's appearance at the Supreme Court in January of 1866 to face charges of 'Riot, assault, and rescue, and such other charge as shall be brought against him.' Dickieson and his friends expected that he would then be released. But Deputy Sheriff Curtis stepped forward with a *ca. sa.* for rent, which Whelan described as 'a very sudden surprise,' given that rejoicing at the prospect of Dickieson's liberation had already commenced.[73] The effect of this new arrest warrant – new only in the sense that the accused and his supporters were unprepared for it, since it was dated 17 July – was that he was re-incarcerated.

The news that Dickieson would not be released shocked and angered the crowd. It also led to an attempt to rescue him as he was taken from the city hall back to the jail.[74] Persons in the crowd threw stones, striking several constables, and the constables used their batons to clear a path through the crowd, apparently landing many blows. The *Examiner* reported the firing of two shots – 'but we suspect they were fired more by accident than intent; and we are happy to say that no one was hurt by them.'[75] *Ross's Weekly* referred to the 'accidental explosion' of a pistol held by a special constable.[76] Despite the opposition, Dodd and his assistants succeeded in lodging Dickieson in jail. The distance was approximately two city blocks: south one block on Queen Street, right turn, and west one block on Richmond Street to Pownal Square. One eyewitness, Solicitor General T. Heath Haviland Jr, later stated that he did not believe the party escorting Dickieson could have taken him fifty yards farther without succumbing to the rescue effort.[77] Although several on both sides received minor wounds, no one suffered serious injury. The crowd dispersed without further incident.

Clearly, the psychological impact of such a demonstration in a town like Charlottetown, with its population of seven or eight thousand, few regular police, and no resident military force to support the civil authority, would be huge. In Ireland, with its 'land meetings' and 'indignation meetings' (a term which had currency in Prince Edward Island during the late colonial period), there was an element of intimidation as well as simply gathering to voice demands or express support or opposition regarding specific matters or policy.[78] Such must have been the case in Charlottetown on 26 July 1865 as well.

The two major points at issue with respect to the crowd are its size and its collective intent. Its actual behaviour is not in question, for all reports agreed that there had been substantial disorder and conflict, centring on an attempt to rescue Dickieson. With respect to size, there was considerable difference of opinion. The low numerical count was Whelan's – between 250 and 350 – and the high one was the *Islander*'s of at least 1,000. Administrator Hodgson reported to London that the number had been 'upwards of 1,000 persons.'[79] In fact, Whelan is alone in placing the size of the crowd decisively under 1,000; and some estimate around 1,000 seems most plausible.

As for collective intent, it is necessary to note that apparently a Tenant League leader from the Lot 49 area had urged the crowd to 'Rescue him [Dickieson]! rescue him!'[80] An inquiry J.W. Morrison, the deputy colonial secretary, made on Hodgson's behalf the following day suggests that this may have been Samuel Lane, one of the founders of the league and someone well known to authorities. The administrator had received a report that Lane was 'very conspicuous ... in inciting by language and gesture the Assembled People to rescue the Prisoner.' Hodgson directed that Attorney General Edward Palmer inquire whether the sheriff had sufficient evidence to justify prosecuting Lane.[81] Such a prosecution was never commenced, but whatever Lane personally may have said or done, the more important question concerns conscious purpose on the part of the crowd as a whole.

Was there in fact a plan to liberate Dickieson by taking him physically out of the custody of the sheriff's party? The *Islander* professed to believe so, but despite the strong powers of logic Pope possessed, the picture emerging from his editorial article on the subject is not internally consistent. Some of the demonstrators were 'from remote districts' and many were 'maddened with rum.' At the same time, 'the mob ... was utterly beyond the control of any leader.'[82] A crowd under the influence of alcohol and uncontrollable would not be an efficient vehicle for undertaking a rescue.

Leaving aside the question of sobriety, what can we conclude about the collective intent of the crowd? In the first place, there was clearly a desire that Dickieson go free. But all evidence indicates that the demonstrators expected that he would be released on bail, rather than have to be liberated by force; therefore, in their calculations before the event, there would be no need to plot a rescue. The information that he would be going back to jail – instead of leading a triumphant parade through the capital – caught large numbers in the crowd by surprise, and consequently whatever they did was improvised, rather than planned in advance. Responding to circumstances as they developed, some were indeed willing to make an effort to set him free or at least to impede the progress of the escort taking him back to the lock-up. Yet, however great the chagrin of the crowd at the outcome of the examination before the court, and whatever loose talk may have been circulating prior to the scheduled court appearance, it is impossible to conclude on the basis of surviving reports that the crowd as a whole engaged in a concerted attempt to rescue Dickieson.

An alternative theory might posit that the crowd originally assembled with the objective of a rescue, but that the constables' possession of arms deterred them from executing their plan. There is no persuasive reason to accept such an hypothesis. In order to do so, one would have to assume that the leaguers expected what in fact surprised them, namely, Curtis's production of the *ca. sa.* The only other assumption which would support the theory of a planned rescue is that they intended to pre-empt the court by liberating Dickieson on the way to an examination which they expected would set him free. But this is predicated on reasoning too irrational and reckless to be entertained seriously, and in fact the account in the *Islander* confirms that there had been no resistance in taking Dickieson from the jail to the hearing.

A general pattern that emerges from conflicting accounts concerning the crowd and its behaviour is that those commentators most favourable to the Tenant League – *Ross's Weekly* and the *Herald* – tended to minimize whatever threatening aspect there was to the situation, and to discount any semblance of a concerted attempt to rescue Dickieson. On the same day Hodgson directed that the attorney general inquire of Sheriff Dodd whether prosecution of Lane would be justified, *Ross's Weekly* claimed that several Tenant League leaders tried to calm the crowd and prevent the attempted rescue by 'rowdies whose minds had been inflamed by liquor.'[83] In its first issue after 26 July, the *Herald* referred to 'a few excited persons' attempting to rescue Dickieson and failing for want of organization; in a later issue, the editor added the adjective 'drunken.'[84] If the disorderly behaviour on 26 July could be attributed to inebriation, then surely the Tenant League did not constitute a serious threat to public order in the political sense. Indeed, *Ross's Weekly* carried the attempt to establish distance between the movement and those who engaged in overt acts of flouting the law beyond disowning the actions of Dickieson and Clow, and repudiating the riotous behaviour on the 26th; it actually complimented Sheriff Dodd, whose conduct 'was marked by that forbearance and yet that firmness which could not fail to elicit the praise of all.'[85]

At the other end of the spectrum, those directly responsible for upholding the rule of law and keeping Dickieson in custody tended to support an alarmist interpretation, to emphasize how close the situation was to being out of their control, and to seek a guiding hand behind the commotion. On the night of 26 July, Hodgson took special 'security' precautions, placing guards at the jail, the magazine, and the building containing rifles and arms on loan from the British government. When he reported to Downing Street a week later, he still had guards at the armoury and the magazine.[86]

Whelan's reaction to the events culminating in the demonstration is especially interesting, for he was in the classic position of a moderate in a highly polarized situation, caught somewhere between the Tenant League, whose tactics he believed would be catastrophic for the tenantry, and the Tories, whose company he had always shunned. When initially commenting on the resistance to the deputy sheriff at Curtisdale, he had volunteered his opinion that 'This is a melancholy piece of business for Dickieson. He will certainly be severely punished, if not ruined for life.'[87] On 26 July he apparently gave an unrealistically low estimate of the size of the crowd – which may have reflected a wish on his part that the movement simply dissipate.

On 28 July, one day after *Ross's Weekly* stated that the explanation for the Charlottetown demonstration was that Dickieson was kept 'in one of the *lowest dungeons*' of the Queens County Jail and denied all visitors,[88] Whelan went to the institution. He investigated for himself Dickieson's conditions of incarceration and the situation with regard to visitors, but perhaps he wished also to gain more insight into the thinking of Tenant

Leaguers. At the jail he talked with Dickieson and Clow, whom he described in the next issue of the *Examiner* as 'both plain farmers, of just about average intelligence, for their class, with none to spare, but evidently not men of the hero type.' He reported that there was no truth whatever in the characterization of the place where Dickieson was kept as a dungeon: 'It is a spacious room, well ventilated, and fairly lighted.'[89]

Whelan learned also that by the morning of the 26th DesBrisay had authorized visits to Dickieson on eight occasions: once a day, on average, for each day he had been there. Such a pattern would be consistent with a report by a grandson in an interview more than a century later that Dickieson had told him he 'never ate a bite of jail food,' and that sympathizers brought food to him.[90] Presumably, they made deliveries every day or almost every day. Although *Ross's Weekly* had stated that Dickieson's own brother had been turned away 'without even the privilege of having one word with him,'[91] in fact a brother had visited him on the morning of the 26th, prior to the court appearance. Whelan recounted that when the same brother came back during the afternoon and attempted to make another visit, *without* a new order from Des-Brisay, 'while groups of League men still lingered about the jail, and before the excitement had subsided,' *then* the jailer had refused. The veteran journalist and politician stated that he conveyed this information in order to refute what he regarded as mendacious reporting, both politically inflammatory and damaging to the reputation of the Island as a civilized community.

The same *Examiner* contained the interesting revelation that Whelan had conversed recently with several Tenant Leaguers who could pay their rent – yet absolutely refused. He found their attitude perplexing. They

are in really independent circumstances, ... [and] estimate the value of their property by several hundreds, and in some cases by thousands – and they have assured us that they would rather part with every shilling they own, or every shilling's worth, and undergo any penalty and suffering, than pay one penny of rent. 'Better,' said we, 'to pay rent, however great the hardship, than pay exorbitant fees to lawyers – undergo degrading and health-destroying imprisonment, and have their property afterwards forcibly taken from them to pay the original bill, more than doubled, perhaps, by costs.' Other debts, they state, they are willing to pay; but they look upon the landlord's claim as a swindle, and they are determined to resist it to the last extremity, no matter what the consequences may be. ... men apparently sensible and rational upon all other subjects ... talk thus.[92]

These men who were 'apparently sensible and rational upon all other subjects' may have included Clow. The *Islander* would report it had been informed that Clow *'purchased* the leasehold interest on no less than three farms, fully aware of the terms on which the lands were let, and after paying rent for years, refused any longer to perform the covenants contained in the leases.'[93] As described in the *Islander*, he was exactly the type Whelan had in mind: 'a farmer in very independent circumstances – and we believe he enjoys the respect of his acquaintances by whom he is esteemed a worthy, honest man. Mr. Clow is quite able to pay his rent.'[94] Yet many Island tenants looked upon rent in a different light from other debts, and were remarkably averse to paying it. At the land commission of 1860, William McNeill, the chairman of a committee presenting a statement of grievances from Lot 23 residents, remarked that 'From the knowledge I have of the feeling of the people, I believe they would rather emigrate than pay rent.'[95]

At the end of July 1865 it was not at all clear where events on Prince Edward Island were going if Islanders continued to support the Tenant League. Would Dickieson be liberated from jail? Would sheriffs and constables be able to function? Would landlords be able to collect rent? These questions must have swirled in the head of Administrator Hodgson when he met the Executive Council on Tuesday the 1st of August. He had requested 'a full attendance' because of the principal item he intended to raise. He brought 'the disturbed state of Queens County' to the attention of members, referring to Sheriff Dodd's statement that he was no longer able to execute writs. His crucial question for Council was whether members believed it 'advisable' to send to Halifax for an armed military force. The councillors concluded that it would be 'highly expedient' to send for two companies of troops, and in his despatch to London Hodgson would report that they had been unanimous.[96]

The Executive Council simultaneously commenced in earnest the process of purging the ranks of minor officials, beginning with inquiries about individuals' levels of involvement with the Tenant League. But members of the Island government, recognizing that the situation in parts of the colony was entirely out of control, had concluded that punishment of the few Tenant Leaguers who held one form or another of government employment or appointment would not suffice to put down the league. Something more decisive and forceful would be needed to assist the sheriff in the countryside: British soldiers. It must have been a difficult pill to swallow, for resort to military authority meant that the civil power had failed. In a colony enjoying a constitutional system of responsible government and a broad franchise, it also entailed grave consequences for the governing political party.

James Curtis and Charles Dickieson

It is necessary to consider the roles played by Curtis and Dickieson. Their minds are difficult to penetrate to any depth: not especially articulate, they spoke for the public record only in cases of conflict, and thus no balanced picture is possible. Nonetheless, they are irresistible subjects.

Curtis, as the deputy sheriff in Queens County during an exceedingly difficult period in the history of the land question, was clearly a man of some backbone and determination. The record even leads one to wonder whether he sought confrontation; Charles Palmer, the lawyer for Dickieson, asserted at his trial that Curtis 'and his associate, Collings, are well known for their turbulent temper; they like the business of hunting and hounding unfortunate men, whenever the opportunity offers.'[97] Yet in Curtis's defence it must be noted that serving unwelcome writs of *fi. fa.* and *ca. sa.* was part of his job, and if he did not do it, he would have to seek other employment. His judgment was sometimes at issue, as in his attempt to arrest Fletcher in the midst of a Tenant League demonstration in Charlottetown on 17 March. Some hostile observers, such as Whelan, occasionally cast aspersions on his courage; significantly, though, he did not do so in the case of the affray at Curtisdale. Instead, in that instance, it was Tenant Leaguers who 'did not appear to have much pluck.'[98] Indeed, after reading the depositions arising out of Curtis's forays into Tenant League territory, it is difficult not to marvel at his tenacity in returning to the fray time after time. He seems also to have appreciated the importance of the rule of law, and to have exercised reasonable restraint in situations which could become exceptionally trying. There is no evidence that he doubted the rightness of what he was doing. An enigmatic character, almost an automaton, apparently not reflective but also not stupid, he was an implacable agent of law enforcement, reminiscent of Javert, the detective in Victor Hugo's *Les Misérables*.

Dickieson stepped onto the stage of officially recorded history when he threw a stone at the deputy sheriff, if one discounts his preceding failure or refusal to pay rent. He emerged as an angry man. The *Islander* newspaper would describe him as having 'a countenance expressive of determination rather than stolid indifference. His appearance would indicate a nature not hardened by crime, but capable of rash and revengeful acts under exasperation.'[99] Eventually he would go to jail for his offence, and he would not show remorse for his outburst of choler on 18 July. When released early in a show of clemency by the authorities, he remained defiant. One hundred and twenty-one years after the events at Curtisdale, Roy Dickieson, sitting in his living room in New Glasgow, Lot 23 on a rainy August afternoon in 1986, remembered clearly his sturdily built grandfather who lived until 1936, when he was almost 103 years of age; as a 17-year-old, Roy had been one of his pallbearers. A powerful man, although not unusually large, Charles was perhaps five feet eight inches in height and 165 pounds in weight. He would not have been easy to subdue. Roy recalled that his grandfather, who 'really enjoyed life,' had been a storyteller. As a centenarian, he had retained a clear mind, and Roy had listened as a teenager to his stories of life in earlier generations on the Island. Each individual narrative was always the same, no matter how many times he retold it. There was no variation in the details to please the audience or mood of the moment – there was absolute consistency.

In the early 1930s Charles Dickieson told his grandson of the Tenant League and how the horns were a signal that the sheriff was coming and that those anticipating legal action should hide their possessions. According to him, on the day of his arrest, Deputy Sheriff Curtis's party 'had been drinking.' One of them struck him in the mouth with a billy, leaving a scar that remained for life; he covered it with a beard. Roy Dickieson recalled that his grandfather never expressed regret over association with the Tenant League, since he viewed the winning of freehold tenure as 'a life goal that had to be met.' He was proud of his role, despite the consequences. He had made an impact on Roy, who was carrying on his grandfather's tradition of activism and commitment to solidarity. As a vigorous member of an agrarian aid group, 'Farmers Helping Farmers,' Roy had travelled to Kenya and Tanzania to assist farmers there, and in New Glasgow he had hosted African associates of the group.[100] The Dickieson family farming tradition in New Glasgow would continue 170 years, from 1820, when James, the father of Charles, arrived from Scotland, until 4 February 1990, when Dana, age thirty-four, a son of Roy and great-grandson of Charles, died in a house fire attempting to save his son Riel, age two years nine months. Dana, too, exemplified some of the traits of Charles Dickieson: an activist who was president of his local of the National Farmers Union, and who had travelled to Nicaragua as a supporter of the farmers' union there, he was also a man of such resolve and courage in the face of personal risk that he received a posthumous Citation for Bravery from the Governor General of Canada.[101]

The Curtis *versus* Dickieson confrontation on 18 July 1865 was a case of the relentless force of the law being confronted by the irresistible force of the desire for freehold tenure among Island tenants. Each man was of an unyielding character; neither was inclined to give an inch. When combined with the deputy sheriff's mandate and the tenant's commitment to the tenant movement which defined their roles, the human material from which each had been moulded made significant friction all but inevitable.

8

The Beginning of Repression

The necessity for troops in this Island arises out of the disputes between the landed proprietors and their tenants.
Administrator Robert Hodgson to Lieutenant Governor Sir Richard Graves MacDonnell of Nova Scotia, 1 August 1865[1]

The presence of a regular military force in this Island will, in itself, have the effect of preventing further resistance to the Laws, and ... I have no reason to apprehend that there will be any necessity for bringing the troops into conflict with the people.
Hodgson to British Colonial Secretary Edward Cardwell, 2 August 1865[2]

The local Government ... have gone as far as they considered the Law permitted them to do, and the Leaguers up to the time of the recent disturbances have been too wary to do anything which would bring them within its reach.
Lieutenant Governor George Dundas to Cardwell, 25 August 1865, confidential[3]

Troops Summoned from Halifax

At approximately 2 a.m. on Sunday the 6th of August, 1865, two companies of British troops under the command of Major Thomas Tydd disembarked in Charlottetown. After the Executive Council meeting on the previous Tuesday, Administrator Robert Hodgson had written to Lieutenant Governor Sir Richard Graves MacDonnell in Halifax. He apprised MacDonnell of the existence and objectives of the Tenant League, an 'illegal union,' and stated that its influence 'throughout the greater portion of Queen's County' had led Sheriff Thomas W. Dodd to report 'his utter inability to execute the process of the Supreme Court.'[4]

As a consequence, Hodgson was requesting military assistance; he cited statements by imperial officials in the 1850s as evidence of his authorization to make this request and, by implication, the obligation of MacDonnell to comply. Prince Edward Island Colonial Secretary William Henry Pope took this message to Halifax in person.

On Wednesday the 2nd, Hodgson wrote to British Colonial Secretary Edward Cardwell reporting a deterioration in the situation since 23 March, when Lieutenant Governor George Dundas had assessed it in a despatch on the day after he issued his proclamation against resistance to the law. Then Dundas had informed Cardwell that he had 'too high an opinion of [Prince Edward Islanders'] good sense' to expect 'any serious difficulty in maintaining the law.'[5] But on 2 August the administrator, as he had done in his despatch to MacDonnell, expressed his agreement with Sheriff Dodd. Although Hodgson did not venture to provide even an approximate figure for the number of Tenant Leaguers, he reported estimates that there were 'several thousands,' and added that in his opinion the sympathizers were 'still more numerous.' His conclusion was that the Queens County sheriff could not do his duty in safety 'unless attended by an armed force.'[6]

Hodgson proceeded to explain carefully why British troops were necessary. The statute governing the local Volunteers did not permit them to be used in cases of civil commotion. The widespread sympathy with the Tenant League meant that the militia was 'not ... available.'[7] His analysis appears to have been shared by the Island government and also to have been accurate. The local authorities were sufficiently uncomfortable that they undertook to disarm all Queens County companies of Volunteers, which meant reclaiming arms from almost seven hundred men.[8] According to Edward Whelan's *Examiner*, 'it was found that the Volunteers in Queen's County were nearly, if not quite all, Leaguers to a man' – which meant that their possession of arms posed a potential threat to law enforcement.[9] In reporting on this operation later, Lieutenant Colonel P.D. Stewart, adjutant general of militia for the colony, stated that during the summer of 1865 he had 'considerable difficulty in some cases in collecting these arms.'[10] There been no regular troops stationed on the Island since 1854, and this lack, combined with the other factors, meant that there was no adequate local means for supporting the sheriff. Therefore, intervention by an outside agency would be required, and hence the appeal to Halifax.

Part of Hodgson's argument in his despatch to Downing Street revealed the potential for friction which would in fact soon develop between the military and the civil power. He expressed his belief that 'the presence of a regular military force in this Island will, in itself, have the effect of preventing further resistance to the Laws, and ... I have no reason to apprehend that there will be any necessity for bringing the troops into conflict with the people.'[11] In other words, although he had stated that Sheriff Dodd needed armed support to discharge his duties, and that this was the reason he was appealing to Halifax, he also professed to believe that actual use of soldiers would be unnecessary. His theory was that knowledge of their proximity would deter resistance and allow the sheriff to proceed; he was making a sharp distinction between the 'deterrent' capacity of a military presence and the active employment of soldiers to support the civil power in the execution of legal processes.

The deterrent function implied that the troops would be on the Island for an indefinite period – at least until the sheriff had time to set some precedents in executing writs of *fi. fa.* and *ca. sa.* successfully in disturbed parts of Queens County. This would be a subtler use of the military than sending soldiers out with the sheriff at once, but could imply a much lengthier commitment. In a general sense, there was precedent for Hodgson's approach within the British North American colonies: historian Scott W. See, in his work on New Brunswick, has noted the avoidance of using soldiers in direct physical confrontations while making successful 'symbolic' use of their presence in order to prevent communal rioting.[12] The complicating factors in the Prince Edward Island case were the ongoing nature of the issues behind the disorders, the existence of an organization dedicated to systematic resistance, and the isolation of the Island during winter.

Hodgson had not made the 'deterrence-versus-actual use' distinction in his despatch to MacDonnell in Halifax. The closest he had come to acknowledging that the use of the troops might not be immediate was in the following passage, where the prospect of a prolonged stay was presented almost as an afterthought to other logistical considerations:

owing to this Colony having been unused to military occupation for several years, no Quarter-masters' or Ordnance Stores are available. ... in consequence of there not being at present any building which can be appropriated as temporary barracks, all camp equipage must be brought with the detachment. *Should the occupation be protracted beyond the fair weather season, means for securing winter quarters can be adopted.*[13]

Judging solely from this despatch, and without knowing the real perspective of Hodgson and the Island government, one would conclude that any lengthy stay of the soldiers in the colony would be owing to resistance they encountered, and had still to overcome. There was no hint that the possibility of an extended posting, with the reference to 'winter quarters,' was linked to a policy of avoiding active use of the troops supporting the sheriff in his duties.

On Thursday the 3rd of August in Halifax, Garrison Orders nos. 1 and 6 indicated that 135 members of the 16th Regiment of Foot were to leave for Charlottetown as soon as possible, supplied with sixty rounds of ammunition per man.[14] From the beginning, the military authorities in Halifax lacked enthusiasm for the mission. The general officer commanding, Major General Charles Hastings Doyle, was absent in Montreal, but his substitute, Colonel J.H. Francklyn, anticipated his attitude perfectly. On Friday the 4th, Francklyn wrote to Hodgson asking, *inter alia*, that he be informed of 'the earliest period' the troops could be returned to Halifax, citing as his authority a despatch from London to Charlottetown in 1855, to the effect that 'troops are not to be detained [in Prince Edward Island] "a moment after the occasion for their services have [*sic*] ceased."' He made this request two days before the soldiers arrived in Charlottetown.[15]

The officers in Halifax were uneasy about the perennial military problem of desertion. Indeed, on 5 June Doyle had made a particular point of emphasizing his concern when formally approving and confirming a sentence by Francklyn in a court martial: 'I have determined to put a stop to Desertion in my Command, and have pledged myself to the men ... that I would try all Deserters by the highest tribunal.' The reference to the 'highest tribunal' meant that he would take deserters before the body with the greatest power of punishment. The soldier whose case had occasioned this declaration was to be imprisoned for five years and branded.[16]

Desertion was in the first place obnoxious to the military establishment because it represented such a definitive rejection of discipline. Its prevalence in British North America meant that concern over prevention was an unwelcome constraint on strategic planning, and indeed the tendency of troops to desert influenced their deployment in the colonies during the American Civil War. Desertion was also an expensive problem, involving costs of apprehension, trial, detention, and replacement that one historian has described as being so great as to be 'incalculable.'[17] These factors – the relationship to discipline in general, the consequences for strategic decisions, and the costs both in money and in time – were present in military thinking wherever soldiers were stationed in British North America.

In addition to general considerations, the military authorities had specific concerns about sending soldiers to Prince Edward Island. There had been a history of problems over desertion when troops had been on the Island previously,[18] and since the colony no longer had a barracks to house the men, controlling their movements promised to be especially difficult. From the perspective of the military leadership in Halifax, the men should be used for active support of the Queens County sheriff as soon as possible, and once they had reasserted the authority of the civil power they should then be returned promptly to their permanent base.

The viewpoint of the Island government was quite different. Premier James C. Pope and his colleagues were anxious to reassert the authority of the Supreme Court and the sheriff without actually using the soldiers. There was a long-standing political taboo on the Island associated with collecting rent 'at the point of a bayonet.' This metaphor was probably borrowed from Irish experience, where it connoted a tenant refusing the demands of a landlord and using all possible obstructions until the sheriff arrived, thereby forcing the issue to the legal limit, causing the landlord maximum inconvenience, and facing significant legal costs himself. Adopting this course was a measure of resoluteness on the part of the tenant.[19] As most commonly employed in Prince Edward Island, the phrase had an entirely different meaning. It emphasized the harshness of the action the political authorities were taking in support of the landlord – an indication of heavy-handedness in enforcement.[20] Resort to 'the point of a bayonet' involved, for politicians, crossing a line beyond which the electoral consequences might be exceptionally negative.

The only way the Island government could escape from the crisis without ever sending the troops out with the sheriff would be to keep them in the colony for many months, available although not actively engaged. For the Pope government, a military presence of long duration, with the soldiers remaining in the Charlottetown area as bystanders but as potential back-up for the sheriff, would be infinitely preferable to sending them into the countryside to assist him in executing writs. This attitude was absolutely consistent with the indefinite response Hodgson gave to Francklyn's request for information on a departure date for the troops: that he had no idea when he would be able to dispense with them.[21] A few days later the *Islander*, edited by the premier's brother, stated that the troops would be retained as long as necessary.[22]

If the problem of desertion was at the root of the eagerness in Halifax to have the troops returned, the best the Island government could do was to try to stem whatever flow of desertions developed, and if the military leadership still desired withdrawal, to attempt to prevent its wish from prevailing. The soldiers of the 16th Regiment rapidly provided proof that the fears of the Halifax authorities were realistic. Within eight days of arrival, two had deserted.[23] These defections would make the officers in Halifax all the more anxious for the return of their men – and therefore also their active use at an early date in reasserting the authority of the sheriff.

The local government did act, but not in the way Halifax desired. On 14 August the Executive Council decided to add £7 to the usual reward of £5 currency paid for apprehension of deserters. In isolation, that measure would have been pleasing to the military, for it reinforced their efforts. But at the same meeting the executive councillors took a decision which indicated that they were not proceeding on the assumption of a rapid departure by the troops: they established a committee to procure a site for a barracks and to arrange for its erection.[24] On the same day, the *Examiner*, Whelan's newspaper, reported that all was quiet. Whelan expected the Tenant Leaguers to 'be less demonstrative than they were before the troops came ... the League ... will not, we think, provoke a collision with the Military.' In turn, as he saw it, 'the gallant detachment of the 16th Regt. can have no excuse or authority for interfering with the League.' He anticipated a waiting game of indefinite duration.[25]

Of course destabilizing developments were possible. Aggressive action on the part of the sheriff might generate resistance of such a nature or on such a scale as to force active use of the military. Or withdrawal of the soldiers might occur because of pressure from Halifax, provoked by the spectacle of some soldiers deserting while none were actively employed in supporting the sheriff. In fact the situation was so delicate that Whelan's words, with the picture of inactivity they portrayed, had an almost incendiary impact at the highest military level in Halifax.

Fuelled by a reading of the *Examiner*, Major General Doyle wrote to the administrator on 22 August asking bluntly whether he intended to use the troops for the purpose for which he had requested their presence. If this was not done, 'in consequence of the numerous desertions [at least five] which have already taken place ... aided, as is well known, by the people themselves,' Doyle would recall them.[26] Hodgson responded by emphasizing the role of the troops in preventing problems, and outlining quite precisely those uses of the soldiers which he believed to be proper and those he considered improper. Rather coldly, he brushed off Doyle's assertion that the inhabitants in general had aided desertion, and went on to state that it should be no surprise that Tenant Leaguers would do so, noting the likelihood that procuring the withdrawal of the troops would be the objective of such activity.[27]

There is ample evidence that the military leadership was strongly opposed to the prospect of soldiers remaining on the Island for an extended period. The day after an advertisement calling for tenders for the erection of a barracks in Charlottetown, with a planned completion date of 15 October, appeared in the *Royal Gazette*, Major Tydd informed Hodgson that Halifax wanted the plan suspended.[28] Hodgson and the Pope government continued with the project. From their point of view, proceeding with the construction of a barracks was an urgent necessity if they were to have it ready for occupation before the weather turned cold. For this reason, Hodgson had informed Cardwell, they had not waited for his approval before undertaking the project, and for the same reason they would not comply with the request from Halifax.[29]

The Island government was facing an awkward dilemma in the days after the arrival of the troops. Actively employing them against Tenant Leaguers was certain to be politically costly, but failure to use them could lead to their withdrawal. Constructing a barracks would at least make it possible to keep them comfortable over the winter, but the decision to commence building went beyond considerations of comfort. Lack of a barracks could contribute to a decision to withdraw the soldiers, whereas construction willingly undertaken at the expense of local taxpayers placed the Island government on moral high ground in requesting their continued presence.

The pressure from the military authorities in Halifax appears to have been constant. Doyle himself showed up in Charlottetown on 7 September, and remained for several days; in a despatch, Hodgson recounted that this had given him the opportunity to explain matters more fully than was possible through written communications, but on Downing Street a sceptical Arthur Blackwood noted that Hodgson did not reveal how Doyle responded to the explanations. Doyle's presence was widely known locally, for it was reported in the Island press.[30] The military's continuing desire to withdraw was a significant ingredient in the situation as it developed over the late summer and the autumn of 1865, and in the calculations of both the government and the Tenant League.

Capture of Joseph Doucette by the Civil Power

If the approach of the local government – discouraging and demobiliz-
ing the Tenant League without active use of soldiers against Island tenants and therefore with minimal political damage to itself – was to
succeed, some vigorous and effective action on the part of the civil
power would be necessary in order to create a 'demonstration effect' on
leaguers. Thus, any such initiative should be understood, like the
increase in pecuniary reward for capturing deserters and the plan to
build a barracks in Charlottetown, as part of the overall strategy of
retaining soldiers on the Island while avoiding actual use of them
against the leaguers. The ultimate authority over both Doyle and Hodgson was the imperial government in London, and Hodgson and the
Pope government had a strong interest in convincing Downing Street
that, while order was being restored without the political embarrassment involved in actively using troops against tenant farmers, the presence of these unused soldiers was nonetheless essential to keeping the
peace.

On Monday the 14th of August, the same day that the Executive
Council increased the reward for apprehension of deserters and set
about to build a barracks, Sheriff Dodd received from Theophilus Des-
Brisay, JP a warrant for the arrest of Joseph Doucette 'for a riot, assault
and rescue.' DesBrisay was also the father-in-law of Colonial Secretary
W.H. Pope, and Doucette had been identified as the man who had
swung the post that broke Deputy Sheriff James Curtis's arm at Curtisdale on 18 July.[31] Little is known of Doucette, although the *Islander*
newspaper would provide the following physical description: 'tall and
muscular, of middle age, and bearing on his countenance traces of an
excitable and turbulent nature.'[32] His home was in Wheatley River, Lot
24, some fourteen miles from the capital. Dodd acted at once, and

knowing that in case of resistance, I could not depend upon any assistance from
those who lived near him, I took with me nine men well armed, under the hope
that by leaving Charlottetown during the night, and arriving at Doucette's
house very early in the morning, I might be able to take him before a large number of people could assemble to rescue him.[33]

The sheriff's party arrived just after daylight on 15 August. Dodd
reported that

Although men had been stationed by Doucette in his out-houses to watch for and give notice of the approach of any force which might be sent to take him, it seems upon this occasion that they had gone to sleep at their post, and we were thus enabled to surround the house before we were discovered.

The moment we were discovered, the men in the out-houses at once began to blow their trumpets, and although we silenced them as soon as we possibly could, ... this signal ... [was] heard and answered from farm to farm.[34]

When the sheriff demanded entry at the door of the house, those inside refused. Curtis, who was one of the party, stated that Doucette even opened a window and, in an act of defiance, blew a trumpet himself. The sheriff then ordered Constable Jonathan Collings to break the door open and, in Curtis's words, as it was 'strongly fastened,' the result was that it was 'smashed.'[35] According to Dodd,

when we entered we found that Doucette had retreated up stairs, and with a number of men armed with large sticks prevented our getting after him. Several times the attempt was made by my men, but the opening of the stairs was so narrow that they were unable to defend themselves, and from the heavy and repeated blows showered upon them, were knocked down several times and received severe bruises.

The sheriff then ordered entry through an end window, but the constable who made that attempt 'received a severe blow on the head which felled him to the ground.'[36]

Rebuffed in the attempt to enter through a window, the sheriff's men tried again to mount the stairs. This time they succeeded after Curtis ordered his son to proceed upstairs with a fixed bayonet on his musket. Others followed Curtis's son,[37] and, by Dodd's account, 'one of them seized Doucette by the hair of his head and dragged him down, and we secured him.' Curtis later stated under oath that there were six or seven persons in addition to Doucette upstairs. But tin trumpets had been sounding more or less steadily, and the sheriff concluded that the best his party could hope for was to escape with Doucette securely in their custody.

Owing to the number of men we saw collecting, I did not think it judicious to endeavor to make any other prisoners, and the information I subsequently received confirms the opinion I then formed, for we had not left the place more than a few minutes before (I have been since informed) a hundred men had collected for the purpose of rescuing any prisoners we arrested or property we had levied upon.[38]

In the report Dodd wrote the same day, he stated baldly that the problem in executing writs 'arises from the active sympathy shewn by all the tenants for each other,' from their 'ingenious' tactics, and from 'their system of terrorism' which, according to him, made it 'utterly impossible to look for any assistance outside of the town.' His words were both revealing and, over time, controversial. The accusation of 'terrorism' touched a sensitive nerve, for he claimed that the Tenant Leaguers 'intimidate the well disposed under threats of burning their premises and taking their lives.'[39] His report would be published as part of an appendix to the House of Assembly *Journal* for 1866, and in the following year Tenant Leaguer Manoah Rowe would commence a prolonged public attack on him for stating that leaguers had used threats of arson and murder.[40] Rowe was not the only one to take up the issue. In 1868 there would be an impassioned exchange in the assembly between former Tenant Leaguer W.S. McNeill and former Solicitor General T. Heath Haviland Jr on the same subject.[41]

But aside from wounded sensibilities, Dodd's letter is important for what he conceded about Tenant League strength. The sheriff, as the law officer on the front line in Queens County, was evidently convinced of the overwhelming support for the league in the countryside; and he was also impressed with its capacity for organization. The very fact that there were several guards – Curtis estimated 'five or six'[42] – in outhouses at Doucette's farm is testimony to the systematic way in which the league was capable of acting in protecting those whom they believed law enforcement officers were seeking.

The events in Wheatley River could be interpreted in contradictory ways. Dodd and Curtis had demonstrated that it was possible for the sheriff, assisted by constables, to arrest a Tenant League fugitive or to execute writs without participation by the military.[43] But the league had not disintegrated with the mere disembarkation of troops and their encampment at Spring Park on the outskirts of the capital. The ferocity of the resistance in Wheatley River suggested that arresting leaguers might be a costly and dangerous undertaking. Dodd stated that in the course of the mêlée, 'all my men had been struck, and some of them seriously bruised.' Indeed, he attributed the capture of Doucette 'wholly to the fact that he and his accomplices saw the fire arms which I had placed in the hands of my Constables.' Curtis reported firing a pistol inside the house, although he said he did not aim it at anyone in particular.[44]

On 16 August, the day after the arrest, Administrator Hodgson wrote to Cardwell again. He more or less took the line that events were unfolding as they should, and reiterated his opinion that the presence of the military would suffice to uphold the law. But he also stated that recalling the soldiers to Halifax before the lengthy winter freeze-up 'would leave the Island for a period of five months at the mercy of the Tenant Union, and practically divest the Supreme Court of the power to enforce its authority.' Therefore, as he had already indicated to Colonel Francklyn, he did not know when it would be safe to withdraw the troops; in fact, he thought such a time was 'in all probability distant.' Under these circumstances, he reported that the local government was planning to erect a barracks 'without delay' and was offering significant rewards for capture of deserters.

Hodgson did not mention that there had already been at least two desertions. Yet, with his decades of involvement in public affairs on the Island, he was familiar with the desertion problem. In the 1830s and 1840s, when he had been attorney general, the public records reported payments for such services as apprehension of deserters and advertising their descriptions. He must also have been aware of the British attitude to desertion, and have realized that Cardwell would have been interested to know that desertions were occurring. Given his long experience, it is difficult to avoid the conclusion that he was deliberately withholding information that might prompt Cardwell to adopt the attitude of the Halifax authorities. Thus, in mentioning the issue, Hodgson put all his emphasis on the ostensible objective of the local government: 'to check, and if possible, entirely prevent the loss which detachments have so often suffered in this Colony from desertion.'[45] But by the following day the total number of deserters had risen to five.[46]

Government Dismissals

We have already noted the removal of James Laird as a justice of the peace, and the declaration by the Board of Education, consisting of government appointees, that it would dismiss teachers who supported the Tenant League actively. Both initiatives were undertaken in late July. At the beginning of August, during the same Executive Council meeting at which members resolved to call for military assistance from Halifax, they instituted several inquiries aimed at persons who held minor positions of emolument or honour under the Island government, and who had reportedly displayed sympathy with the league. The most prominent of these was Joseph Wightman of Lot 59, Kings County, a former Liberal assemblyman and a justice of the peace. A well-educated Lowland Scottish immigrant active in several sectors of the economy, he was a notable member of his community.[47] He had been the prime spokesman for a delegation from southern Kings which appeared before the land commission of 1860. In his opening statement to the commissioners, he had declared that 'The people whom we represent have a bitter antipathy towards the rent-paying system. They might dispose of almost the last article they possess to become freeholders.'[48] The Executive Council noted a report that Wightman had presided 'recently' at a meeting called to form a branch of the Tenant League, and on the following day, Charles DesBrisay, the clerk, wrote to him to inquire.

Wightman replied immediately, confirming that he had been called to the chair when the Tenant League met at Montague Bridge, Lot 52 (at the site of the present town of Montague). But, he reported, when he heard the league's constitution explained, he objected to a rule not upholding the supremacy of law. This would no doubt be the 'tenant's pledge,' by which members were bound 'to resist the distraint, coercion, ejection, seizure, and sale for rent and arrears of rent.' Since Wightman did not approve of the strategy of the league, he had taken no further part in its proceedings and planned to attend no more meetings. When the Executive Council met on 14 August, members accepted his explanation and ordered that the correspondence with him be published in the next issue of the *Royal Gazette*. The publication would have the dual function of publicizing the distancing of a veteran Liberal politician from the league, and flushing out any further evidence of involvement by Wightman, should the councillors wish to test the validity of his account.[49]

Less prominent persons proved more difficult to deal with. On 1 August the Executive Council also noted reports that two justices of the peace, Hugh Carr and William Campbell, and one commissioner for the recovery of small debts, John Balderston, had attended Tenant League meetings. Carr resided in eastern Prince County, and Balderston and Campbell in western Queens. The Council ordered that DesBrisay ask whether they had attended league meetings 'at any time,' and whether they were presently members of the league.[50]

Balderston, a close neighbour of George Clow and William Large in North Wiltshire, Lot 31, was entirely unapologetic about being a Tenant

Leaguer and added that he had been present at the founding convention of 19 May 1864. His letter was argumentative and defiant, reaffirming his commitment to solidarity with his neighbours in the struggle to 'release ourselves from serfdom, oppression and an unjust rent.'[51] Campbell replied, but not to the satisfaction of the Council. A second letter of inquiry sent to him, dated 29 August, received an answer, in the phrasing of the Executive Council Minutes, 'of a very extraordinary character, and couched in most insulting language.' When Carr replied, he acknowledged that he had presided at a meeting convened to form a branch of the league, had encouraged its formation, and had contributed funds. The Executive Council dismissed Balderston, Campbell, and Carr on 19 September; Campbell had attempted to pre-empt dismissal by resigning, but the Council rejected that manoeuvre.[52]

Another justice of the peace the Executive Council dismissed on 19 September was Robert Gordon from western Prince County. On 14 August the Council had received a report that he and Cornelius Richard O'Leary, health officer for the Port of Cascumpec, had recently participated in a public meeting called to establish a branch of the Tenant League, and ordered that they be directed to state whether they had done so 'at any time,' and whether they were members. Gordon had been delegate and spokesman for Lots 4 and 5 before the commission of 1860, and was a member of a family with a history of confrontation with landlordism, specifically the Cunard estate.[53] No one familiar with the Gordons would have wagered in favour of a submissive response, and Robert proved true to the tradition established by his father, John Gordon Sr, who had carried on a long and bitter feud with the formidable James Horsfield Peters, then the land agent for Sir Samuel Cunard.[54] There was a biblical certitude about the Gordons when they took a stand, a fact reflected in the fate of two brothers of Robert, both Presbyterian missionaries slain in the south Pacific island of Vanuatu; the death of one in 1861 had prompted the other to follow him to the same island three years later, after publishing a eulogistic volume.[55]

Robert Gordon showed comparable spirit in dealing with the Executive Council. In his initial reply, dated 30 August, he declined to give a full answer to the inquiry DesBrisay had addressed to him, and delivered a brief homily on the rights of the subject in relation to the Crown. He did state, nonetheless, that 'I am not a member of the tenant union.' Dismissed on 19 September, he protested to London, and almost two years later he was still asking for more information on the reason for his removal. In his memorial dated 10 January 1866 he stated that at the meeting he attended he had voted against formation of a branch of the league, that he had not joined it, that he did not contribute money to it, and that he did not attend additional meetings of the organization.[56] The government did not pursue the inquiry into O'Leary's conduct, although there is no record of his having replied, and although there is evidence that he was in fact a Tenant Leaguer.[57]

On 30 August 1865 David Lawson, a member of the Central Committee of the league, was dismissed from his largely honorary post as captain of a company of 'Volunteers.' The Volunteer movement had flourished in the early 1860s, and many Islanders had enrolled. As already noted, the local legislation governing the Volunteers explicitly prohibited their use in cases of civil disturbance; indeed this limitation on their role had been an important pre-condition for the success of the movement. Thus there was never any possibility that Lawson would be called upon, as an officer in the Volunteers, to assist the sheriff. But it was widely recognized that there was overlap between the respective memberships of the Volunteer movement and the Tenant League, and the Queens County Volunteers had even been asked to surrender their weapons. The adjutant general of militia, Lieutenant Colonel Stewart, reported that on 22 August Lawson, who lived on Lot 34, eastern Queens County, sent him 'a highly insubordinate letter, subversive of all order and military discipline,' presumably in response to the request for a return of arms. Stewart dismissed Lawson, who did not protest; or if he did, no documentation survives.[58]

Other cases were also investigated. Veteran Liberal assemblyman and Justice of the Peace Francis Kelly of Fort Augustus, in the heart of Tenant League territory in eastern Queens, was asked whether he had subscribed funds to the league. His reply, supported by a joint affidavit signed by two others, was deemed sufficiently satisfactory that no further action was taken. Another justice of the peace, James MacDonald, also of Fort Augustus, was judged to have been guilty of 'an error in judgement' with respect to the incident involving intimidation of Constable Bernard McKenna on 22 July, but not to have done anything warranting dismissal. He had replied to a request for an explanation with a letter, accompanied by affidavits.[59]

The harm the dismissals of minor officials did to the individuals involved is open to question. These positions – justice of the peace, commissioner for the recovery of small debts, Volunteer captain – provided nominal remuneration at most, and therefore it is unlikely that anyone's livelihood was imperilled. Did removal for Tenant League activism undermine prestige in the community? This is doubtful. Both Balder-ston and Campbell would later be elected to the legislature. Balderston's victory on 19 December 1866 was particularly noteworthy: in an election for a Legislative Council seat representing all of western Queens County, he won 62.7 per cent of the votes as a Liberal candidate in a constituency which had elected a Conservative in 1863.[60] The net effect on the community standing of these men does not appear to have been negative, and the disciplinary impact the dismissals had on the general population of western Queens would seem to be questionable also.

Another repressive initiative was the proposal by the Board of Education on 27 July 1865 to deprive teachers who supported the Tenant League 'in future' of their salaries and to remove them from the roll of licensed teachers.[61] The most important function of the Board was the licensing of teachers, and such action would effectively dismiss them. Consequently this was a serious matter and it did involve a person's livelihood. But it is not clear whether the Board had the right to do what it proposed to do. Justices of the peace, commissioners for the recovery of small debts, and officers in the Volunteers all had roles in either the administration of justice or the defence of the crown, and a case could be made that the 'tenant's pledge' put them in direct conflict with that commitment. The connection in the case of the teachers was more tenuous. The Board made its position known late in July, but it does not seem to have pursued anyone other than Alexander McNeill, secretary of the Tenant League. Others who might have been singled out, such as John Bassett of Lot 22, a vocal supporter of the league, remained in their places, and no record of threats to their positions as individual teachers survives.

On 3 August McNeill replied to the Board of Education's statement of policy, enclosing the Mount Mellick resolution which in effect demonstrated that he had the support of parents in his school district. The Board countered tartly that it 'could not ... be any longer responsible for the payment of his salary.' On 14 September McNeill asked for an investigation of the accusation against him. Two weeks later he had a hearing in person before the Board. In a sense the meeting was inconclusive, because although he stated that he no longer held an office in the Tenant League, he refused to sign a written declaration renouncing all connection with the organization. The Board threw the issue into the lap of the Executive Council a week later by reporting on its dealings with McNeill and stating that it had declined to authorize payment of his salary – £20 was owing – without the sanction of the Council.[62]

For its part, the Executive Council seems to have doubted the legality of, in effect, dismissing teachers for Tenant League activism. In its minutes around this time there is evidence of concern lest successful appeals of other dismissals reveal abuse of authority which would force reversal of decisions announced publicly for the purpose of deterrence. Before deciding to discharge minor officials, the Council appointed three lawyers to a committee to make recommendations: Attorney General Edward Palmer, who was not a member of the Council, and executive councillors John Longworth and Solicitor General Haviland.[63] Clearly, it wanted to proceed with caution, for publicity regarding the imposition of discipline, if such was the outcome of an inquiry, was integral to its intent; therefore there could be no mistakes which would be embarrassing, politically damaging, and counterproductive. News of the dismissals of Balderston, Campbell, Carr, and Gordon, decided by the Council on 19 September, was published in the *Royal Gazette* the following day – in all probability before notification reached the four.[64] This was the same issue of the official newspaper that announced to the public the sentencing of an Islander to six months of hard labour after conviction for attempting to induce a soldier to desert.

Wary of making a misstep, the Council referred the report of the Board of Education to Palmer for an opinion on the Board's power to withhold the salaries of teachers accused of association with the Tenant League.[65] Palmer, a member of the Island bar since 1830 and of the legislature since 1835, and always a stickler for procedure, replied on 16 October that such action would have to be taken under the rubric of 'gross misconduct or neglect of duty,' and that due process would have to be observed.[66] Eventually the Council would authorize payment of McNeill's £20 and ask Palmer, further, whether the Board had been justified in withholding his salary.[67] Palmer replied on 2 December that he believed the Board had been justified, but admitted that there was room for legal argument; in other words, there was the political risk of defeat over a legal challenge should the matter ever go before a judge.[68]

On the same date the secretary of the Board stated, in response to a query from the Council, that McNeill's claim for £20 related to past services and 'is apart from his present position – he is now virtually under suspension – the Board having resolved no longer to recognize him as a licensed teacher while he refuses to comply with its requirements.'[69] More than two additional months would pass before, on 7 February 1866, the secretary would inform the Council that because McNeill had disavowed his connection with the Tenant League 'under his hand, in the presence of the Board,' it would again recognize him as a licensed teacher.[70] Later that month at least four Island newspapers would report McNeill's disavowal.[71]

Throughout the confrontation with McNeill, the Board of Education appears to have been at least as zealous as the Executive Council in pressing him to comply with their conception of a teacher's role in society, which included being an exemplar in upholding the rule of law. Since they did not pursue others, like Bassett, their actions suggest a desire to make a conspicuous example of McNeill specifically. Given that the Council feared the political consequences of a legal defeat in the harassing of McNeill (which Palmer believed was possible), it is conceivable that the Board, an unelected body appointed by the Island government, was consciously stepping forward to do the work as the Council's surrogate. In effect they were striking at the head of the Tenant League rather than the body, but the 'demonstration effect' would be for the benefit of the membership as a whole. If this was the intent, it may be judged a success. Under pressure, McNeill had renounced his commitment to the league, lending support to the contention of those, like Whelan, who had argued that the Tenant League leaders were not the men to carry through the struggle to the bitter end, regardless of personal consequences, and that they were opportunists who would lead others to their destruction while protecting their own hides.

The example of McNeill must have disheartened at least some of the rank and file, and was probably more damaging to morale than the dismissals of Balderston, Campbell, Carr, Gordon, Laird, and Lawson combined, for it suggested a lack of spirit at the highest level of the organization. The *Herald*, generally sympathetic to the league, carried the news of McNeill's reinstatement under the heading 'Alas! How are the mighty fallen,' and remarked that this 'suggestive communication ... requires no comment.'[72] In the summer of 1866 the *Islander*, consistently hostile to the league, would carry a news item which implied that McNeill's rehabilitation as a respectable member of society, taking an active part in public issues other than the land question, was complete.[73]

Was the Tenant League Illegal?

The apparent reluctance of someone so steadfastly conservative as Attorney General Palmer to encourage taking punitive action against Tenant Leaguers who were not involved directly in some aspect of law enforcement or defence of the crown suggests fundamental questions.

Was the league actually illegal? Indeed, *what was* an 'illegal' organization? When Administrator Hodgson requested that troops be sent to the
Island, he referred to '[t]his illegal union.'[74] Thus Hodgson, normally
chief justice and formerly the attorney general for more than two
decades, appeared to have no doubt about the illegality of the Tenant
League. Dundas, when endorsing Hodgson's application to Halifax for
soldiers to support the civil power, referred to the league as 'an illegal
association.'[75] W.H. Pope, who was a lawyer as well as editor of the
Islander newspaper and colonial secretary, wrote in his editorial capacity
that 'Persons by assembling and associating under [the 'tenant's
pledge'], thereby, in our humble opinion, commit an indictable offence,
and render themselves liable to be indicted for a conspiracy to subvert
the laws.'[76] This was quite different from arguing, as he had done
shortly after the formation of the Tenant League, that adherence to the
'tenant's pledge' would lead leaguers to commit acts by which they
would be breaking the law. In other words, Pope was now making the
case that membership in the league was itself an offence; and he stuck to
this opinion throughout the period of the Tenant League's promi-
nence.[77]

But Palmer, as the lawyer responsible for providing sound legal
advice to the local government, framed his words in a more guarded
fashion when writing to the Executive Council regarding McNeill, the
secretary of the organization. He commenced as follows: '*Assuming* that
the Tenant League or Union is an unlawful combination of men, and
that the members of it are liable to be indicted and punished criminally
for a violation of law ... ' He went on to spell out the need, even if the
league were illegal, for the authorities to observe proper procedure in
undertaking such measures as withholding the salary of someone paid
from the colonial treasury, if they – the Island government – were to act
'constitutionally.'[78]

In public discourse on Prince Edward Island, the terms 'illegal organization' and 'secret society' appeared frequently with respect to the
league. Among journalists, Whelan was especially prone to use them
more or less interchangeably with 'oath-taking society.' Such expressions had meanings rooted in Irish history, and since it was Ireland that
provided Whelan and some other Islanders, such as his fellow-journalist
and political rival Edward Reilly, with much of their political frame-
work and vocabulary, the associations and assumptions they made in
using the terminology require scrutiny.

'Secret society' indicated that members might be sworn not to reveal to non-members the range of their actions or principles, or their methods of governance, or their passwords, or the names of their members; and this code of silence with respect to outsiders has made the secret societies all the more difficult for historians to explain.[79] One historian of Ireland has pointed out that they fell into three broad categories: 'those with socio-economic purposes (usually agrarian conspiracies), those with sectarian ends, and those with essentially political objectives.'[80] Examples of religious and political groups would be the Orangemen, based on Protestant solidarity, and the Fenians, committed to the political independence of Ireland.

An instance of a prominent agrarian secret society, the type most relevant to analysis of the Tenant League, would be the Whiteboys of the pre-famine period. According to historian Paul E.W. Roberts, they engaged in 'agrarian terrorism ... primarily aimed at redressing the economic grievances of the poor and thus mainly directed against the middle class.'[81] The poor were small farmers and rural labourers, two social groups without a clear boundary between them and with much in common in their daily lives. The rural middle class consisted of such persons as shopkeepers, publicans, millers, and medium and larger farmers. The agrarian secret societies met in secret, and usually made their raids only by night. They delivered warnings, and if these nocturnal messages were not heeded, they would punish by means of fire, torture, mutilation, or murder. In order to reduce the possibility of identification, they might use persons from another district to do the actual deeds. Compulsory enlistment of peasants into the organization also helped to maintain the veil of secrecy; those who refused to join a secret society or who defied its orders were treated as enemies. In the event of apprehension of a member for a crime, other members could, through intimidation of witnesses, make it extremely difficult to obtain a conviction.[82]

'Oath-taking' referred to the pledging of members of an organization to certain actions or principles, although in some instances in Irish history the oaths might focus on the need for solidarity as such and in all circumstances – rather than a programmatic goal, as usually understood. A case in point would be the summarized version of the initial oath taken by the Caravats, a regional outgrowth of Whiteboyism: 'to be true to each other on every occasion, *particularly quarrels at fairs.*' Yet the Caravats were an exceptionally class-conscious group, for, unlike mainstream Whiteboys, they made no effort to enlist sympathetic members of the middle and upper classes; they treated those classes as a whole as enemies. Roberts has depicted the Caravats as 'a kind of primitive syndicalist movement whose aim was apparently to absorb as many of the poor as possible into a network of autonomous local gangs, each exercising thoroughgoing control over its local economy, and the whole adding up to a generalized alternative system.' Caravats referred to the policies they wished to enforce as their 'laws,' and they presented themselves as a sort of shadow government with a special mandate to protect the interests of the poor.[83]

Caravatism, with its combination of class exclusivism and an oath emphasizing the overriding need to support one's fellows in fairground brawls, underlines the extent to which it was possible to confuse a socio-economic organization with a gang of toughs displaying a propensity for faction fights. In fact, the Caravats and the opposing group they generated, the Shanavests, whom Roberts describes as 'an unprecedented middle-class anti-Whiteboy movement,' were exceptionally violent. They fought pitched battles involving hundreds and sometimes thousands of combatants at a time. As many as twenty people may have died at one fair alone, and Roberts has estimated that the total death toll over the course of approximately five years was in the hundreds. The carnage continued until a special commission suppressed the feud between the two groups.[84] Such a mixture of ruthless violence and socio-economic struggle, with strong political overtones, but also featuring behaviour indistinguishable from banditry or vigilantism, helps to explain the distaste of many Irish for the whole phenomenon of oath-taking groups.

'Illegal organization' was a term commonly used with respect to 'oath-taking' or 'secret' societies. Members of such groups might be, as a matter of course, involved in actions which were illegal or which at least obstructed legal processes. In such instances, their very opposition to the laws was sufficient explanation for their oaths and their secrecy, and hence the label 'illegal.'

The term 'secret society' evoked especially negative reactions from two important sources: clergy and government. In Ireland, the clergy provided part of the popular sanction against this sort of body, since Roman Catholic priests anywhere objected strongly to the notion that any part of a believer's life was exempt from the religious obligation of confession.[85] Yet according to historian T. Desmond Williams, 'Secret societies were probably more influential in the shaping of ordinary politics in Ireland than they were in other countries. With one or two possible exceptions, such as in Tsarist Russia, [they] appear to have been more "normal" ... than elsewhere.'[86] Thus, given their importance as well as their nature, they were certain to arouse concern among the Roman Catholic clergy who ministered to the majority of the Irish population.[87] Within British political culture in the United Kingdom and overseas, the very fact that secret societies were usually Irish, and were uncharacteristic of other areas in the British Isles, added to the suspicion the term elicited, especially in official circles.[88]

Prima facie, the terms 'illegal organization,' 'secret society,' and 'oath-taking society,' assuming their interchangeability, suggest that members or at least officers in a body so characterized might be subject to prosecution by virtue of their relationship to it. In fact, one cannot assume that 'oath-taking' means 'secret' means 'illegal' in any precise sense. Political opponents of such controversial groups as the Tenant League or the Orange Order might assert these equations as part of their propaganda warfare against them. Perhaps Whelan, who was also a vigorous and unrelenting opponent of Orangeism, was particularly likely to do so in part because he was a native of north Connaught, an area where, like south Ulster, according to one Irish historian, as late as the 1850s 'the habits of primitive collective violence endured.'[89]

But although the Tenant League of Prince Edward Island had a 'tenant's pledge,' nothing bound members to secrecy. Indeed, the Central Committee of the league regularly advertised in the Island press the time and place of its monthly meetings, which were held in Charlottetown. Charles Dickieson emphasized this fact in a letter to a newspaper in August of 1866. The leaders of the league had met every month in the capital, yet the government had never attempted to arrest or prosecute them in this connection.[90] In the words of Manoah Rowe, another prominent Tenant Leaguer writing on the same theme in later years, 'the displeasure of the Government is one thing, and the violation of law is quite another affair.'[91]

Any case arguing that the Tenant League was an illegal organization and therefore, by extension, that membership in it constituted a breach of law would have to rest on interpretation of the 'tenant's pledge.' That commitment bound members 'to resist the distraint, coercion, ejection, seizure, and sale for rent and arrears of rent.' Rent and arrears of rent were legally collectible debts, and it is arguable that the offence of membership would consist in taking the pledge to resist legal proceedings to collect these debts. What legal opinion of an authoritative nature do we have regarding the legality of the Tenant League? On 12 October 1865 Colonial Secretary Cardwell, a practical and clear-headed man who had ample experience with contentious agrarian issues as the former chief secretary for Ireland (1859–61), raised this question in an internal Colonial Office memorandum when he asked 'to see the actual proof that the society is illegal.'[92]

On the next day Sir Frederic Rogers, the person at the Colonial Office who was most learned in the law, replied. He had impressive scholarly credentials, for he had been a professor of law, indeed a Vinerian fellow, at Oriel College, Oxford University. He was also reasonably free from any hint of special pleading on behalf of tenant organizations. As well as being employed by the Colonial Office, he was married to the daughter of a man who had been land agent for the former Selkirk estate.[93] In an unusually messy, stroked-through passage, he expressed doubt that the Tenant League was illegal in the sense of being 'punishable by Law.' Moreover, he added, 'I do not understand on what exact grounds it cd. [could] be so.' At the same time, he believed that the Queen's Representative was justified in dismissing servants of the Queen – such as justices of the peace – who had continued to belong to the league after the proclamation by Dundas on 22 March 1865 when he had denounced it (although not by name) for its involvement in agreement to resist the payment of rent, given that such resistance led to breach of the peace.[94] In summary, although Rogers would not advise Cardwell to discourage the Island government from dismissing justices of the peace and others from office, neither was he persuaded that membership in the Tenant League could be construed as illegal. A memorandum by the chief clerk of the Colonial Office, T.F. Elliot, on 1 March 1866, possibly influenced by Rogers's memorandum, revealed that he too was uncertain whether the league was, properly speaking, an illegal organization.[95]

What could the Island government do to punish the leadership of the Tenant League through use of the law rather than the troops? Their latitude was limited. We have already noted that after the pro-Dickieson mass demonstration in Charlottetown on 26 July 1865 Administrator Hodgson had inquired whether Samuel Lane, a Tenant League leader, could be prosecuted for inciting the crowd to rescue the prisoner. Hodgson was apparently impatient or dissatisfied with whatever response he received. At his direction, on 5 August J.W. Morrison, the deputy colonial secretary, wrote to the crown law officers, Palmer and Haviland, submitting a 'case' for their opinion.

Hodgson's hypothetical case was as follows. The Tenant League had marched in Charlottetown more than once (on 17 March and 26 July), and although there had been no violence or intimidation beyond the fear created by their numbers, noise, and shouts,[96] he wanted to know whether these demonstrations were legal or illegal, and whether, if illegal, they could be legally 'dispersed' and their participants arrested and charged. Finally, he wished to know, if the civil power was inadequate, whether the military could be legally used to assist in dispersal and arrest.[97] He made these inquiries on the eve of the arrival of the troops, and although such eventualities never came to pass, it is a fact that after the soldiers arrived the Tenant League assumed a more defensive posture. Members no longer marched in Charlottetown, and thus the occasion for arresting Tenant Leaguers in the act of demonstrating there never arose. No record of the opinion that the crown law officers rendered survives, but Hodgson's inquiries indicate the range of options which at least some in the local government were contemplating, such as challenging the right of leaguers to demonstrate in public.

John Ross, publisher of *Ross's Weekly*, was one supporter of the Tenant League whom elements in the Island government seem to have been especially eager to punish. Writing to Cardwell from Scotland on 25 August 1865, Dundas reported that he had repeatedly urged his 'Law Advisers' – Palmer and Haviland, the attorney general and the solicitor general – to prosecute Ross 'for the seditious advice which he was giving to the Tenantry, but they [the crown law officers] with every desire to proceed against him, assured me that it would be impossible to get a Jury to convict him.' The lieutenant governor portrayed both Liberals and Conservatives as having been sound from the outset concerning the illegitimacy of the league, and he outlined the perspective of the local government: to get at whomever they could through dismissals.[98]

Yet even in dismissing employees, the attorney general had reminded the Executive Council of the obligation to act 'constitutionally,' by which he meant 'according to the principles of right and to the general rules applicable to the administration of justice by the laws of England, that is, that the party accused should first be made distinctly acquainted with the nature of the charge alleged against him.' In other words, the government or such quasi-governmental bodies as the Board of Education were bound by law, and not above it. A specific charge had to be spelled out, and the accused 'should have a fit opportunity and sufficient time to be heard and to adduce evidence on his behalf before any order [be] made against him.'[99]

On 7 October 1865 the Executive Council asked the crown law officers to examine an article entitled 'Constitution of the Tenant League' in *Ross's Weekly* of 14 September. In the view of the Council, the article 'appears to be of a libellous and seditious character.' If the crown law officers concurred, they were to 'adopt the most prompt and efficacious proceedings against the parties whose names are appended thereto, or against the Publisher, as they may be advised.'[100] The officers, when reporting to the Council on 13 November, indicated that the appropriate charge was seditious conspiracy. They stated that they would proceed 'against the parties whose names appear at the foot of the article as being present on the 5th September last' – as soon as they had sufficient evidence in order to go before a grand jury. This would require, they reported, testimony by non-accomplices. Such a charge was never taken to court, presumably because they were unable to obtain enough of the proper kind of evidence.[101]

The Council responded by digging deeper into the history of the Tenant League agitation and referring *Ross's Weekly* of 16 March to the same officers, to determine whether prosecution of the publisher for sedition or libel was warranted, and directing, once again, that if they concurred they should 'adopt the most prompt and efficacious' measures. That issue had included an article entitled 'The Tenant Union and the Courts of Law,' in which a sheriff or bailiff acting in certain circumstances was characterized as a 'raider.'[102] The reply of Palmer and Haviland, as read to the Executive Council on 12 December, was that the article in *Ross's Weekly* of 16 March did create sufficient grounds for prosecuting Ross on a charge of libelling the administration of justice on the Island. They proceeded to recommend the best procedure for gaining a conviction, which involved delay. The recommendation of such a course of action was grounded in the assumption that pro–Tenant League sentiment still ran high. The most prompt means of proceeding was not, in their view, the most efficacious.[103]

The question of whether the league was an illegal organization and whether membership in it constituted an offence is distinct from the matter of the sympathies of Island jurors. Nonetheless, these two issues cannot be separated entirely, because in trial by jury common Islanders would ultimately determine the outcome. It was widely accepted that jurors on the Island tended to favour tenants over landlords whenever they had any room for discretion. Therefore, taking any doubtful case involving landlord-tenant relations, or resultant confrontations, before a jury was risky for the Island government, especially if acquittal could be politically damaging. Experienced lawyers like Palmer were probably well aware of the challenge and the danger facing them should they undertake prosecution in borderline cases where the charges involved membership in a body which was not clearly illegal. As a politician, Palmer would also fear the consequences of conspicuous defeats, for acquittals of prominent leaguers would mean major propaganda victories for the league. This would worsen the situation of law enforcement officials to an unpredictable extent, and could make what was already difficult unmanageable – something no prudent Island government would want to risk.

There is an inescapable conclusion arising from any reflection on the first actions of the Prince Edward Island government after calling upon Halifax for military assistance: tentativeness characterized almost every action undertaken. The administrator summoned troops, but the civil power did not use them to support the sheriff, who continued to encounter resistance from determined Tenant Leaguers. The government commenced dismissals, but with caution. The legal advisers of the government knew that it would not be easy to prosecute leaguers successfully, and the government itself feared the cost of failure. Were the league, its words, or its actions illegal? The Island government was reluctant to let Islanders be the judge, and the opinion of Rogers indicates that there were sound legal as well as political reasons for their wariness. Curiously, the views of common Islanders may have been closer than those of Hodgson to British legal orthodoxy. If a lawyer employed by the Colonial Office could not comprehend the grounds for treating membership in the Tenant League as illegal, it is highly questionable whether any prosecutor could reasonably expect to convince a jury on Prince Edward Island, drawn as it would be from a political and social environment hostile to landlordism. Under the circumstances, prosecution of Tenant Leaguers for anything other than the most overt offences was a double-edged tool which involved a real risk of exposing in court the weakness of the government's position in a law-enforcement situation which was already delicate.

Tentativeness characterized the authorities; boldness characterized the Tenant League. The situation remained unresolved.

9

Military Desertion and the Tenant League

In a public house ... in Charlottetown, ... the defendant Green came in, and, after treating all the soldiers to drink, threw his purse on the table and said, according to the testimony of [Private] Macgory, 'that any man who wanted to desert he was the man to exchange the clothing. He was the very boy who would do it for him. ... there is plenty of money, and as long as the money lasted he would see any of them off the Island.'
Report of the trial of Henry Green, shopkeeper, for attempting to induce desertion[1]

Inform me whether it is your intention to attempt to enforce the Law in those cases for which the services of the troops were required, for if such be not the case, ... in consequence of the numerous desertions which have already taken place from the ranks, aided, as is well known, by the people themselves, I shall ... recall the detachment from Prince Edward Island.
Major General Charles Hastings Doyle to Administrator Robert Hodgson, 22 August 1865[2]

Military Desertion and Its Context

As the Island government hesitated and hoped that the Tenant League would fade away, instead the troops called from Halifax began to melt away. The problem was desertion. Together, Tenant League encouragement of military desertion and the soldiers' response constitute an extraordinary chapter in the history of the league and indeed the history of British North America. Here we pass beyond issues of agrarian insurgency, leasehold tenure, and specifically Prince Edward Island history, and into questions of military discipline, British recruitment policy, and migration history.

By 19 September 1865 there had been seventeen desertions. Only three deserters were ever captured. Two, Michael Fearon and Michael Nash, both age twenty-seven and Irish, were in the Kings County port of Souris, probably ready to leave Prince Edward Island, when apprehended on 21 September. They had been at large more than two weeks. The third, Thomas Blair, age twenty-six and also Irish, appears to have been on board a vessel in Georgetown harbour, Kings County, when he was taken in early October; expenses regarding his case included a payment for 'boat and crew.' He had disappeared sometime between 12 and 19 September.[3] None of the seventeen deserters was native to the region or to any part of British North America. Thirteen were Irish, three were English, and one was Scottish. In their non-colonial origins, these soldiers were typical of enlisted men: since they were vulnerable to being shipped to distant locations, and since the imperial government aimed to build up the permanent civilian population in the colonies, recruitment was done in the British Isles.[4] The seventeen deserters ranged in age from 22 to 36 years and averaged 26.2. Most were privates, although two, both from Northampton, England, were above that rank, a lance-corporal and a corporal.[5]

The total number of enlisted men dispatched from Halifax to Charlottetown had been 130.[6] The per centage who had deserted was 13.1; if allowance is made for the three captured, it was 10.8. Either figure was, by any standards, a high rate of desertion. In a careful study, historian Carol M. Whitfield has calculated an annual desertion rate of 1.77 per cent from the rank and file of the military establishment in the Nova Scotia military command for the year 1865. The two companies on the Island formed a small part of that body, whose monthly average total that year was 4,124.67.[7]

The high number of desertions in Prince Edward Island during August and September of 1865 was clearly anomalous within the command as a whole. How is it to be explained? Ideological sympathy should not be discounted entirely, but it would require considerable faith in the power of ideas, particularly the idea of trans-national class solidarity, in order to believe that so many were won over to the Tenant League cause, and so quickly. Indeed, the fragility of such notions as class solidarity in a supra-local sense becomes especially apparent when one examines closely such a group as the Caravats in Ireland. One consistent goal of this group of defenders of 'the poor' was to drive out 'strangers,' by whom they meant persons of the same class but from a different, although probably neighbouring, county, and whose offence was that their presence lowered wages and inflated rents.[8] In the case of the Tenant League, the loyalties of the soldiers had not been tested by heart-wrenching confrontations with leaguers. They had not taken part in arrests or process serving, and therefore had not had to make a choice between the authorities and the tenantry.

An explanation of how such an exodus from the ranks of the two companies on Prince Edward Island could occur so rapidly has to begin with a review of some general considerations surrounding the phenomenon of desertion in British North America. In reality, desertion was in the minds of some soldiers from the time they joined the ranks – and long before they knew anything about the Island's land question or, for that matter, local political or social conditions anywhere they were likely to be sent. The fact that regiments going to or already in British North America recruited in the United Kingdom represented an opportunity for some people in the British Isles. Many British commoners were emigrating to North America in the nineteenth century, and for those who were too poor to raise the passage money, entering the army for service on the other side of the Atlantic was a rational emigration strategy. After arriving, when a chance to desert presented itself, such recruits would seize it. Once safely out of grasp of the army, they could seek out friends already in North America or begin earning money to bring over family members. The usual destination was the United States, and proximity to the American border was a factor in desertion; there was no provision for extradition of deserters unless they had committed serious crimes. Whitfield comments that

The army was ... something of an unofficial and reluctant emigration service. Commanding officers consistently accused the Irish of emigration enlistment. This accusation is not surprising since a large proportion of the rank and file were Irish and accordingly figured prominently in any activity of soldiers.[9]

This analysis of the context in which desertion in British North America occurred fits well with the situation on Prince Edward Island, including the predominance of Irishmen among the deserters there. American fishermen who frequented ports like Georgetown and Souris by the hundreds, especially during storms, provided a link with the United States; in fact, Americans could be found in all fishing areas of the Island.[10] The disturbed state of the colony in 1865 meant that there was an organization with a strong motive for enticing soldiers to desert. In other words, the Tenant League crisis presented an extraordinary opportunity to those who had always thought of desertion as an option.

Moreover, summer, the very season when the troops arrived on the Island, was traditionally the season of choice for desertion. All in all, circumstances could hardly have been more propitious both for the soldier contemplating desertion as part of an emigration strategy and for the organization promoting it in order to further its own ends.[11]

It was no secret that desertion had been a problem before on Prince Edward Island. In the years surrounding the withdrawal of troops in 1854, the *Journal* of the House of Assembly had included official correspondence on the subject, and the published documents made it clear that there was disagreement among previous lieutenant governors on whether the inhabitants had been actively encouraging desertion. For example, the last two before the withdrawal of the garrison differed: Sir Donald Campbell (1847–50) suspected that Islanders had been doing so 'for the purpose of ... procuring cheap labour,' but Sir Alexander Bannerman (1851–4) thought not, and stated explicitly that 'I think my predecessor laboured under a wrong impression.'[12] Like the rest of informed Islanders, the Tenant League leaders were well aware of the official concern over desertion. They must also have realized that rapid removal of the soldiers was a real possibility, for the press discussed the prospect openly.

The Island government knew it faced a delicate situation, and was anxious to stop desertion. By 30 August it was advertising that it would pay £30 currency for the apprehension of any deserter or for any information leading to the same result – an increase of 150 per cent since the most recent augmentation, just sixteen days earlier.[13] This was six times the amount of the reward at the time the troops had landed, and in contemporary terms it was a significant sum. Thirty pounds represented exactly one-half the maximum annual salary the Island legislature specified for a male school teacher with the second, or higher, class of licence, teaching in a rural district.

The Role of the Tenant League

The strategy of the Tenant League appears to have been to promote as much desertion as possible in an effort to goad the military authorities in Halifax into withdrawing their men. There is no known surviving document in which league leaders declared this to be their policy, and in fact creation of such a written statement or record of intent would have been extraordinarily imprudent. If it fell into the wrong hands, it could be presented in a court as evidence of a conspiracy to violate the imperial statute for punishing those who made any attempt, direct or indirect, to persuade a soldier to desert.[14] But such an interpretation of league intentions is consistent with much contemporary commentary in local newspapers and with a critical reading of official correspondence. '"Bring your soldiers to the Island, and we will very soon induce the whole of them to desert," was the threat uttered by a Tenant Leaguer,' reported the *Islander* newspaper.[15] The league was also gambling that the Island government led by James C. Pope had no desire to use the troops against tenants because of the adverse political consequences which would almost certainly follow. Desertion of soldiers in significant numbers, combined with non-use of the two companies to support the sheriff, would, the leaguers hoped, create sufficient pressure from Halifax that London would order removal of the detachment.

The leaguers did not wish to provoke physical confrontation with the troops through, for example, a mass demonstration at the Spring Park encampment site. Nor did they hold any public parade or demonstration or march or rally or display or protest or picket in Charlottetown – which was wise, given their strategy, for we know that on the eve of the arrival of the troops Administrator Robert Hodgson thought in terms of using soldiers to assist the civil power in challenging the right of Tenant Leaguers to hold public demonstrations. Any physical conflict with the troops would undermine the arguments made by the Halifax authorities in favour of withdrawal, because once the soldiers were actively engaged, then the officers could no longer complain that their men were not being used, and that the sole process effectively at work was attrition of numbers through desertion. Indeed, desertion was the only activity in which the Tenant League wished to see the soldiers engaged as long as they were on the Island. In the meantime, while waiting for the troops to be withdrawn, the leaguers were, as they proved in the case of Joseph Doucette's capture, willing to exchange blows with the sheriff's men. But any clash with the soldiers would mean initiating a struggle they could not win.

The Tenant League went about promoting desertion audaciously. Our best evidence on this point comes from the trials of two men convicted of attempting to induce desertion. The first was Donald McLeod, convicted of trying to persuade William Glynn, a soldier whom he found in Charlottetown under the influence of alcohol, to desert. In his testimony, Glynn, who, according to a military officer, 'had no leave of absence,' stated that he. 'had a couple of glasses, but was not the worse of liquor.'[16] Glynn's denial that he had been intoxicated should be read in light of the fact that he had a vested interest in making such a declaration. 'Habitual drunkenness,' sometimes defined in terms of the number of occasions one had been recorded as being drunk within a year, was a significant offence under military law. Whitfield, in her study of the British soldier in the British North American colonies, reports that conviction resulted in 'some sort of physical punishment, flogging or imprisonment, and a loss of pay.'[17] On 15 September Glynn had met McLeod, who was driving a 'truck.' McLeod

asked him if he was going to the barrack [Spring Park Camp], to which he replied yes. Prisoner [McLeod] told him to jump on the tail of the truck. He did so. Prisoner then spoke of the number of men who had deserted from the Regiment, said it was likely the deserters would not be found. Prisoner said ... that he would drive him – Glynn – 17 miles into the country if he wished, when he would see someone whom he, Glynn, knew – mentioned the name of a man – 'Goodman,'[18] who had deserted. He also said there were four others who had deserted, and was about to name the place where they were, when the stoppage of the truck prevented him.[19]

The 'stoppage' arose from the arrest of McLeod after he inquired directions from Ellen Duffy, in the words of the reporter for the *Royal Gazette*, 'a young and very intelligent person.' Once she perceived the soldier on the truck, she seems to have realized what was happening and to have simultaneously misdirected McLeod towards the camp and given the alarm.[20] The *Examiner* newspaper reported that McLeod was 'driving at a rapid rate' when apprehended, implying that he was making an effort to escape.[21] At the trial, the soldier stated he 'believed that the prisoner intended to drive him into the country the seventeen miles.' The arresting policeman, Thomas Brennan, testified that although McLeod was sober, the soldier was 'the worse of liquor.' According to Brennan, McLeod was defiant. Admonished on the way to the lockup, he declared 'that the Horse and Truck were his own, and that he would take fifty of them [the soldiers] if he could.'[22]

On 19 September Justice of the Peace Theophilus DesBrisay sentenced McLeod to six months' hard labour in the Queens County Jail. DesBrisay informed him that he could apply to the Executive Council for clemency, but there is no record that he did so.[23] A fairly detailed report of the case appeared in the *Royal Gazette* on 20 September – which meant that the arrest, two hearings, the sentencing, and publication of the newspaper report all occurred within six days. Little is known of McLeod other than that the *Examiner* reported him to be from Lot 22, a hotbed of Tenant League support.[24] The *Herald* hinted that it had been a case of entrapment, but no evidence supporting this theory has emerged.[25] The degree of publicity was highly unusual for a case tried in magistrate's court, and it was doubtless intended to have a deterrent effect.

By way of contrast, a prosecution against Jane O'Halloran, apparently a resident of Souris, for harbouring deserters received little publicity. Yet the trial did occur. The colonial government paid John Lawson, a veteran member of the Island bar, £8 8s. 8d. for his services in prosecuting the case in Georgetown, and disbursed £5 15s. 2d. to justices of the peace there for expenses incurred in the case.[26] One justice of the peace stated in the Island press that 'four of her [the accused's] family were taken there as witnesses against her.'[27] Thus it is clear that the Island government prosecuted the case vigorously. But no evidence that O'Halloran was convicted appears in newspapers or government records, and hence one may infer the strong probability of an acquittal. The government and its allies against the Tenant League had no interest in publicizing a prosecution that failed; and a policy of hushing up unsuccessful prosecutions against civilians for aiding deserters would be complementary to its policy of advertising successes.

In terms of newspaper coverage, the most publicized aspect of the case concerns an oath that Alexander Halloran, a son of the accused, took in Souris on 16 October, attested by a justice of the peace. He published it, among paid advertisements, in at least three newspapers.[28] By this solemn declaration he denied a rumour that was circulating to the effect that he had given information leading to the capture of Fearon and Nash at his father's home.[29] The fact that Halloran went to such lengths to contradict the gossip would appear to illustrate something about public attitudes with respect to assisting these searches; it certainly signified a desire on his part not to have the reputation of being an informer. Many years later, a founder of the Tenant League would refer in the legislature to the statutory £5 reward for apprehension of deserters as 'encouraging immorality,'[30] a choice of words indicating an alternative standard of morality. It is also quite possible that Halloran's action was motivated by fear of retribution.

Whatever Halloran did or did not do to assist the authorities, on the day after he took his oath the Executive Council authorized the Island's colonial secretary, William Henry Pope, to disburse £60, 'Being a reward for the apprehension of two Deserters from Her Majesty's 16th Regiment.'[31] Government correspondence later in the year reveals that the

£60 had been paid through a justice of the peace in Souris to the 'person' (singular) who provided the information leading to the capture of the two deserters.[32] Someone in the area was collecting; and after the capture of the third deserter, in Georgetown, the Executive Council authorized payment of £30 through someone in that area.[33] In none of these documents is the name of either informer revealed. Nor are there any details that would assist in identification.

The temptation to inform when the reward for doing so was £30 must have been significant for many Islanders. For a tenant farmer in arrears to his landlord, collecting a £30 reward would mean being able to pay off several years back-rent, even if the lease were drawn for sterling rents. In such circumstances, popular suspicions that an arrest of deserters resulted from a calculated act of betrayal with a pecuniary motive must have been rife. Sowing the seeds of distrust and the fear of double crossing among those involved in assisting deserters, and thereby crippling the operation, would have been one of the effects, and probably one of the objectives, of increasing the reward by a multiple of six (from £5 to £30) over the course of less than three weeks in August. The knowledge that persons who appeared to be accomplices or sympathizers or simply neutral bystanders could earn £30 for informing would almost certainly have induced greater caution and hesitation.

Yet the £30 reward, the McLeod sentencing, and the report of the sentence proved to be imperfect as deterrents to Tenant Leaguers or those working to assist them. On 3 November the *Islander* reported the conviction of Henry Green, a Charlottetown shopkeeper, on the charge of attempting to induce desertion. At a hearing before T.H. Haviland Sr, the mayor of Charlottetown, Green was convicted of trying on two separate occasions, more than two months apart, to persuade Private Peter Macgowan, the servant of Major Thomas Tydd, the commanding officer, to desert. No date was given for his arrest or the hearing of his case, nor was it made precisely clear how he came to arrested. But the details that were published revealed a great deal.

The first attempt is noteworthy for occurring a mere two days after the arrival of the troops in Charlottetown, and the second occurred well after the McLeod trial. On 8 August Green approached five soldiers, including two non-commissioned officers, who were drinking in a tavern. Macgowan was among them.

After treating all the soldiers to drink, [Green] threw his purse on the table and said, according to the testimony of [Private] Macgory, 'that any man who wanted to desert he was the man to exchange the clothing. He was the very boy who would do it for him.' According to that of [Private] Lamb, Green threw his purse on the table and said, 'that if any man wished to change his clothing there was his purse at his command, and he, Green, would see them safe off the Island. Did not care what creed or profession they were, he would stand to them.' Macgrory stated that Green took out his purse and threw it on the table, saying 'there is plenty of money, and as long as the money lasted he would see any of them off the Island.' ... James Keenan, who was called on the part of the defendant, stated that he saw Green take out his purse and throw it on the table, saying 'that he did not care for creed or profession; ... a man's principles was all he wanted.' All the witnesses agree in stating that Green was drunk.[34]

One of those Green approached on 8 August, Lance-Corporal George Mumby of Northampton, England, age twenty-three, did in fact desert, and was never apprehended.[35] Mayor Haviland noted, *inter alia*, the similarity between Keenan's account on behalf of the defendant, and the testimony of three prosecution witnesses. He put little stock in the defence that Green was too intoxicated to know what he was saying or doing. 'The language made use of is connected and perfectly rational, and appears to be the result of a previous determination. ... The proverb *In vino veritas* is not without an apt signification in the present case.'[36]

On 11 October, according to Private Macgowan, he met Green on the street between 5 and 6 p.m. Green was as direct as he had been previously, and

asked him if there was any word of Private Moffit [*sic*][37] – one of the men who had deserted; to which he replied there was no word of him. Green then said there will not be, for it was I [who] put him away, and I buried his clothes [his uniform] by tying a stone to them to sink them. Green then said that if he, Macgowan, was willing to go, he, Green, would put him, in the same way.

Several defence witnesses testified, but Haviland found their testimony to be either 'very curious' in terms of credibility, or irrelevant, or actually working against the defendant. Green received a sentence identical to McLeod: six months' hard labour.[38]

The inescapable inference, if one credits the sworn testimony in the McLeod and Green cases, is that the Tenant League worked hard and took significant risks to induce desertion. It specially targeted soldiers who had been drinking, and its approach was sophisticated, based on analysis and forethought about the need to create a supportive infrastructure in order to ensure success. The deserting soldier was in urgent need of civilian clothing and a place to go *to*; he almost certainly also needed money. Without a change of clothes, a British soldier 'stood out like a red thumb,' in the words of one archivist.[39] This was in fact the intent of an army rule forbidding soldiers to own civilian clothing. Prevention of desertion was also the rationale for a long-standing tendency among officers to disapprove of any saving of money by the enlisted men; a deserting soldier without funds had less chance of success.[40] Tenant Leaguers offered a haven, civilian clothes, transportation, money, and the prospect of a link to the United States. Their appeal must have been seductive.

Yet the discipline a deserting soldier faced – lengthy imprisonment and possibly the lash or branding with the letter 'D' for 'deserter' or the letters 'BC' for 'bad character' – was daunting, and it was impossible to be unaware of the sanctions.[41] In fact, the penalties were so severe that a sober man might not want to risk them, and at least one War Office official, in commenting on the Prince Edward Island situation, made the link between 'the facilities for desertion and drunkenness.'[42] According to the *Herald*, Nash, one of the deserters arrested at Souris, received lashes: it reported twenty-five in its 4 October issue, and stated a week later that it had been mistaken and that in fact the number was fifty. Unfortunately, the *Herald* is not an entirely reliable source on these matters, for the same brief paragraph in the 11 October issue, which published the ostensible correction, also contained an erroneous report that two more deserters had been captured.[43] Fearon, the other soldier apprehended at Souris, had already been branded, apparently for an earlier attempt at desertion.[44] In addition to the penalties if caught, a deserting soldier, even if successful, had to live with the fact that 'the regiment informed the deserter's family and home parish that the man had committed a felony.'[45]

Consequently, the Tenant League had to be well prepared if it was going to make a serious attempt to induce enough desertions to cause withdrawal of the troops. It had to counteract the measures of the British army and the local government to prevent desertion. This required making careful plans in terms of offering potential deserters safe haven, accomplices, clothing, money, and transportation, as well as moral support; and it had to be sufficiently organized to render success probable. Members involved in enticing soldiers to desert made a point of referring to comrades who had already done so successfully. In a group of 130 enlisted men, individuals would be known by name, and as a result the query from a stranger 'Have you heard of Sean Doe recently?' was engaging. But if the first deserters had been captured, news of that, with the attendant awe-inspiring punishments, would have been disastrous for the Tenant League strategy. The evidence indicates that the leaguers who were involved in inducing desertion knew from the beginning exactly what they were doing, how to do it, and what the stakes were; and they were determined. Their success was remarkable.

The End of the Desertions

Eventually desertion ceased. The reasons are difficult to identify with certainty. As noted, successful desertion in Prince Edward Island at this time seemed to involve the participation of Tenant Leaguers as well as the deserting soldiers themselves. Therefore, any analysis should take into account the considerations facing leaguers as well as those confronting soldiers. The convictions of McLeod and Green may eventually have deterred some Tenant Leaguers; and as the autumn wore on, other events began to erode league morale and commitment. The capture of three deserters may have had a dampening impact on the soldiers, and the changing of the seasons would have been a factor.

Furthermore, the Island government took a number of exceptionally proactive preventive actions which may have had some effect. Since the last days of August, at the request and expense of the colonial government, Mayor Haviland had been providing five night watchmen to patrol the 'outlets' from Charlottetown. The standard pay was 5s. per night, and the shift extended from sundown to midnight. At a meeting on 7 October, the Executive Council authorized payment of £10 5s. to Frederick Curtis and others 'for Services performed as Detectives in keeping nightly watch to prevent desertion'; it also voted £7 15s. and £7 10s., respectively, to two other groups, led by Francis Bell and Hezekiah H. Pollard, for the same. Ten days later, the Council provided for payment of £12 15s. to Edward Shea and others 'as Detectives in watching nightly on various Public Highways.'[46]

The government did something else, which was virtually unheard-of. It furnished 'plain clothing' for soldiers to wear when on military search parties seeking deserters, so as to enable them to travel and make progress without being detected.[47] The 'Detailed Public Accounts and ... Vouchers of Expenditures' that it presented to the assembly on 18 April 1866 included a payment of £31 10s. for seven suits of clothing, 'including hats and neck scarfs,' which Major Tydd reported on 11 September 1865 to be 'in use by men ... under my command.'[48] Like Halloran's widely published oath, the wearing of civilian apparel to disguise military search parties indicates that the general population, at least in some areas, was not inclined to cooperate. For rank and file soldiers, the irony of the situation must have been apparent. Ordinarily forbidden to possess civilian clothing in the interests of preventing desertion, now, for the purpose of pursuing deserters, they were provided with it free of charge and ordered to wear it! Researchers who have published on the subject of desertion in British North America seem unaware of these exceptional measures that the Island government took, and provide no comparable examples. The closest parallel would seem to have occurred in New Brunswick in 1849, and it stopped far short of the actions taken on the Island: 'A small detective force of civilians was formed ... to watch out for escapees on roads adjacent to the border [with the United States].'[49]

It is not known whether any members of search parties, outfitted with civilian clothing through the agency of the Island government so as to blend in with the civilian population, took the opportunity to desert themselves. Scholarly research indicates that members of military parties searching for deserters elsewhere in British North America might seize the occasion to abscond. Historian Peter Burroughs has even written that 'It was not uncommon for military policemen and carefully chosen look-out patrols or pickets posted on major roads to commit the very crime they had been specially selected to prevent.'[50] The Island records are silent on the matter of desertion by soldiers who had been sent out looking for deserters, but it is a fact that there was reluctance on the part of officials to report embarrassing information to superiors; for example, the fiasco of the *posse comitatus* is an aspect of the official record which does not appear in any of the despatches which Lieutenant Governor George Dundas sent to Downing Street. Thus the silence is not proof that the provision of civilian clothes – which actually increased the known risk of desertion by the searchers – never backfired. Indeed, the unreliability of soldiers in this sort of situation may have been the reason that the Island government also paid for non-military patrols of the roads at night.

There is no way of knowing precisely how many active searches for deserters were undertaken, especially in the Charlottetown area, since operations there would not have involved extraordinary expenditures that would find their way into the public records. For missions outside Charlottetown, there is a better chance of documentation surfacing. In fact, the Executive Council Minutes record the voting of money to various persons for assisting or participating in these endeavours. On 19 September 1865 the Council ordered that 'Larter and Weldon' and Harry Binns be paid £2 16s. 6d. and £1 10s., respectively, 'For hire of horses and Wagons in pursuit of Deserters.' The 7 October meeting approved payment of £20 6s. 2d. to John H. Gates for the hire of horses and wagons by the military 'in pursuit of Deserters'; the magnitude of the sum suggests either an expedition on an unusually large scale (or perhaps of long duration), or a number of expeditions. The same meeting approved payment of £2 5s. to William Hooper for hire of horses and refreshments for men 'in pursuit of Deserters.' All of these persons appear to have provided services to military search parties.

Ten days later the Executive Council voted £1 to Pollard, who had acted as night watchman on the roads, and Angus MacLeod, both civilians, '[f]or travelling to Lot 67 – in pursuit of Deserters.' Their quarry, according to a warrant in the 'Detailed Public Accounts and ... Vouchers of Expenditures,' was Alexander McPherson, age thirty-six, the sole Scot among the deserters, who was described as having a 'broad Scotch accent.' He and one other soldier were the first to be reported as having deserted; and as the oldest of the seventeen to desert, he may have conformed to an ethnic stereotype of the shrewd Scot which military authorities expressed, whereby the Scots reputedly planned their desertions carefully, waiting for the right moment, and were rarely captured. In contrast, the English or Irish deserter was supposedly more impulsive, less calculating, and likelier be apprehended.[51] Pollard and MacLeod did not retrieve McPherson as a result of their trip to western Queens County, which they made at the direction of Mayor Haviland; perhaps it is worth noting that Lot 67 is not the most obvious destination for a deserter who planned to leave the colony, since it is one of only two landlocked townships on the Island.

At the same Executive Council meeting on 17 October, the Georgetown mail courier received £2 3s. 9d. for transporting troops to and from Georgetown. Several months later 'Georgetown Mail' collected an additional £2 12s. 6d. 'For fare of Seven Soldiers to and from Georgetown, in October 1865 – in the matter of deserters.' The occasion was most likely the apprehension of Private Blair in that town and in that month. Although not many deserters were captured, money for services and provisions was certainly being expended in the effort. At the 12 December meeting, councillors authorized payment of £10 5s. 6d. to Colonial Secretary W.H. Pope, a member of Council, for unspecified 'expenses incurred in apprehension of Deserters.'[52]

The precautions the authorities took to guard the roads out of Charlottetown at night and to enable search parties to travel incognito undoubtedly put more pressure on everyone with a stake in desertion. The Island government adopted these measures in addition to offering substantial rewards for information, and commencing work on a barracks to house the soldiers. In one of Hodgson's despatches dated 30 August, he informed British Colonial Secretary Edward Cardwell that the Island government was about to spend 'little short of £4,000 sterling' on construction of 'permanent' military quarters. That translated into £6,000 local currency, and although Hodgson described the sum as 'The entire cost about to be incurred by the Colony,' in fact the warrants listed under the heading 'Victoria Barracks' in the Public Accounts for the financial year ending 31 January 1866 would total £8,248 3s. 3d. In another despatch dated 30 August 1865, Hodgson stated, with respect to stopping desertion, that Major Tydd 'has been informed of the willingness of the Government to adopt any additional measure which he [Tydd] may suggest.'[53]

Although the desertions did end, the Tenant League policy has to be considered a notable success in terms of tactics and execution. This is especially apparent when one notes that the men were drawn from a military command where, in terms of its own history, the annual desertion rate had 'hit one of its low points' during the years 1861–4, fluctuating between 0.78 and 1.70 per cent. That rate was low not only by the standards of the Nova Scotia command; it was much lower than the contemporary rate in Canada, where, during the same years, the rate had ranged from a low of 3.85 to a high of 8.35 per cent.[54] Therefore, by any realistic comparative yardstick, it would appear that the league was not dealing with a group of military men whom one might expect to be unusually prone to disaffection – just the reverse, in fact. Yet of the 130 enlisted men who came to the Island 17 deserted in the space of exactly six weeks, from 8 August to 19 September; the rate, if annualized, works out to well over 100 per cent.[55] The key to explaining the high desertion rate during August and September of 1865 in Prince Edward Island lies in the enticement and support the Tenant League offered, in short, the opportunity to desert with a reasonable expectation of success.

In the larger strategic sense, the league policy of encouraging desertion as a means of forcing withdrawal of the troops from the Island failed. They stayed for almost two years. But the failure to bring about a return of all soldiers to Halifax is attributable to factors other than the quality of the campaign to promote desertion. That effort began with an appreciation of the conflicting tugs between the government in Charlottetown and the military command in Halifax, both of which ultimately had to gain and retain the approval of higher authorities in London for their actions. It was also rooted in a realistic comprehension of the concrete situation of the potential deserter, and it was carried out with skill and daring. Without a sophisticated understanding of the inherent tension between Halifax and Charlottetown, the league would not have undertaken the operation, and without competence, nerve, and the realization of what the prospective deserter needed and wanted to hear, it would never have succeeded to the extent that it did.

Although the soldiers remained on the Island, the Tenant League strategy of promoting desertion did create a great deal of difficulty for the local government and caused much friction between it and the military authorities in Halifax.

10

Resistance, Outrage, Escalation, and Coercion

A constable or Bailiff cannot travel on the public road without being abused by a Lawless Mob of those Tenant Union men, boath [sic] with their tongues and those tin horns.
Bernard McKenna to the Administrator of the Government in Council,
4 September 1865[1]

Deponents solemnly swear that they firmly believe that had they endeavoured to execute any of the writs ... they ... would have been killed by the mob by which they were surrounded.
Affidavit of Deputy Sheriff James Curtis and Bailiff Jonathan Collings,
22 September 1865[2]

One large portion of this community has lately required military coercion to compel obedience to the laws of the land.
Robert P. Haythorne to the Editor of the *Islander*, 7 November 1865[3]

Charlottetown, London, and a Military Presence

After soldiers of the 16th Regiment of Foot arrived in August and as some deserted in that month and some more deserted in September, the attitude of the Prince Edward Island government remained constant: keep the troops in the colony, but avoid using them against tenants if at all possible. As already noted, the perspective of the administrator, Robert Hodgson, who was in charge of the government in the absence of Lieutenant Governor George Dundas, was virtually indistinguishable from that of the Executive Council led by James C. Pope. Hodgson and the Council saw themselves as attempting to keep the lid on a dangerous situation. In their view, the military presence was to function strictly as backing or protection of last resort for the civil forces, but the overriding imperative was that it must not be withdrawn.

The absence of British soldiers from 1854 to 1865 had made Charlottetown the exception to the rule among colonial capitals in British North America. One historian, writing of the colonies, has commented on 'the political impact of the garrison, a result of its high public visibility.'[4] The Pope government could interpret its own position – having the military there as a reminder to Tenant Leaguers that the armed might of Britain was available, if necessary, to support the sheriff – as simply a variant on the theme of 'political impact ... [as] a result of ... visibility.' For the government in Charlottetown, keeping the troops in the colony seemed consistent with the norm elsewhere. They certainly did not think they were asking anything unreasonable in wanting two companies of soldiers to remain over the winter of 1865–6 or longer, particularly in light of the volatility of the local situation. Beyond the immediate need, there was precedent: troops had been stationed on the Island until urgent military requirements during the Crimean War had led to their withdrawal. Moreover, when soldiers had been present previously, potential aid to the civil power in disputes arising from the land question had been the major rationale for retaining them whenever removal had been suggested.[5] On the other side of the Atlantic Ocean, Colonial Secretary Edward Cardwell and the British authorities generally tended to construe the attitude of the Island government as reflecting a covert desire to have the British taxpayer provide for a local police force. Such a reading of Island intentions would not seem far-fetched to them because in Britain the army had served as the police until 1829.[6]

The Pope government believed they strengthened their case for keeping the troops when they promptly set about constructing barracks at substantial financial cost. They also accepted without question a bill for £300 sterling (or £450 PEI currency) to cover the expense of transporting the men from Halifax to Charlottetown, and undertook to pay the costs of troop movements within Prince Edward Island.[7] They would not balk when asked to pay field allowance to officers for the period they lived in tents for lack of barracks,[8] or when asked to pay field allowance to officers when they were engaged in supporting the civil power,[9] or when asked to pay the lodging expenses of officers who could not be accommodated in the new Victoria Barracks.[10]

There was a scarcely disguised punitive element in the attitude of British authorities towards the Island over the matter of clawing back military-related costs. They blatantly used demands for payment as a means of pressuring the Island government in the direction of an early withdrawal of the troops – or *vice versa*.[11] The British did evoke protest from the government in Charlottetown when they demanded that it pay maintenance expenses for the two companies while they were stationed in the colony. The Islanders had no shortage of arguments to rebut the British position. In the first place, since the imperial government provided for the pay and allowances of British soldiers elsewhere in British North America (such as in Halifax), these expenses were not extraordinary costs occasioned by the disorders on the Island, and therefore not attributable in any sense to, for example, a failure on the part of local authorities to provide sufficient policing. Secondly, the leasehold system that caused the turmoil had originated in, to use the term of the Executive Council, 'Imperial error.' The Pope government registered its objection through the lieutenant governor, with William Henry Pope recording his dissent; already at odds with most of his colleagues over Confederation, William was committed to the proposition that Islanders should pay dearly for the defiance of the law by the Tenant League, and was probably unconcerned about prospects for personal re-election. The government also objected to being charged for the expenses of transporting troops between colonies when the British began to rotate different regiments in October of 1865 for reasons unrelated to the situation on the Island. In the end, it was excused from payment of the ordinary maintenance expenses of the two companies; but it was required to comply with the imperial request for payment of the expenses of rotation. The penalty for non-compliance, the Island government was informed, would have been removal of the troops.[12] The local government would have to pay again after another rotation of regiments in April of 1866.[13]

Would those soldiers ever go into action on Prince Edward Island? That depended upon the success of the sheriff's men in re-establishing their ability to serve legal processes resulting from landlord-tenant frictions. The place where the tin trumpets of the Tenant Leaguers were likeliest to challenge the authority of the sheriff was Queens County, especially the western portion, where tenancy was the norm in all but one township (the closest to Charlottetown), and where the most serious recent breaches of the peace had occurred. It was almost inevitable that the approach of the Island government would be tested first in western Queens; and it was probable that its success or failure would be determined there, for the only other area where resistance was likely to be so formidable in the late summer of 1865 was the Tracadie estate in eastern Queens.

Resistance to Deputy Sheriff Curtis on Lot 65

On Monday, 18 September, Deputy Sheriff James Curtis set out in a horse-drawn wagon for Lot 65, to serve a number of writs on behalf of Colonel Bentinck Harry Cumberland and his wife Margaret, a daughter of Edmund Fanning, lieutenant governor at the turn of the century. The township, in western Queens, was located across the Charlottetown harbour from the capital. Although it had been well populated for at least a full generation, at the 1861 census less than 20 per cent of land occupiers were freeholders. All recorded leases specified an annual rate of more than 1s. and up to 2s. sterling per acre; apparently the practice on the Cumberland estate was to stipulate either 1s. 6d. or 2s.[14] The Cumberlands probably owned about one-third of the township, and although they were absentees, they had lived on the Island for many years before returning to England in the early 1850s. They employed the same land agent as the two sisters of Mrs Cumberland who owned much of Lot 50, namely, Charles Wright; he was also the other major proprietor on Lot 65.

The Cumberlands had long had a reputation as hard-line, diehard landlords. In 1840, the visiting absentee proprietor Sir George Seymour had noted that then-Captain Cumberland was 'inclined to be violent' when speaking of the lieutenant governor, Sir Charles FitzRoy, and the local officials, for he believed that they were insufficiently supportive of the proprietors.[15] Over the years, the Cumberland estate had been involved in various controversial assertions of proprietary rights and power, including the case of Neil Darrach, an elderly unilingual Gaelic-speaking and illiterate tenant whose imprisonment became a symbol of harsh treatment, indeed a *cause célèbre*.[16] Writing from Guernsey in June of 1864, Colonel Cumberland had urged the Colonial Office to disallow the Fifteen Years Purchase Bill because it would only whet the appetite of tenants for further concessions. Measures like this, in the colonel's view, accounted for the Tenant League agitation.[17]

Curtis and his bailiff, Jonathan Collings, succeeded in serving a writ on James Gorveatt at his home, evidently taking him by surprise. According to the joint deposition of the two law enforcement officers, Gorveatt remarked that 'You will not go along as quickly as you came down,' and immediately set out to make good his prediction.

he ... got on his horse and [went] through a road at the rear of the farms in that vicinity blowing a tin trumpet until he reached the South shore settlement. When deponents [Curtis and Collings] overtook ... Gorveatt he remarked ... that they had better turn back. ... deponents proceeded a short distance further and met about half a dozen persons on horseback, all blowing horns ... the numbers of persons increased until about thirty persons were assembled all on horseback. ... they surrounded deponents and refused to allow deponents to proceed ... James Curtis ... requested them to allow deponents to proceed upon their way which they refused to do. ... after some time they proceeded slowly along the road, but ... were so surrounded that [they] were unable to move any faster than the mob would permit them ... during all this time trumpets were blown and every attempt was made to create a serious disturbance. ... deponents ... endeavoured to persuade the mob to make way for them but with no effect.[18]

Eventually, after two hours or more of this tumult, Curtis and Collings reached Rocky Point, at the extreme northern tip of Lot 65, just two miles south-west of the capital. They took a ferry to Charlottetown, and they may have had little choice about their method of leaving the township, for in a letter written the following Saturday, High Sheriff Thomas W. Dodd would describe them as having been 'hemmed in on Rocky Point Wharf by a large body of men.'[19] While on the wharf, just before departing, Curtis gave a writ to Patrick Doolan, thus effecting personal service. Whether intentionally or not, this demonstration of determination to serve the legal processes in his possession, regardless of immediate circumstances, provoked the crowd. 'Doolan ran up to the mob who were standing a short distance from deponents, and ... they commenced blowing their horns and moved in a body down the wharf towards deponents ... [who] got into the ferry boat and shoved off.'

Curtis and Collings left their horse and wagon behind in a stable at Rocky Point. Overnight, the stable was broken into. The deputy sheriff and his assistant reported that when they returned the next day they found the horse's tail shaved, and the wagon in the water. According to newspaper reports, the horse's mane had also been clipped. Curtis and Collings were unable to make any progress in serving more writs, for on the road they met, by their count, sixty-seven men, many of them armed with sticks. Shouting at them, the crowd refused to let them pass, and it was apparently only with difficulty that they managed to escape a second time by taking a ferry back to Charlottetown. The Tenant Leaguers of Lot 65 evidently made a powerful impression on the deputy and the bailiff, for they solemnly swore that 'they firmly believe that had they endeavoured to execute any of the writs against any parties in the said Township that they ... would have been killed by the mob by which they were surrounded on the two several occasions.'[20]

Other sources, including Dodd's letter, suggest that Curtis and Collings had left some particularly unpleasant aspects of the story unstated in their joint deposition. Dodd informed W.H. Pope, the Island's colonial secretary and his usual conduit for official communication with the administrator in Council, that

From information afforded by my deputy, I understand there were other circumstances connected with this outbreak, which though not necessarily set forth in the affidavit show a very demoralized tone of feeling, and an utter disregard, not only for the laws, but also for the decencies usually observed when the remains of the dead are being finally removed.[21]

On 18 October, Lady Georgiana Fane, a proprietor resident in England, wrote to Cardwell conveying a story that the deputy sheriff's dog had been killed; it is possible that her statement may connect with Dodd's reference to death.[22] The *Islander* newspaper declared that 'the poor faithful dog,' who apparently had been left behind at Rocky Point with the horse and wagon on 18 September, had been 'brutally ill-treated' – without further elaboration.[23] On the basis of the reports from Fane and the *Islander*, it is not entirely clear whether the dog was killed or mutilated, perhaps in some way similar to the horse. But, regardless of what precisely happened to the dog, the wording of Dodd's letter suggests that *human* remains were involved. There may have been a macabre part of the story that Curtis and Collings did not report in their affidavit, one that so violated community taboos that it did not find its way into the newspaper press. It is an intriguing detail with no parallel in the known history of the Tenant League.

Another element in the series of events at Rocky Point, about which there is certainty in at least some respects, made it unique among the recorded incidents of confrontation surrounding the Tenant League. This was the maiming or killing of animals. Animal maiming, while universally condemned, played a role in nineteenth-century rural protest that has been little studied or understood. Judging from statistics on crime compiled by police in the United Kingdom, there was no simple relationship between the incidence of maiming or killing of animals and cruelty to animals. The English and Welsh were likelier than the Irish to be cruel to animals, but less likely to maim or to kill them.[24] One crime was directed against the animal, the other against its owner. Since few perpetrators were ever identified and convicted, and therefore little of a definite nature can be clear concerning motivation, animal maiming remains, in the words of historian John E. Archer, 'the dark crime of the countryside.'[25] But one thing was clear: when an animal was mutilated rather than stolen, material gain was not the point. Rather, the crime represented retaliation against the owner for transgression against a code of behaviour.[26]

In Prince Edward Island, animal maiming had occurred in connection with the Escheat agitation of the previous generation, and so it was not entirely unknown in the colony.[27] Writing of the practice in East Anglia (Norfolk and Suffolk counties), the part of the British mainland where it occurred most frequently, Archer has declared that, although 'never a common crime,' it was

vengeance crime 'par excellence.' It was a more personal act of violence by the maimer on the victim than any other protest crime. One can view it almost as a form of symbolic murder, ... In short maiming could be an extreme form of psychological terror which could leave the victim appalled and fearful for his own safety.

Archer has also noted that persons occupying positions of authority in rural life, such as constables, were 'obvious targets.'[28] These comments seem both reasonable and applicable to the events at Rocky Point: there was an extreme antagonism towards Curtis, such that passing beyond normal bounds of protest in an attempt to instil mortal fear might seem, to some members of the community, justifiable. The treatment of the horse and, if the reports of Fane and the *Islander* were correct, the dog was probably a factor that undermined the resolve of Curtis and Collings to press forward on 19 September. They did subsequently swear jointly that they had feared for their lives.

A constituent part of the behaviour associated with animal maiming, according to Archer, was a tendency of the persons whose animals had been maimed to be reticent about reporting the facts. This fits with the failure of the deputy sheriff to report anything about the dog in his deposition, and the probability that he under-reported about the horse. The fact that he reported as much as he did may have been a reflection of pressure by government officials who were not on the front line of law enforcement and who wished to increase public distaste for the Tenant League. There would be rational reasons for Curtis to remain silent about an outrage against an animal of his, especially if he were genuinely horrified. He would not want to do anything, through recounting unpleasant details in an affidavit which would become part of the public record, to stimulate or inspire imitative behaviour when he went elsewhere to serve writs. Indeed, the more repugnant the facts of the case, the greater the incentive for someone in Curtis's position to remain silent, for whatever happened had occurred in the course of performing a duty in a situation – attempting to deliver writs in a Tenant League area – that was certain to recur, and recur soon.

In addition to being a highly intimidating crime against the person of the owner by transference, the attack on Curtis's animal(s) was also in a literal sense a property crime. Furthermore, animals were a form of property that was rarely insured, a consideration that made animal maiming, Archer has suggested, 'perhaps, a more successful method of protest in narrow economic terms than incendiarism.'[29] Again, Curtis, who lived in a rural area, on Lot 33, along the Malpeque Road, a main thoroughfare north of Charlottetown, would be inclined to shun unnecessary publicity so as not to suggest to ill-disposed persons this means of venting vindictive feelings against him. In 1865 many people would have motives for retaliating against him, and he would not wish to draw possibilities to their attention.

This is part of what Archer has referred to as 'a powerful psychological terror' that maimers exercised over their victims.[30] Animal maiming was 'intended to show that the powerless were indeed capable of striking back, often with terrifying and bloody effect.' Conversely, the act emphasized the vulnerability of the apparently powerful victim, and in a particularly personal way, especially if he lived in a rural area. It could, therefore, be an exceptionally effective crime in instilling fear: fear for one's property, and since that property was a living creature, fear, by redirection, for oneself. Historian David J.V. Jones has noted a case in north Wales in 1870 where 'one extended family was alarmed when the tongues of their horses were cut, and horrified when one of the humans later received the same treatment.'[31] In only one respect does the incident at Rocky Point in September of 1865 deviate from Archer's conception of animal maiming as a crime: whereas he describes it as 'individualistic,' the circumstances in this instance indicate the likelihood that it was a collective act – a community effort to instil fear and horror, a fact which would probably give it an even greater impact, suggestive more of a lynching than a murder by an individual.[32]

In forwarding the Curtis-Collings affidavit to W.H. Pope, Sheriff Dodd requested 'to be informed if Military assistance will be afforded me, if applied for in cases of emergency – because I find it will be impossible for me to induce men to risk their lives in the execution of such duties.'[33] He had become more pessimistic since late August, when he had expressed confidence in his ability to serve writs for rent if authorized to hire sufficient constables.[34] The pressure on the local government was building. The rapid gathering of a crowd on 18 September after the serving of Gorveatt with a writ was clearly not a one-time-only occurrence which could be classified as an entirely spontaneous outpouring of anger unlikely to recur. The inhabitants of Lot 65 had given a repeat performance, definitely premeditated, and even more menacing, the next day.

The animal maiming may have constituted the crossing of a threshold in terms of creating a sense of personal threat among law enforcement officers, for in official reports the references to the level of danger became more dire. On Tuesday the deputy sheriff and the constable beheld the results, on Friday they swore that they believed their lives had been in danger, and on Saturday the high sheriff informed the Island's colonial secretary that he was in the untenable position of trying to persuade men to imperil their lives in order to deliver writs. Maiming have well have had the psychological impact, as Archer suggests, of an act of symbolic or transferred murder.

On the same day that Curtis and Collings disembarked from the ferry to survey the damage and to attempt once again to serve legal processes, the Executive Council read a communication from Bernard McKenna of Lot 48, a constable, concerning harassment of persons in his position:

a constable or Bailiff cannot travel on the public road without being abused by a Lawless Mob of those Tenant Union men, boath [sic] with their tongues and those tin horns. ... I have been three days at different intervals serving summonses for the Small Debt Court and there has been each day 100 or 150 men and boys, belonging to the Union. They were not like men. They were more like Devils.

McKenna named another constable, William Farquharson, as being part of the mob on one occasion. He was particularly perturbed by the din the trumpets created: 'a person travelling the road with a fractious horse would be in danger of being killed or hurt, by the beast running away.'[35] The complainant was the same constable who had been coerced on 22 July into signing a pledge to desist from acting as constable for a year or until the Tenant League disturbances were over.

Around Wednesday, 20 September, according to Fane, there was a near-incident on Lot 65. In the evening hours, a Prince Edward Island correspondent of hers 'passed through a mob of about thirty men on horseback, to whom someone was distributing penny trumpets. He was told that they were "waiting" for the Sheriff, whom they believed to be coming on behalf of a Gentleman named Wright.' This was probably Charles Wright, proprietor of part of Lot 65, and land agent for the Cumberland estate. Because of illness, the sheriff appears not to have shown up, and thus a confrontation did not take place; certainly no record of such an incident has come to light. Fane's letter was also interesting because of a hint she gave as to the intelligence system of the Tenant League: 'The people must have learnt that a Writ was out through some Lawyer's Clerk in Charlotte Town.'[36]

In some respects, Fane is a source who must be used with caution, and not only because she was an ocean away from Prince Edward Island. The tone of her correspondence was sometimes alarmist, and indeed was often coloured by a deep suspicion of the Charlottetown élite. This was one of a series of letters she wrote during the summer and autumn of 1865 to the Colonial Office. There they were received with some dread and the knowledge that, in the words of Arthur Blackwood, who lived a few steps from her on Upper Brook Street, she had 'the pen of a ready writer' and anything more than the briefest response would 'elicit volumes of corresp[ce] from her.'[37] Sir Frederic Rogers, the acknowledged expert on Downing Street concerning the Island's land question, viewed her as a member of 'the no surrender party in P.E.I. who w[d] [would] like the Imperial Gov[t] to carry on war with the Tenants.'[38]

But Fane was intelligent, she had spent time on Prince Edward Island, she had correspondents there, and her theory about the intelligence network of the league is plausible. Sympathetic law clerks may well have been a source of information on outgoing writs from the Supreme Court. Once leaguers in the capital knew the destinations, they could dispatch messengers to give timely warning to the inhabitants of the districts where the sheriff's men were going. This would be consistent with a pattern the Welsh historian David Jones has noted among the Cardiganshire peasantry around 1820. Planned resistance was 'co-ordinated by special signals and messengers,' and was also contained in scope, like the Tenant League disturbances, rather than spreading to random targets.[39]

A few days later there was more evidence of activity by those opposed to landlordism in western Queens County. On the morning of

Saturday, 23 September, what was apparently intended to be a threatening letter was discovered nailed to the front gate at the house of the largest resident proprietor on the Island, Robert Bruce Stewart, in Strathgartney, Lot 30. Stewart, who by his own estimate had nearly eighty thousand acres in the colony, was an outspoken defender of proprietary rights, with a reputation as a severe and unbending landlord. The note had a heading in English, 'Take Warning,' and an indecipherable five-line text. It was signed 'Enonymous [sic] Cestpirce R.'[40] There had already been arson (at Glenaladale, 27 May) and animal maiming (at Rocky Point, earlier that week in September) in incidents associated with the Tenant League agitation in 1865, but this is the only solid evidence of the third staple of 'agrarian outrage,' namely, the threatening letter, to emerge in research on the league.[41] Jones, writing of the British countryside, describes the threatening letter written for the purpose of intimidation as 'overwhelmingly' a crime of protest, reflecting community anger arising from a sense of collective grievance, rather than a private quarrel; it was certainly not a tool used by criminal gangs or families.[42] Stewart was a correspondent of Fane, and perhaps it was knowledge of this incident that led her to declare to Cardwell later in the autumn that 'If the Island continues in its present disturbed state, the lives of the Resident Proprietors will not be safe.'[43]

The note on Stewart's gate has survived as part of a private collection held by one of his descendants. There is no evidence that he reported the incident to the local authorities, whom he distrusted, and therefore it appears not to have received publicity at the time. The exact message of the text, which bears no resemblance to English or French, remains a mystery. Historian Deborah Stewart, who had 'full access to family documents,' states simply that it is 'written in Gaelic,' without indicating why she believes this to be so, and whether she means Irish or Scottish Gaelic. She also reports that the landlord, a native of London, England who had never lived in Scotland, did not speak Gaelic; and given the extent to which the Gaelic tradition was oral, it is highly unlikely that he could read the language if he did not speak it.[44]

If one assumes Gaelic to be the language of the text, the most obvious theory to pursue is that the writer was Irish. When census results were tallied, natives of Ireland consistently outnumbered natives of Scotland on Lot 30, and the threatening letter was a much more prominent weapon in the Irish than the Scottish arsenal of protest. Eric Richards has reported finding only one such letter in his research on the Highland Clearances, whereas, according to W.E. Vaughan, the threatening letter was 'the most commonly recorded agrarian crime' in Ireland.[45] But Celtic Studies scholar Mairin Nic Dhiarmada is definite in stating that the text of the note is not in Irish Gaelic, for it has none of the signs or markers of written Irish at that time. In making this point, she cites accents and the formation of letters in particular; to her, the text looks very much like the product of an illiterate person. As a consequence, on the basis of the text, there is no reason to link the note to a person of Irish origin.[46] Ronald Black, a senior lecturer in Celtic Languages and Cultures at the University of Edinburgh, has stated that he believes that the text is a poorly spelled attempt to convey the first verse of a famous song by a popular Scottish Gaelic poet. The song is satiric, with an edge that could be interpreted as threatening.[47] But others fluent in Gaelic have concluded that it is not Scottish Gaelic; one present-day speaker of Scottish Gaelic has reported that the words in the note resemble no words in that language and, independently of Nic Dhiarmada, has suggested that it is the work of a semi-literate person.[48] A reasonable hypothesis in the circumstances is that, whatever the author of the text intended to convey, someone else may have added the English words 'Take Warning' at the top. Thus there may have been two different writers at work: the writer of the heading, literate in English, and the writer of the body, not literate, the former possibly wishing to clarify the meaning of the latter, whose spelling evidently conforms to that of no known language.

We cannot be certain that we are aware of all the confrontations and instances of intimidation of one sort or another that happened at this troubled time. Our lack of certitude can be illustrated by examining the purely serendipitous way in which we have come to know what we do about events in the area of Bagnall's Hotel, a landmark inn located in the district of Hazel Grove, Lot 22, halfway along the road between Charlottetown and Princetown, the old county seat of Prince County, on Wednesday, 27 September. Our sole source is the personal diary of land agent and surveyor Henry Jones Cundall, who had formed part of the *posse comitatus* in April. He was on a surveying trip, approaching Bagnall's, and he heard 'the blast of a Tenant League trumpet.' When he was within sight of the inn, he saw a crowd of eighty to a hundred men and boys, some on horseback and some on foot, who had positioned themselves 'across the Princetown Road in order to obstruct the Sheriff's deputy.' The trumpets continued to blow, and when he went inside Bagnall's, he found that the clamour had 'unpleasantly frightened some ladies who with their gentlemen friends were on a pleasant drive to see the country.' Cundall himself was known to the leaguers, for he recorded that 'When Curtis or Collins [sic] or myself appeared outside a louder blast was sounded.' The commotion continued until the deputy sheriff and the bailiff left Bagnall's 'when the whole crowd rag tag & bob tail pursued them with their trumpets.'[49]

This is all we know of the incident. Curtis and Collings were doubtless on some official business, and they were most probably prevented from completing their work. If Cundall had not been present or if he had not kept a diary that survived after his death, we would know nothing of the episode; and he had no responsibility to report on what he saw happen between the sheriff's men and the Tenant Leaguers. Those who were there as law enforcement officers apparently made no report, or at least made none that survived. It would not be surprising if several other occurrences such as this and the harassment of McKenna went unrecorded.

As September of 1865 drew to a close, order had emphatically not returned to the countryside, especially not to western Queens, where the Curtisdale affray had occurred in mid-July. On 29 September, W.H. Pope's *Islander* called for the vigorous prosecution of twenty or thirty of those who had formed the riotous assemblies on Lot 65, hoping for twelve-month sentences of imprisonment to drive home the point that such conduct was intolerable. Pope suggested that Samuel Fletcher's evasion of the law for the previous six months had created the impression that such behaviour could go unpunished. If Pope was serious, common sense indicated that no force other than the military would be able to effect the arrests of so many Tenant Leaguers. Use of the troops in the townships was a step that members of the government dreaded taking, but of all the leading Conservative strategists, W.H. Pope was the least inclined to shrink before public opinion or disapproval.

Already, two days earlier, when Hodgson wrote to Cardwell enclosing accounts of the resistance Curtis and Collings encountered on Lot 65, there was an indication that the administrator's position on the use of troops was shifting. He reported that he had sent the evidence to the attorney general, asking that warrants be prepared for those participants in the 'outrages' who could be identified. If such warrants were issued, Hodgson stated, it was 'probable' that the sheriff would require the assistance of soldiers to execute them.[50] In this admission – which must have been an admission as much to himself as to anyone else – Hodgson was conceding that the policy he had favoured was failing. He wrote the despatch the same day that, according to Cundall, Curtis and

Collings were yet again surrounded and escorted along the road by a noisy and hostile crowd.

Thwarting the Civil Power on Lot 22

On Monday, 2 October 1865 Curtis led three bailiffs on an expedition to Lot 22, in western Queens County, to the north of Lot 65. Their objective was to serve writs on four tenants of Laurence Sulivan, former deputy secretary-at-war for Great Britain, and a resident of Fulham, on the outskirts of London.[51] The Sulivan estate was one of the largest on the Island, and historically the family had been reluctant to sell freehold title. Laurence Sulivan had consented first to the land commission of 1860 and then to the Fifteen Years Purchase Bill of 1864. The agent for his estate and the Cunard estate had announced in the spring of 1865 that the owners would grant no more new leases, but would sell freehold at rates of at least 20s. per acre. These changes were significant, for they represented recognition that past policies were no longer tenable. During debate in the Legislative Council on the Fifteen Years Purchase Bill, Donald Montgomery, president of the Council, had stated that many Sulivan tenants had hundred-year leases with thirty or forty years expired, and that 'At present they have not the right to purchase at any price.'[52]

The consequence of three generations of disinclination to allow conversion to freehold title was that, according to the 1861 census, on each of the four townships where Sulivan owned most of the land the percentage of freeholders was unusually low: 6.8 per cent on Lot 9, 11.8 on Lot 16, 9.3 on Lot 61, and only 5.1 on Lot 22, the one with the largest population. Indeed, the family's resistance to giving in on this issue was evident in the fact that on two townships there were more squatters than freeholders. The squatters' survival as such was due to inattention on the part of the proprietors rather than any calculated policy of toleration, for often the Sulivans did not have land agents on the Island. They saved money that way; they also gained a reputation as being archetypes of proprietors who were not developers. Despite the alterations in policy during 1864–5 concerning the right to purchase freehold title, which represented progress in principle, the prices were too high to make much of a positive impression on the tenantry.

Rents on the townships dominated by the Sulivan family were not onerous. The 1861 census indicates that all leaseholders had rents set no higher than 1s. sterling. But land occupiers on those townships suffered

TABLE 10.1

Tenurial status in 1861 of land occupiers in individual townships where Laurence Sulivan owned most of the land, expressed in percentages and absolute numbers

	Freeholders		Tenants		Squatters		Total number of occupiers
	%	No.	%	No.	%	No.	
Lot 9 (Prince)	6.8	3	81.8	36	11.4	5	44
Lot 16 (Prince)	11.8	16	83.1	113	5.1	7	136
Lot 22 (Queens)	5.1	13	88.9	225	5.9	15	253
Lot 61 (Kings)	9.3	14	83.3	125	7.3	11	150
Totals	7.9	46	85.6	499	6.5	38	583

Note: Calculations based on totals derived from the 1861 census in PEI, Assembly, *Journal*, 1862, app. A

one significant disadvantage in particular. On each, there was a large degree of tenurial insecurity, although the reasons varied according to the township. On Lot 9, there were high proportions of tenants at will, tenants with short leases, and squatters. On Lots 16 and 22, among leaseholders, 74.8 and 96.9 per cent, respectively, had short leases; the overall percentage within the colony was 15.7. Although there were no short leases on Lot 61, more than one-quarter of the tenants had no leases at all.[53]

The final major distinguishing characteristic of the Sulivan estate was its good connections with the centre of political power in Great Britain. Sulivan had become a privy councillor in 1851 upon his voluntary retirement after twenty-six years as deputy secretary-at-war, and was the close friend and brother-in-law of Henry Temple, Viscount Palmerston, the British prime minister, who had been a member of parliament for fifty-eight years and a minister of the crown for forty-eight. Sulivan and Palmerston had known each other since their days together as students at St John's College, Cambridge University, where they formed what Palmerston's biographer Kenneth Bourne describes as his subject's 'principal adult friendship with a man.'[54] The closeness between them antedated Sulivan's marriage to Palmerston's younger sister in 1811, and long survived her death in 1837. They also worked together at close quarters in government, since Palmerston was secretary-at-war from 1809 to 1828.[55] He brought Sulivan with him into the War Office as his private secretary in 1809, and promoted him rapidly; by 1825 he held the deputy secretaryship at £2,000 per annum. Sulivan retained the position under a dozen ministers in the twenty-three years between the departure of Palmerston and his own retirement from the War Office. The favours went in both directions. Sulivan loaned the prime minister money. When he made his will on 19 July 1865, among the gifts he left to a daughter was 'all money secured to me by way of mortgage upon the estates of Viscount Palmerston.' The prime minister – who would predecease him – was named as an executor of the will.[56]

For whatever reasons (and these must have included his respect and affection for his long-time friend, colleague, and brother-in-law), Palmerston had strong, indeed dogmatic opinions on the Prince Edward Island land question that he expressed forcefully, persistently, and insistently. He was the landlords' political bulwark, the statesman-politician who was willing to break with established principles of non-interference in the internal affairs of self-governing colonies, and tell his colonial secretary that he had a personal interest in the matter. He believed the

Island land question was important both for its own sake and for the precedents it could set. It was a cause on which he remained determined until the end of his life; the evidence suggests that his deepest prejudices were involved. In the words of E.D. Steele, the historian of the Irish land reform movement during the period, 'While Palmerston lived, neither Prince Edward Island nor Irish tenants could obtain the concessions of principle they sought.'[57]

The Sulivan estate was thus notable for its size, its historic intransigence on the issue of conversion to freehold (which was just beginning to weaken), its lack of interest in development requiring investment, and its exceptional linkage to the pinnacle of political power in London. The possibility that Sulivan could exert decisive political influence in the making of crucial decisions was known on the Island, where some saw his political weight as comparable to that of Sir Samuel Cunard. In the spring of 1864, the *Vindicator* newspaper wrote

Tenants, Beware! – Col. [John Hamilton] Gray [the premier] said a few days ago, in the House of Assembly, that it was useless for the tenantry to refuse the payment of rents, for one of the proprietors, Laurence Sulivan, being a near relative of the Premier of England, could easily obtain from the Home Government a sufficient number of troops to enforce the payment of rents at the point of the bayonet.[58]

Many years later, Louis Henry Davies would refer to the Fifteen Years Purchase Bill as 'Laurence Sulivan's Act.'[59]

Despite an earned reputation as a delinquent proprietor in terms of authorizing land agents to act on his behalf, Sulivan did not have a name as a harsh or exceptionally exacting landlord. Indeed, in later years, one long-time resident of Lot 22 would testify in the legislature to the leniency of the estate.[60] But the landlord was very much an absentee. One tenant who said he had resided on Lot 22 for sixty-four years stated before the land commission of 1875–6 that 'I am not aware of the proprietor doing anything to assist the tenants.'[61] The Sulivan family may have concluded, while watching for three generations the spectacle of such well-equipped proprietors as Sir James Montgomery losing money on their Island estates, that holding on to their pounds, shillings, and pence was the prudent policy. Or simple lack of attention to the property may be the explanation; although the Sulivans had owned land on the Island since the 1760s, no Sulivan is known to have visited the colony by 1865. The disorganization of the affairs of the estate is reflected in the declaration by George W. DeBlois that 'When I was appointed Agent, in 1860, I found everything in confusion. ... There never was an estate in such a condition.'[62] It is possible that his attempt to bring order to the chaos – which presumably would entail pressing tenants in arrears to pay some back-rent and demanding that squatters attorn – was a factor in igniting resistance on the township in 1865.

When settlers from Lot 22 had testified and presented a memorial to the commission of 1860, their complaints were of a general nature concerning the land question, although they also focused on the lack of a land agent authorized to grant leases, for, by their calculation, a period of some thirty years in the early decades of the century. In addition, they said they were 'involved in large arrears of rent, ... from sheer inability ... to pay.'[63] This condition was not universal on the township, and in 1875 a settler who had arrived in the 1840s with £100 capital reported that he had never been in arrears.[64] The Lot 22 census-taker for 1861 declared that with certain exceptions 'The quality of the land ... will compare favourably with any land in Queen's County.'[65] Little is known about the four tenants against whom Curtis and his bailiffs had writs. It may be significant that they all had Scottish surnames; the census-taker had commented on the poor agricultural skills of settlers on the township who had come from Raasay, an island situated between the Isle of Skye and the Scottish mainland, and he had noted that they had made little progress in twenty years.[66]

It was to this township that Curtis and his assistants went on 2 October.[67] The deputy sheriff reported that upon arriving on Lot 22 with Bailiff Andrew Cranston, he 'heard trumpets sounding in different directions.' He estimated that twenty or more persons immediately appeared on the road, some on horseback, some on foot, and most 'armed with sticks and other offensive weapons.' Curtis described the *modus operandi* of the Tenant Leaguers: 'one portion of the Crowd went ahead of deponent, the others surrounded and followed him, they appeared to be aware of the object for which deponent was there, they began to abuse deponent using very violent language.'[68] In fact, he reported that they threatened him with death if he attempted to execute the writs he had with him. He decided to turn back, and as he and Cranston retreated to join the two other bailiffs, who were travelling a considerable distance behind, in Cranston's words, 'we were pelted with sticks, stones, and mud.'[69] By the time they met the bailiffs, Collings and Montague Irving, Curtis estimated the crowd had grown to sixty or more. Collings testified that he could hear the sound of horns half an hour before the scene came into view. When the crowd saw the other two bailiffs, both of them heard someone call out 'here is Collings the bugger.' The party of four, travelling in wagons, proceeded to Bagnall's at Hazel Grove. But the crowd was not finished with them. Richard Bagnall, the host, reported that the tin trumpets blew steadily while they were inside. According to Irving, they remained approximately fifteen minutes.

When Curtis and his bailiffs emerged, the crowd once again followed and harassed them. It was still increasing in size, although the evidence is contradictory with respect to numbers: Collings estimated one hundred or more, Cranston, less experienced and perhaps more subject to panic, some two hundred. There was no difference in the testimony concerning its spirit. Irving related that 'a young man' on horseback fired a gun, and that as a consequence Collings became involved in a struggle with him over the gun; the Tenant Leaguer prevailed.[70] The leaguers were becoming more aggressive, and the deputy reported that he was 'several times struck with stones and clods of Earth.'[71] Guns were visible in the crowd, which remained hostile as it followed the sheriff's party of four along the road. In vain, Curtis called upon the assembly to disperse. That having failed, he requested James Pound, one of the crowd, to assist in identifying individuals. Sworn accounts by all four of the sheriff's party indicate that Pound appeared willing to do so, but the crowd dissuaded him; the key person in restraining him appears to have been J. Alexander Bovyer, the schoolmaster at Hazel Grove. Cranston gave sworn evidence that 'one person in the crowd said that they would send the Deputy Sheriff to Hell and he would have a greater crowd after him there than they have here.'[72] Collings reported seeing a gun fired from within the crowd, and watching a Tenant Leaguer on horseback come alongside Irving, in a wagon, and blow a trumpet in his ear.

The four returned to Bagnall's, with the Tenant Leaguers following them all the way there, still throwing stones. Bagnall believed they had been away forty-five minutes. They ate dinner, and during the meal the crowd surrounded the inn, waiting for them to emerge. More leaguers arrived. The noise from trumpets and shouting did not abate, and in the uproar, according to Collings, Curtis's name was called 'frequently.'[73] Bagnall testified that he heard two or three gunshots. The *Islander* newspaper reported that the 'disorderly mob' alarmed other guests at the inn, 'among whom, we understand, were several strangers'; again, as with Cundall's account of the scene on 27 September, the suggestion surfaces that the Tenant League was disrupting normally enjoyable social occasions which had nothing to do with the land question.[74]

Following dinner, Curtis and his bailiffs attempted to proceed on their way. The crowd soon made that impossible, and in the commotion surrounding the departure of the deputy sheriff's party and the Tenant Leaguers from his premises, Bagnall believed he heard a pistol shot coming from one of the bailiffs' wagons. Cranston reported that the party attempted to go to the district of Mill Vale, also on Lot 22, to the north and west of Hazel Grove. But 'the crowd ... blocked up the road both on the front and on the rear. The crowd asked Curtis what was his business in Mill Vale. Curtis said that he was not going to writ. The crowd shouted, We will not believe you. They then pelted us with sticks and stones.'[75] Collings swore in an affidavit that Curtis 'was struck several times in the head and back.'[76] Eventually, in the face of death threats, they decided to give up and go home – still without having served a single legal process. The deputy sheriff reported that he turned back in fear for his life. Bagnall testified that he 'Heard several ... say that they would not allow Curtis to go on any rent occasion and they would not believe him if he said that he was going on other business.'[77] It was clear that the deputy was *persona non grata* in an absolute sense on Lot 22. Many in the crowd had advised him and his bailiffs to escape with their lives while they could.

There was a unique factor in the reported events on Lot 22. In sworn testimony, Curtis and Cranston stated that one in the crowd, William Harris, *alias* William Harris Corney, had blackened his face; Harris would confirm this, and state he did so because he had heard that Curtis had a warrant for him. His evidence, in which he mentioned the occasion 'when the nine Constables came out,' indicated that he had been one of the crowd that gathered when Sheriff Dodd's party stormed the house of Joseph Doucette in Wheatley River on 15 August.[78] This is the only recorded incident of a person associated with the Tenant League using a disguise. Ethnic origin may have been a factor; Harris is a common Welsh surname, and blackened faces seem to have been used frequently in rural or isolated areas of Wales during collective protest that involved breaking the law or attempting to impose an alternative standard of justice.[79]

Harris, in the information he gave to two justices of the peace, showed himself to be unique in the recorded history of the league in another way. No one else charged with significant offences associated with league activity is known to have incriminated others, presumably in an effort to curry favour with the authorities. On the day before the commencement of the Hilary Term of the Supreme Court for 1866, when the Tenant League cases were to be tried, he named no fewer than six persons from the crowd on 2 October. According to him, one had wielded a gun, another had struck Curtis on the back of the neck with a stone, and a third had hurled a stick at the deputy. Two others had hidden behind a bush with intent 'to waylay Mr Curtis and his officers,' and Harris said he had heard a shot fired from that bush. He reported that a sixth person had both thrown a stone at Curtis and uttered 'serious threats,' saying 'that he would shoot Curtis as soon as he would a dog.' At the same time, Harris, who had been charged with riot and assault, insisted that this own activity had been limited to shouting and horn-blowing.[80]

William Harris (Corney) was also without parallel in yet a third respect: he is the only person who is known both to have played a prominent role in a league action and to have engaged in ordinary criminal behaviour at approximately the same time. On 6 September 1866, less than a year after the Lot 22 confrontation, he made a formal written confession of guilt in a case of larceny.[81] This aspect of Harris's role is particularly noteworthy because it is so untypical of Tenant Leaguers, and because the almost complete freedom of the organization from criminal hangers-on underlines its serious-mindedness and clear-headedness as a social movement. It was involved in calculated defiance of the law, and it was willing to go further than passive resistance, but it did not cross over the line into indulgence in common criminality, a problem that plagued the reputation of Irish secret societies. In his use of disguise, his willingness to provide evidence against fellow-participants in a Tenant League action, and his criminality, Harris was a truly exceptional case.[82]

As noted, Curtis and his bailiffs had failed to serve any of their legal documents. The crowd that had gathered and persistently dogged them could count the confrontation as a victory in this respect. The ambush had in fact been even more successful than the effort to thwart the deputy on Lot 65, where he had at least delivered two writs. There had also been a degree of escalation in the tactics used: more shots had probably been fired than on any other occasion during the history of the Tenant League. In this incident, as in all others involving the league, there is no indication that anyone discharged a firearm in an attempt to wound. The purpose of the shooting appears to have been simply to add to the din that the tin trumpets created.

Although the encounter had been a success for the Tenant Leaguers in terms of their immediate objective of preventing the deputy sheriff from serving his writs, and although there had been unprecedented willingness to use firearms in the presence of and in defiance of law enforcement officers, the incident was in reality the beginning of the end for the organization. It triggered the use of troops to assist the civil power. Why? Why precisely then? Was it simply a matter of 'one incident too many,' especially given that there had been a somewhat similar occurrence, although apparently of lesser magnitude, at the same place – Bagnall's – a mere five days earlier? Or was it the use of firearms? Did the repeated firing of guns mean that the leaguers had passed over an invisible line between tolerable and intolerable conduct? One factor was that a witness to part of the encounter was Judge James Horsfield Peters of the Supreme Court. Peters came across the scene of bedlam entirely by chance while on his way to court in St Eleanors, Prince County, in the company of Attorney General Edward Palmer and other members of the bar. Writing on the following day from St Eleanors, Peters described it as 'a great disturbance of the public peace.' There was a crowd of two hundred blocking the road.

As I drove up I heard one of them call out to 'clear the way and let the man pass in peace as this was not Curtis.' The way was cleared at once, but they lined the roadside, and commenced blowing tin trumpets, (which each appeared to have), making such a noise that any traveller, whose horses were not very steady, would have been much endangered. I drew up and told them that their conduct was most illegal, that they had no right to make such a disturbance and conduct themselves in that disorderly manner in the highway. Curtis the Deputy Sheriff then came up to my carriage and said he had taken their names. I told him to do so and to take care that he did it so that he could identify them. They became perfectly quiet while I was speaking to them, but I had not proceeded twenty yards, when a tremendous noise of trumpets took place, and for a mile along the road, I met parties on horseback blowing trumpets and riding furiously towards Bagnall's, which appeared to be the rendezvous.[83]

Peters was a singularly formidable individual, in fact a living legend in the history of the Prince Edward Island land question. A native of New Brunswick, he had married a daughter of the great landlord Cunard, and moved to the Island in 1838 to manage his father-in-law's properties there, 'my only instructions being to treat them as if they were my own.'[84] In the course of ten years, he made a colony-wide reputation for himself as an exacting, businesslike, and remarkably efficient land agent. He had probably been the most important agent in the entire history of the land question. For many Islanders, Peters personified the rent-paying system, and because of the level of hostility directed towards land agents generally and himself particularly, he had travelled armed with two brace of pistols, a detail that added to the legend. The danger was real; on one occasion, fire apparently set by an arsonist consumed a barn close to the house where he was sleeping in western Prince, and although he was not harmed, his horse perished in the blaze. Appointed to the Supreme Court while in his thirties, he would be a member of the bench for more than forty years. He was widely acknowledged to have been an excellent lawyer and to be a judge of exceptional ability. Yet, handicapped by lasting unpopularity resulting from his ten years as agent for the Cunard estate, he would be passed over three times for promotion to chief justice.

The administrator of the government, Hodgson, was Peters's nominal superior when serving on the bench. Peters was younger, but widely regarded as abler; he even had more experience as a judge. The two had served together for ten years as crown law officers, Hodgson as attorney general and Peters as solicitor general. When Peters, as the assistant judge on the court, applied to the Colonial Office in 1852 for the vacant chief justiceship, he stated 'a fact which no one in this Island will contradict – that when Mr. Hodgson and myself were at the bar ... I had so decidedly the lead over him ... that where he received one retainer as Counsel I received Twenty.'[85] Peters was also not a man to shrink from controversy or from identification with unpopular policies or decisions. He told Hodgson precisely what he should do in the circumstances of October 1865:

You have the means of enforcing the law. If the Sheriff is not supported by a sufficient force, while in the execution of his duty, each successful resistance will encourage others, and the lawless spirit which appears at present confined to certain districts will soon extend through the whole country, then only to be suppressed by a loss of life, which makes me shudder to contemplate. Mercy, therefore, to these misguided men, and others who may by their success be encouraged to follow their example, should induce your Government to use the means at their command with the energy and promptness which the case requires.[86]

It is impossible to know exactly what Hodgson and the Pope government would have done had Peters not come across the scene near

Bagnall's on 2 October. But his involvement was almost certainly a significant factor. He was highly respected by those on Downing Street who knew him, and it was also known on the Island that he had never been afraid to speak out. If tragedy such as loss of life in a confrontation between Tenant Leaguers and the sheriff's men occurred *after* he had warned the administrator in writing that the time had come to act, then Hodgson – and his constitutional advisers – would be unable to defend inaction in the face of questioning from London with the plea that no one of stature and with relevant experience had apprised them of the gravity of the situation; that consideration would have weight with someone as ambitious for appointments and honours as Hodgson. The chance presence of Peters during the confrontation of 2 October 1865 and the fact that he took the trouble to put his analysis on the record may have tipped the balance in Charlottetown away from caution and towards use of the military.

In any event, Sheriff Dodd wrote to Hodgson on Friday, 6 October, four days after the events on Lot 22, stating that military assistance was necessary to enable him to do his job.[87] On the same day, at the command of Hodgson, W.H. Pope, as the Island's colonial secretary, wrote to William Swabey, a justice of the peace, concerning the events at Bagnall's. Hodgson, as administrator of the government, had directed that two military officers and twenty-five men accompany the sheriff and his assistants. He had ordered that the letter be written to Swabey because he wished the military force to be under the control of a magistrate who would be present to judge whether there was sufficient danger to the civil officers to require active intervention by the soldiers.[88]

On the next day, Saturday, Swabey left Charlottetown with Dodd, eight bailiffs, and two officers and twenty-five men of the 16th Regiment, all going to Bagnall's Hotel in Hazel Grove, Lot 22. It was exactly two months and one day since the soldiers had arrived on Prince Edward Island. The editorial article in the next issue of the *Islander* declared grimly that 'The time for trifling has gone.'[89] Events had moved past the stage of feints, manoeuvres, and posturing; the military would be putting the will of the Tenant Leaguers to the test. This was not the risible *posse comitatus* in search of Samuel Fletcher.

Execution of Writs with Military Support in Tenant League Areas

The expedition to Lot 22 that commenced on 7 October represented the first use of troops to support the Queens County sheriff against the

Tenant League. Hodgson, as administrator of the government, had instructed Swabey, the justice of the peace in charge, through W.H. Pope, 'that you take especial care that nothing short of well founded and just apprehension of loss of life or serious bodily injury to the Sheriff or his Officers induce you to use the Military force at your disposal.'[90] As in the case of retaining twenty-five special constables to escort Charles Dickieson to and from the court house in Charlottetown on 26 July, he emphasized the imperative need for those who controlled the reinforced civil authority to exercise as much patience and forbearance as possible.

It is not known precisely why Hodgson chose Swabey, a Liberal supporter, as the justice of the peace to lead the expedition. But it is conceivable that Swabey's Liberalism was a factor; as someone not compromised in the eyes of tenants by being a landlord or a land agent or a person politically identified with the local government, he may have been a reassuring presence to at least some tenants. From the perspective of an administrator probably desiring a bi-partisan gloss for the policy of using soldiers to assist the sheriff, the device of using a Liberal to head the campaign when the government was Conservative would make good sense. It would also make sense from the perspective of a Conservative government that had wished to avoid ever sending troops into the rural areas; if something went terribly wrong and serious injury or death resulted, and if the soldiers at the centre of it were at least theoretically under the control of a Liberal, then perhaps not all the odium would fall on the political party in power. For Swabey, there was a considerable commitment of time and the reward of a substantial sum of money: twenty-five days' work over the next several weeks, for which he received from the government £3 per day, totalling £75.[91]

The soldiers returned to Charlottetown on the evening of Friday the 13th. The operation appears to have been as close to a complete success as anyone in authority could have hoped. Several persons accused as rioters had been arrested, and others had surrendered voluntarily; according to the *Herald* of 11 October, three of these had already turned themselves in to the civil authorities to Charlottetown, presumably to avoid the ignominy of being taken prisoner by the military. W.H. Pope predicted publicly that the outcome for persons convicted of riot-related offences would be 'long months of imprisonment.' In case anyone was so foolish as to attempt resistance to the sheriff, accompanied by troops, the result would be 'the use of the bayonet or minnie ball.'[92] Hodgson reported to Downing Street that, in addition to apprehending rioters, the sheriff was able 'to serve many Writs on refractory tenants.'[93] There does not appear to have been any active resistance, although feelings were high and anxiety was deep if there was any basis for a report in the *Herald* concerning Bagnall, the innkeeper in Hazel Grove: 'it is said that Mr. Bagnall has become so apprehensive since the soldiers have been located in his barn, that he has tried, but without success, to effect an insurance on his premises.'[94]

Swabey, Dodd, two military officers, and twenty-five soldiers departed on Wednesday, 18 October, for Lots 65, 29, and 30 to serve legal processes. The proprietors on Lot 65 were the Cumberlands, Wright, and the heirs of Sir Samuel Cunard; on Lot 29, they were Fane and Viscount Melville; and on Lot 30, the landlord was Stewart, a resident. These, with Sulivan, formed a group of some of the largest, most vocal, and best-connected landlords. After completion of this mission, Hodgson reported to Cardwell that 'Although displaying a very angry and hostile spirit, there was no attempt to resist the Sheriff, supported as he was by the military.'[95] In fact, there is no evidence that the soldiers fired any shots in anger, the bloody projections of Pope notwithstanding. The tenants apparently conformed to a pattern Richards has discerned in cases of resistance during the Clearances in the Scottish Highlands: when the sheriff, after failing to do the job by ordinary means, showed up with a supporting military force, the people offered no opposition. 'The news of impending [military] intervention was usually enough of itself to lead to a collapse of resistance. Troops intervened on ten occasions but were never actually engaged in physical hostilities.'[96] Soldiers would not be engaging in physical hostilities in rural Prince Edward Island during the autumn of 1865 either.

On 27 October, a changing of the guard began in Charlottetown. The men of the 16th Regiment departed for Halifax to replace men of another regiment who had been sent to Jamaica to deal with a rebellion. Two companies of the 15th Regiment in Fredericton arrived in Prince Edward Island and marched into the new Victoria Barracks, which was now ready for habitation.[97] On Thursday, 2 November, Dodd and his assistants, accompanied by a military party of forty-three, all under the direction of Swabey, began a ten-day sweep through eastern Queens County. They went first to the Tracadie estate, Lots 35 and 36, and then to the townships where the formal leadership of the Tenant League had been concentrated: Lots 48, 49, and 50. This was the largest, longest, and last of the expeditions into the countryside to enforce the writs of the Supreme Court with the assistance of the military. Unlike the forays into western Queens, this mission apparently did not include an attempt to arrest suspected rioters. The group set out to enforce writs for rent, and only writs for rent; the symbolism of doing this in the very area where the Tenant League leadership lived must have had an impact on the rest of the Island, and certainly on the general membership of the league. On 22 November, Hodgson reported that the inhabitants had offered nothing more than passive resistance, except for the destruction of three small wooden bridges, presumably to impede the progress of the expedition. The sheriff's party replaced them.[98]

In the course of the trek through eastern Queens, the troops may have engaged in some vandalism or at least permitted themselves some careless behaviour at the expense of one Tenant League leader. John Mooney resided in the district of Ten Mile House on Lot 35, where the soldiers made their first stop. Many months later, after much of the furor had subsided, he demanded £3 'for damage sustained by the burning of 300 longers [unfinished wooden poles] by the Troops, ... and for damage done to his grass.' The Tory councillors at first declined to allow the account, but later softened their stand, apparently recognizing that there was some justice in his claim; they authorized payment of £1 10s., one-half the amount he had sought.[99]

After returning from the rural districts of eastern Queens, the sheriff had informed Hodgson that he was able to serve 'a large portion' of his writs.[100] But on the same day that Hodgson put that information in a despatch to Cardwell, the former proprietor Robert P. Haythorne wrote to a newspaper of hearing that in the Tracadie area 'several families ... left their homes and encamped in the woods' as part of an effort to avoid the sheriff and his military escort. These included 'old men and old women, young mothers and young children.' The weather at the end of October and the beginning of November had apparently been uncommonly severe in Prince Edward Island in 1865. Haythorne asked, in his letter to the editor of the *Islander*:

Do any of your readers believe that the recollection of the camp at Tracadie will ever be obliterated from the minds of those who shared its hardships? Will not elderly people always date the commencement of their rheumatism from those bitter days? ... I cannot regard the fact of families thus abandoning their homes, as otherwise than disgraceful to a civilized community.

In his view, the episode 'shews the unextinguishable dislike of the people to the leasehold system, and affords another argument for the application of a remedy.'[101] Haythorne later related the same incident, although with less specificity, when he was a Member of the Legislative Council.[102] In the autumn of 1867 Edward Whelan's *Examiner* commented on the story in a tone that suggested acceptance of it as fact.[103]

Following completion of the last law-enforcement mission of the soldiers, Hodgson gave the Colonial Office his interpretation of the impact of the three expeditions. 'I am inclined to believe that ... [they] have had a good effect; and that although the animosity of the tenantry on the rent question has not subsided, the dread of incurring heavy costs has induced many to come forward and endeavor to effect a settlement of their arrears.'[104] Hodgson was referring to the legal costs in rent cases, and as a judge and former lawyer and land agent, he knew well the importance of these costs for Island tenants. When Peters had been a practising lawyer, he wrote that in actions against tenants for rent 'I found that to obtain even a Judgment by default the costs would amount to between £8 and £10'[105] – which of course in practical terms for most Island farmers was the equivalent of rent for one or two years. Furthermore, in the words of George DeBlois, a long-time agent, testifying under oath in 1875, 'When the arrears are put into the hands of the Solicitor they are collected by him, and the costs are paid by the Tenant. That is invariable in my experience.'[106] Therefore, once a farmer in the situation of many Queens County tenants in the autumn of 1865 was convinced of the capacity of the sheriff to deliver writs, it made sense to come forward voluntarily before being actually served with a legal process. Hodgson's assertion was almost certainly correct. The logic of the costing of the legal system dictated pre-emptive attempts to settle arrears before the writ, whether in the form of a *fi. fa.* or a *ca. sa.*, arrived under military escort. But conforming to that logic entailed breaking with the Tenant League strategy of withholding rent until the land question was settled. One could not obey both the law and the Central Committee of the Tenant League on this matter. If the Central Committee prevailed, the law suffered, and *vice versa*.

Evidence in the papers of John McEachern, a tenant farmer in the district of Rice Point, Lot 65, provides a concrete example of a tenant attempting to settle accounts with his landlord before being served with a legal process. There is no reason to believe that McEachern, a self-styled 'moderate Conservative,' was ever an active supporter of the Tenant League. But he appears to have been in arrears to his landlords, the Cumberlands, and he did watch developments surrounding the league on his township closely. In a diary entry for Thursday, 19 October, he wrote that the sheriff had 'passed here' that day, accompanied by soldiers. There was a gale the following day, but on Saturday, when the weather had cleared to some extent, he went to Charlottetown 'to settle with Chas. Wright,' the land agent for the Cumberland estate.[107] He was thus apparently conforming to the pattern Hodgson noted subsequently, whereby tenants in arrears, anticipating a visit from the sheriff supported by soldiers, acted before it happened to them.

In a retrospective diary entry dated 1 January 1866, McEachern stated that in the autumn of 1865 landlords had 'distrained upon tennants [sic] that never attended Union [Tenant League] meetings, which causes much dissatisfaction among many of the former supporters of the present Government.'[108] The wording gives cause for reflection on the patterns of behaviour and thought underlying such discontent. Some tenants who did not support the Tenant League were nevertheless apparently willing to use its activism in making, in this case, Lot 65 a 'no-go' area for Curtis and Collings, as a shelter for themselves against the demands of their landlords. Furthermore, in a rather peculiar exercise of their reasoning powers, they seemed to think that they should be entitled to avoid rent paying and the legal penalties for not paying rent because they were not Tenant Leaguers. Agents and proprietors could be excused for impatience with such a point of view. In their eyes, a non-paying tenant who distanced himself verbally from the Tenant League while behaving like a Tenant Leaguer by withholding his rent must have looked rather similar to the non-paying Tenant Leaguer – only somewhat less consistent. There was no immunity from legal action over arrears simply because a tenant professed to be politically conservative; and there could not be, or the result would be the anomalous situation of tenants being allowed to act like Tenant Leaguers with impunity if only they said they were Conservatives.[109] In the case of McEachern, he appears to have conformed also to the oft-remarked-upon phenomenon of the Prince Edward Island tenant farmer who was able to pay his rent but chose not to do so, for when he saw soldiers he was able to settle with the land agent.

Not everybody on Lot 65 was a John McEachern – a Conservative who did not support the league, but was willing to use its militancy to personal advantage in an opportunistic way. There is significant evidence of a continuing degree of passive resistance to the forces of the sheriff and, conversely, support for the league, within the community. Passive resistance, with 'its withdrawal of collaboration,' as Richards has put it,[110] had the potential to irritate or embarrass the authorities, although it could not stop them in the autumn of 1865. A letter from

J.W. Morrison, the deputy colonial secretary of Prince Edward Island, to Major St.George M. Nugent, assistant quarter master general in Halifax, dated 6 December 1865, reveals some surprising details. A captain in the 16th Regiment had complained that on one occasion no prior arrangements had been made for the accommodation of soldiers sent into the countryside. Responding to letters dated 30 October and 20 November, Morrison explained what had happened, from the perspective of the civil authorities. Hodgson, as administrator, personally arranged with Wright, a justice of the peace as well as a Lot 65 proprietor and the land agent for the Cumberland estate, for billeting the troops on his premises.

But when the party arrived via steamer from Charlottetown about noon on the appointed day, 18 October, 'they found that Mr. Wright's Domestic Servants and Farm laborers had all deserted his service from sympathy and connection with the Tenant League, [and] that he had been unable to procure others.' Thus the building where the troops were to stay was not ready. As a consequence, 'the sheriff and his Civil force immediately proceeded to prepare it,' with the voluntary assistance of some soldiers. By three o'clock in the afternoon, the work party had readied the accommodation and cooked a meal, and all had eaten. After the meal, they set about serving writs. Morrison closed his account of the events of that day with the reassuring statement that 'it was a bright sunny day in October and ... no privation beyond a little delay in getting dinner cooked, could have been suffered by the Troops on that occasion.'[111] The refusal of collaboration by Wright's servants did not prevent the serving of writs, but it must have made a point to the soldiers and the sheriff's party generally. There was also embarrassment for the civil authorities, whose plans had gone awry, and there was probably irritation as well. Extra work had to be done by persons who had no reason to expect these duties, and explanations had to be offered, even many weeks later. Certainly the passive resistance at Wright's farm on Lot 65 put a strain on communication between the civil and the military forces.

In the end, the authorities prevailed. On 10 November the *Islander* reported that even the tenants of Fort Augustus and the Monaghan Settlement were coming to an agreement with John Roach Bourke, the agent for their landlord, the Reverend John McDonald. The report appeared in print eight months to the day after the inhabitants of those districts repelled Curtis when he went alone to look for James Callaghan and others. The Tenant League was in trouble in its heartland. Even so, it remained important for local people to dissociate themselves from the suspicion of assisting the sheriff: on 1 December, James Doyle of Lot 48 took an oath, attested by a justice of the peace, to the effect that he did not convey or lead the sheriff and his constables to various tenants in the Johnstons River district, straddling Lots 48 and 35, and did not provide the sheriff's forces with information. He published it in the *Islander*, in a manner reminiscent of Alexander Halloran's oath in October to the effect that he had not given information leading to the arrest of deserters. The wording of Doyle's denial was significant: 'I am innocent of the charge laid to me' – innocent of the charge of aiding a sheriff in his duty![112] Tenant League sentiment was still strong enough to provide an alternative set of norms. Doyle's concern for his reputation or his safety or both had led him to make the highly unusual and unorthodox solemn declaration, and make it public through advertising it in a newspaper.

Yet passive resistance and public disapproval could not carry the day against the sheriff and the landlords, supported by the presence of British troops. The soldiers, when used, proved to be effective instruments of coercion, and their presence at the side of the civil power made the resolve of Tenant League stalwarts crumble. The effective power of the league, the ability to withhold rent, forming the very centre of their program, had been negated.

11

Collapse and the Courts

The excitement which prevailed at the time of the disturbances last summer, and for some considerable time after, has now wholly subsided.
Examiner, 8 January 1866

Stern duty compels me to pass a very severe sentence on you.
Judge James Horsfield Peters addressing Charles Dickieson, 24 January 1866[1]

Disarray of the Tenant League

The successful use of the military to assist the sheriff in Tenant League districts appears to have dealt a virtual death-blow to the league as a formal organization. When troops were on the Tracadie estate, William Henry Pope's *Islander* declared 'we trust this is the last time that it will be necessary to send [them] out ... even the "invincibles" of the League must now be convinced that obstructing the Sheriff and his officers, while in the discharge of their duty, is a useless exhibition of valor.'[2] In fact, it proved to be the last time in the history of Prince Edward Island that soldiers were used to protect the civil authority in enforcing the law.[3]

The leadership of the Tenant League more or less disappeared from public view as an organized force, although there was no attempt to arrest or prosecute any member of the Central Committee. On 7 September 1865 *Ross's Weekly* had given notice that the league constitution was being revised. The point of the exercise was apparently to shift the emphasis away from rent repudiation, with its concomitant defiance of the law, and towards purchase of estates. No complete copy of the revised constitution is known to exist, although parts are accessible in

Edward Whelan's *Examiner*; first, three passages appeared on 25 September, and then more of Article 11, which dealt with the payment of rent, on 9 October. Whelan professed to be disappointed that the revision had not gone far enough. In his view, there had been no real change in the policy of rent resistance, merely a juggling of words. The *Examiner* cited the following from the Preamble: 'further compliance with the terms of *leases* ... would only tend to protract the agitation, by preventing a speedy settlement of this long-vexed question.' Article 11 included the statement that 'He [the tenant] shall pay neither rent nor arrears of rent.'[4] But that would not be the end of the changes. On 7 November, while troops were in eastern Queens County where, coincidentally, most of the leadership resided, the Central Committee stated that it would allow members to pay rent in certain circumstances. At the same time, it declared that it would attempt to assist financially 'those who have been put to expenses by their connexion with the Tenant Union ... as far as the funds will allow.'[5]

One practical problem in attempting to understand the response of the league as an organized body to the events of the autumn is the fact that almost all the relevant information which survives has been filtered through hostile witnesses. Little appears to have been heard from George F. Adams or other recognized leaders. Furthermore, *Ross's Weekly* is not available for either 1865 or 1866, except for a few copies or parts of copies that survive only because other newspapers reprinted certain articles for their own purposes or because 'evidence' from it found its way into, for example, the Colonial Office files in support of explanations why the local government was or was not prosecuting the publisher.[6] The major reason for this lack of a substantial file of *Ross's Weekly* after 1864 is probably a fire in Charlottetown on the morning of Sunday, 15 July 1866 which caused widespread devastation and destroyed the office, equipment, and supplies of the newspaper. In an advertisement printed the following day in the *Examiner*, Ross announced suspension of publication 'for a time.'[7]

Among prominent persons associated with the Tenant League, only Ross wrote anything approaching a retrospective account of the years in which it flourished, and his reliability even in dealing with contemporaneous incidents had been relatively poor. Adams, who was capable of providing a good and readable account of the fate of the league, did not do so, and he seems to have disappeared to the United States, for he died in Philadelphia on 26 November 1879.[8] Alexander McNeill, who had renounced office in the league before the end of September 1865, continued as a district schoolmaster, and lived until 1899. He retained an interest in reform causes, and would be an advocate of female suffrage, but he is not known to have written about his experience with the league.[9]

Some prominent Tenant Leaguers carried on after the autumn of 1865 as they had before the coming of the league. They were certainly not in hiding. At the Easter Term of the Supreme Court in 1866 Samuel Lane appeared as a witness against John Roach Bourke in a civil case, and Adams himself – identified as an innkeeper – gave bail for Samuel Sabine, a tenant, in a suit initiated by Lady Louisa Wood and Maria Matilda Fanning, his proprietors, for £28 10s. 3d. back-rent.[10] The date and circumstances of Adams's departure from the colony are unknown. He was certainly on the Island on 4 September 1866, when he chaired a meeting of the Central Committee and was described as 'President.'[11] Yet he may have been preparing to leave: he advertised an auction sale to be held on 18 October, because he and his father were 'withdrawing from the Mail and Stage Business.'[12] It is possible that such news prompted Richard J. Clark, a merchant in Orwell, a nearby district, to swear out an affidavit of debt against him on 6 November; in that document, Clark described him as a merchant in Vernon River. Adams was probably still on the Island when Clark took his legal action, but the trail runs cold a few months later, after the *Examiner* reported that a political meeting had been held 'at Mr. Adams' Hotel, Vernon River' on 8 February 1867; even that news item does not indicate whether he was still in the colony.[13] Nonetheless, it is clear that he was on the Island on 4 March 1868 following the death of his father several months earlier, for on that day, as the administrator of his father's estate and in the presence of Robert Stewart, another former leaguer, he sold the leasehold interest in fifty acres to Sylvester Kehough of Lot 50 for £135 currency.[14]

For a period, the Central Committee continued to meet on a monthly basis. The last known accounts of these meetings appeared in August and September of 1866 in the *Herald*, a newspaper much more sympathetic than most of the press. At the September meeting discussion apparently centred on 'future political action,' and giving Ross a new start.[15] Although defiant individuals sometimes spoke out subsequently in a manner suggesting they were speaking for an organization, no hard evidence survives of any organized league activity after the 4 September 1866 meeting of the Central Committee. This does not mean that Tenant League sentiment died; in fact, there is ample reason to believe that it did not. But the league had been a body committed to direct action of a particular kind in pursuit of its land reform agenda, and not simply a group of like-minded persons. The use of soldiers had destroyed the credibility of its master-tactic, and consequently it withered.

An important letter from Bourke appeared in the *Examiner* of 2 October 1865. It was significant in undermining the hopes associated with the Tenant League. For several months, rumours had been swirling about that Bourke, a small landlord who owned part of Lot 37, was about to sell to his tenants through the good offices of the league. In the letter, he stated that he had sold some of his property to tenant-occupiers but that the league did not deserve the credit; he strongly implied that he had used Adams for his own purposes in the transactions. In fact, he wrote, 'the lies of Adams and Ross' had prompted him to decide to sell no more land 'as long as the League exists, let the consequence be what it may.' He was responding to the 21 September issue of *Ross's Weekly* and he specified that 'a voluminous statement of falsehoods' by Adams had provoked him.

Bourke declared, furthermore, that he would not advise other landowners to sell in collaboration with the league. That was a noteworthy announcement because he was land agent to such proprietors as Viscount Melville and the Reverend John McDonald. He emphatically denied a report in *Ross's Weekly* that a purchase had already been negotiated for Melville's portion of Lot 29.[16] A story on the same matter in *Ross's Weekly* of 24 August had already prompted him to write to the *Islander* several weeks earlier, denying there had been a purchase, describing such a transaction as improbable, and criticizing *Ross's Weekly* for publishing 'such unblushing untruths.'[17] He had not displayed any particular ill will against Adams in his letter to the *Islander*. But when writing to the *Examiner* of 2 October he was much more sweeping in the way he closed the door to any future cooperation with the league and Adams, probably in part because of his repeated problems with *Ross's Weekly*.

It appears from Bourke's letter to the *Examiner* that Ross had distorted a written statement by Adams about the Lot 29 property. When Bourke had confronted Adams about the published story that there had been a purchase, Adams

positively denied giving Ross any such statement, and said that the statement he gave could be seen in writing there [at the office of *Ross's Weekly*]. I [Bourke] next saw Ross, who referred to the file, and read the statement given by Adams, which was nearly correct. Ross then read what he said in the paper; but trying to qualify it, said that he inferred from what Adams said that there was a purchase effected. No person but himself could think so.

Lest there be any doubt what he thought of Ross, Bourke declared that 'there was nothing said or done that could induce the most illiterate person to believe or think that there was a purchase effected, but quite the reverse.'[18] Bourke's anger with Adams seems to have arisen from an indication by him in *Ross's Weekly* that Bourke had gone to Lot 29 knowing that tenants there were hoping to effect a purchase, a story which, coupled with the 'news' in the same newspaper that an agreement to sell had already been reached, could put him in a bad light with his employer, Melville. In fact, it is conceivable that Bourke was being so categorical in order to avoid the censure of the landlord, a military man whom a visitor, the son of a former proprietor, had described some years earlier as 'more gruff & boozy and prejudiced than ever.'[19] If reports reached Melville that Bourke, as his land agent, was selling his property, they could cost him his job; Sir George Seymour, the father of Melville's visitor, had had precisely that problem with an agent, and he had fired the agent as a consequence.

Bourke vehemently denied Adams's account of the reason the two of them made a trip together on 21 August to a meeting at the Melville Road schoolhouse, in the very district where the initial disturbance associated with organized rent resistance had occurred late in 1863, prior to the formation of the league. He referred to 'Adams's deception,' and implied a motive by stating that 'It is well known' that Adams desired to become an assemblyman. He also reported that 'more than once' Adams had said to him that the leaguers were 'a parcel of d——d fools,' and that McNeill and Lane 'both often made him sick to hear them – that he used to go out from them, and take his pipe and smoke.' In future, Bourke wrote, he would take no notice of what either Adams or Ross said.[20]

For all Bourke's bluster against Adams, he did not explain why he had been travelling in the company of the known leader of the league to meet the tenants of Melville. In his letter to the *Islander*, he had been slightly more forthcoming; one reason was convenience, since he and Adams lived in the adjoining districts of Millview and Vernon River in eastern Queens, and he outlined the practical considerations, which had to do with hiring a horse and carriage. But still it seems that he went essentially because Adams suggested that he go, and that Adams's reason was that the Melville tenants on Lot 29 wanted them to go; the fact that Bourke would do so, even though, according to him, Adams told him that he did not know why the tenants wanted to see them testifies to the influence Adams had with him as recently as late August, even after the arrival of troops.[21] Whatever the explanation was, it suggested that Adams had formerly been acceptable to Bourke, a land agent and proprietor, as a potential go-between. In the context of the escalating confrontation between Tenant Leaguers and the forces of the Queens County sheriff over the following several weeks, Bourke's consorting with the leader of the league could be seen as lending credibility to the organization, and this consideration was probably one reason to establish distance.

Bourke's letter was wide-ranging and included, for example, his interpretation of the purchase of the Haythorne estate on Lot 49. A Tenant Leaguer did not have to accept every part of the letter as entirely accurate or candid in order to find it disheartening. The most important aspect of the matter, in a practical sense, was that Bourke had gone on the public record emphatically declaring that he would not be selling his land on Lot 37 to his tenants according to the league's prescription. Many had probably been hoping that he would agree to such a purchase. The additional information that he would not use his influence to assist other accommodations to the league program was bad news, particularly if anyone had hoped for a breakthrough on Lot 29, where the rate of tenancy had been the highest in the colony at the last census. Instead of a purchase, the inhabitants of the Melville estate on that township could expect a visit from soldiers in October.

Tenant Leaguers were receiving profoundly discouraging news in October and November of 1865. The local government's decision to use the troops meant that leaguers would no longer be able to prevent the sheriff from serving legal processes, and therefore rent resistance could not continue to be the centrepiece of their strategy; the price was too high. Bourke's message, delivered through the press, was that he at least would not facilitate their objective of purchase through negotiation. He followed up with a notice in the *Islander* of 27 October, advising tenants of several non-resident proprietors for whom he was land agent to pay up or face 'harsh measures.'[22] As noted already, by early November the tenants of the Reverend John McDonald, previously known as militant rent resisters, reached an agreement with Bourke concerning payment of rent on his estate. This understanding became public knowledge without delay, and would have further demoralized leaguers.

As prospects for realization of the Tenant League agenda became bleaker, there was, not surprisingly, conflict, acrimony, and recrimination. Fairly detailed evidence exists regarding one case at the local level, and in fact it is the only instance for which significant documentation concerning internal disciplinary procedures of the league has survived. It occurred within the Tenant Union branch at Campbellton, Lot 4, in western Prince County. The organization was not generally strong in Prince, but Lot 4 was a part of Edward Cunard's estate where tenants had formed 96.3 per cent of land occupiers at the 1861 census, and thus the potential beneficiaries of the league program included almost every farmer on the township.[23]

On 10 October members of 'the Campbellton Tenant Union' dealt with an accusation against James Sullivan, its treasurer. He was charged with breaking the code of solidarity, specifically 'by making a private arrangement with the Land Agent.' The 10 October meeting was called to deal with the matter. The official report indicated that the turn-out was 'very large and respectable.' Sullivan was informed of the charge, which he answered. Apparently he did not dispute the facts, but did disagree with the interpretation the branch placed upon them. Six resolutions relating to Sullivan passed, all unanimously. The branch expressed 'contempt at his conduct,' expelled him 'as a rotten member,' and asked that he turn over the books and money belonging to the league.[24]

Two days after the meeting, *Ross's Weekly*, when publishing the resolutions, put on record its view that the Campbellton branch appeared to be rather severe in its treatment of Sullivan, who had evidently been in contact with the newspaper. This development infuriated the local branch, which held a second meeting, on 17 October, and passed six more unanimous resolutions. The branch condemned Sullivan, 'a self-convicted traitor,' for, among other things, attempting to deceive the editor of *Ross's Weekly*. When the report of this second meeting appeared in the *Herald* near the end of November, a 'P.S.' pointed out that Sullivan had broken ranks by paying his rent 'long before the November resolution of the Central Board rendered it quite legal.'[25] The branch was not going to forgive him retroactively. Events had gone too far for that. On the night of 24 October, he had been 'burnt in Effigy by the League,' according to the diarist Henry Jones Cundall, who was in the area on surveying business and who had recently dined with him. Feelings at the local level, where solidarity had been the paramount value, could be exceptionally bitteras the organization crumbled. Cundall reported that the symbolic burning 'much annoyed' Sullivan.[26]

Later that autumn, Sullivan published a letter in which he explained his point of view. He had attended Tenant League meetings, and contributed £1 to the league, but, 'not wishing to place myself in a position which would embarrass my judgment,' had never signed the 'tenant's pledge.' In fact, he said, he had been 'not a *member*, but a *sympathiser*,' and he suggested that this rendered his expulsion from the Campbellton branch moot. He did not state explicitly whether, in his view, he had been treasurer.[27] The dispute dragged on in the press. Apparently the *Patriot* newspaper of 6 January 1866 published a letter by R. McDonald defending Sullivan, or at least attacking the action taken against him. That provoked a response by a member of the Campbellton branch.[28]

There is a much broader question about internal relations within the league during the period of its disintegration: was there significant dissension between the leaders and the mass membership at the colony-wide level? Even in early September, one anonymous writer to the *Examiner* asserted in a lengthy letter that the rank and file had been left in the lurch and their actions disowned after they had acted on the advice of the leadership: 'Adams prated about the divine right of resistance. Stewart publicly threatened to make a martyr of a bailiff, and John Ross got some rascal to say that the Sheriff and his officers were no better than raiders.'[29] This interpretation of the leadership as having instigated ordinary members to challenge the civil authority physically was certainly consistent with the position of Whelan. On 2 October 1865, as events were leading to the dispatch of troops to rural Queens County, the *Examiner* had declared that the 'selfish and cunning' leaders 'are in comfortable circumstances, and well able to meet their obligations to their landlords.'[30] The implication was that they, in contrast with many Island tenants, could afford the visit of the sheriff with his *fi. fa.*, and that there would be no dire hardships for them personally.

Whelan of course was a hostile witness regarding the leadership, who had defined him as an enemy of the Island tenantry, just as he had portrayed them as false prophets. As recently as 23 September 1865, Tenant Leaguers from five townships in eastern Queens, meeting at James Callaghan's in Fort Augustus, had excoriated him as 'low and contemptible ... for his wanton attack on the Tenant Union,' and had called on members to withdraw their financial support from his newspaper.[31] This was an escalation in an already bitter conflict. Earlier in September, Whelan had launched a blistering attack on the Tenant League leadership, in which he singled out Adams, Lane, Stewart, and John Mooney by name. He denounced them for 'fraud and deception' in claiming to be involved in purchases from proprietors on Lots 29 (Melville), 34 (the Reverend James F. Montgomery), 37 (Bourke), and 49 (Robert P. Haythorne); in two of these cases (Melville and Bourke), no transaction had occurred. 'They [the Tenant Leaguers] have not purchased even one acre on any one estate in this Island,' he wrote. He had gone on to accuse Adams and Stewart of using Tenant League funds for their private purposes; according to Sullivan, Stewart was the general treasurer. Whelan damned the league itself as 'the swindling organization.'[32] Whatever the probity of the leadership concerning finances, it is a fact that none of the leading members, except Ross, faced charges in the Supreme Court for actions taken in support of Tenant League goals – and in this respect there was certainly a disparity in suffering between the leaders and the led as collective entities.

One issue on which critics of the league had long accused the leaders of misleading the mass membership was the availability of troops for protecting the civil authorities as they went about their work enforcing the rights of property. Writing from Scotland in August of 1865, Lieutenant Governor George Dundas informed British Colonial Secretary Edward Cardwell that, according to Administrator Robert Hodgson, 'a rumour had been industriously circulated to the effect, that Troops had been applied for, and the application refused.'[33] Once the soldiers were on the Island, it was often said that the Tenant League leaders told their followers that they would never be sent out into the countryside.[34] No surviving document produced by a Tenant Leaguer confirms in a positive way that the leaders did use this assurance in encouraging defiance of the sheriff's forces; but since it is likely that the medium would have been word of mouth rather than writing, the absence of such written proof is not a strong reason to disbelieve the allegation that the leadership used this argument. If the leaders did assure members that they had not to fear that the troops in Charlottetown would enter rural districts in support of the sheriff (save, of course, in such obviously extraordinary circumstances as the death of a law enforcement officer in the course of duty), it is likely they would have lost credibility when events falsified their assurances.

Yet regardless of what the leaders may have said beforehand, the government's successful use of soldiers to assist enforcement of writs in Tenant League strongholds – whether near the homes of the formal leadership in eastern Queens, or simply in centres of strong mass backing, as in western Queens – must have done significant damage to the league. Haythorne, who lived on Lot 34, would soon write, in a thinly veiled reference to the Tracadie estate, that 'not far away, I hear of wholesale distraints'; he mentioned seizures of crops, horses, and cattle.[35] The power relationship in the countryside had changed. It had been demonstrated that the sheriff could indeed call upon soldiers to assist him or his deputy. No one repeated the defiant and threatening scenes enacted a few weeks earlier. Surveying the situation after the Supreme Court appearances at the end of October and the beginning of November, the *Herald* gave the gloomy prognosis that 'The lawyers, we think, will be the greatest gainers by those wretched troubles.'[36]

The local newspapers expressed definite opinions about the demise of the Tenant League. At the beginning of October, before the use of troops in rural Queens, the *Examiner* had believed that the league was far from collapse.[37] But by 16 October, following the expedition to the Sulivan estate on Lot 22, it announced that 'Their [the Tenant Leaguers'] spirit is broken.' On 20 November, after completion of the third and last military mission, the *Examiner* stated that the Tenant League was 'now happily defunct.' Two weeks later, comparing the league, 'a vile fraud,' to the Fenian brotherhood, Whelan's newspaper declared that it was dead.[38] On 29 December, the *Islander* published an article reporting that the league, 'as an illegal association, is dead.' The press had pronounced, and they appear to have done so accurately.

The league was very much on the retreat at the end of 1865. *Ross's Weekly* of 14 December published a Central Committee resolution appealing for money to assist in the legal defence of those to be tried in the Supreme Court in January.[39] On 19 December, with Dundas, who had returned to the Island a few days earlier, present, the Executive Council directed Attorney General Edward Palmer to prepare a bill of indictment against the publisher of *Ross's Weekly* for an article that had appeared on 16 March. At the same meeting, the Council rescinded the authorization of 21 July concerning employment of special constables.[40] This was close to an acknowledgment that the emergency was over. From the point of view of the authorities, the remaining unfinished business was the trials before the Supreme Court.

The Accused Tenant League Rioters before the Supreme Court

By the middle of October 1865, accused rioters from Lot 22 had been listening to bailiffs and others giving sworn evidence against them before justices of the peace. At the end of that month and the beginning of November, they were bound over before the Supreme Court until the

Hilary Term in January of 1866.[41] So great was the sense of moment when the accused Tenant Leaguers met the Supreme Court in January that Whelan predicted 'it is likely to be the most important session that has occurred in this Island.'[42]

The first step in the proceedings involved the grand jury, an institution distinct from the body charged with convicting or acquitting accused persons, known in legal terminology as the petit jury. The procedure surrounding this initial stage in the prosecution would prove decisive in determining how broadly punishment would be meted out to leaguers. Concern for the integrity of the role of the grand jury in the justice system became absolutely, but probably unexpectedly, central to the fate of most who had been charged.

By law, in Queens County the grand jurors for a term of the Supreme Court were chosen by ballot from a list of one hundred qualified persons. The sheriff had the responsibility of drawing up the array or list of one hundred; he did so in the previous June or January, and conveyed it to the prothonotary or chief clerk of the court, who conducted the ballot in open court. At least twenty grand jurors had to be summoned, and the maximum number was twenty-three. Once more than twelve had assembled, they could chose a foreman for the term. At least sixteen members had to be present in order for the grand jury to transact substantive business.[43] In cases such as those involving the Tenant Leaguers, the prosecutor preferred or brought before the grand jury written accusations known as bills of indictment. Its function was to determine whether there was sufficient evidence to justify a trial, and it decided by majority vote. If it concluded that there was reason to put the accused on trial, it wrote 'true bill' on the back of the bill, which became an indictment upon presentment to the court. Then the case proceeded. The other option in dealing with a bill of indictment was to write 'no bill' on the back; that determination meant that the case halted. The work of the grand jury in such cases was complete once it had decided whether the evidence warranted a trial; and given the nature of its function, it heard evidence for only the crown.[44]

The petit jury decided the guilt or innocence of those persons indicted, on the basis of evidence from both sides. The twelve petit jurors had also been selected by ballot, although from a different list, with two hundred names. They could convict only if they were unanimous.

On Tuesday the 9th of January 1866 the term commenced, with Chief Justice Hodgson and Assistant Judge James Horsfield Peters presiding.

The chief justice delivered the Charge to the grand jury of twenty-one members, beginning by focusing in some detail on the Tenant League campaign of rent resistance and explaining the nature of the alleged offences. He did not mention the organization by name, referring simply to 'a widespread and illegal combination amongst the tenantry to resist the payment of their rents' which had apparently resulted in riotous assemblies and other crimes.

Hodgson lucidly explained the distinction between a riot, a rout, and an unlawful assembly. The three offences were situated on a continuum, with unlawful assembly and riot at opposite ends, and rout occupying an intermediate position. A riot was a tumultuous meeting of three or more persons assisting each other against any who opposed them in executing a common purpose with violence, whether the purpose was itself lawful or unlawful. A rout was a similar meeting which made a motion to execute such a purpose, but which did not actually execute it; and an unlawful assembly was a meeting for such a purpose which, if executed, would make the assembly a riot, but which the meeting neither executed nor made a motion to execute. The key element in every riot, according to Hodgson, was the presence of

some such circumstances either of actual force or violence, or at least of an apparent tendency thereto, as are naturally apt to strike a terror into the people, as the show of arms, threatening speeches, or turbulent gestures; ... it is not necessary, in order to constitute this crime, that personal violence should have been committed. Causing terror is sufficient; for if the conduct of the assembled people was such as to make persons of ordinary courage – whether men or women – feel a sense of being less secure than in the usual peaceful state of society, the offence is complete.

As an illustration of how to determine participation in a riot, Hodgson stated that possession of a trumpet, or other distinguishing accessory, 'such as used generally by the rioters, is strong *prima facie* evidence.'[45] The grand jury met until the 20th, a Saturday at the end of the following week. This was an uncommonly long period, and Whelan believed members had set a record in meeting so long.[46]

On 10 January, the grand jury brought in a presentment against Deputy Sheriff James Curtis 'for using undue force and violence' against Joseph Doucette as one of the sheriff's party storming his house in Wheatley River, Lot 24 on 15 August.[47] This was a case of the grand jurors taking the initiative to raise the *possibility* of a charge, and their action represented a dramatic turning of the tables. They were declaring that there were indications the deputy sheriff had committed a crime and, in effect, they were asking the attorney general to gather evidence concerning his behaviour on the occasion in question. Nevertheless, ten days later they would return 'no bill.' In the meantime, evidence would have been gathered, and the grand jury would have concluded that there was insufficient reason to proceed.[48]

Next the grand jurors dealt with the behaviour of Tenant League supporters. On the 12th, a Friday, they brought in a 'true bill' against nine persons alleged to have been part of the Lot 65 disturbance on 18 September. This was known as the case of 'The Queen *v*. James Gorveatt *et al*.,' and the charges were riot, assault, and conspiracy. On Saturday, the grand jurors brought in a true bill against James Devine and fifteen other persons on similar charges arising out of the Lot 22 incident on 2 October. They also returned a true bill against Charles Dickieson, Doucette, and three others (two of whom were absent) for assault and rescue at Curtisdale on 18 July. On the following Tuesday, they indicted Doucette for assault on constables at his home on 15 August.

Those four indictments – relating to the resistance the authorities encountered on Lots 22 and 65 and at Curtisdale and Wheatley River – were the most important the grand jurors brought in. But there were at least three other bills of indictment or presentments relevant to the Tenant League, and in each case which definitely involved leaguers the decisions of the grand jury could be considered favourable to them. In the case against Ross for libel, the foreman wrote 'No Bill For Self and Fellows.'[49] The grand jurors brought in a presentment against Curtis and George Swan, a constable, for assault and false imprisonment of three persons, arising out of an incident which apparently occurred on 5 October. The location of the alleged detention was Curtis's house,[50] and the duration one hour, described in the bill of indictment as 'a long space of time.' The grand jury subsequently brought in a 'true bill' but struck the words 'beat' and 'wound' from the standard phrase 'beat, wound and ill treat' in the presentment.[51] The grand jurors also delivered a presentment against Curtis for an assault on Elizabeth Devine allegedly committed on 9 October, and later returned a 'true bill,' in this case striking from the usual formulation the word 'wound.'[52]

Finally, there was a case involving assault on a constable executing his duty on 2 May 1865. This was Bernard McKenna, the same constable Tenant Leaguers had harassed on 22 July at Fort Augustus, and the grand jurors indicted three men, all surnamed McKenna. Although the surviving evidence does not indicate the circumstances or place of the offence, it is certainly a fair possibility that the Tenant League was a factor on that occasion too, for when one of the accused was convicted, Francis Kelly, an assemblyman and also a justice of the peace who lived in Fort Augustus and who had a slightly ambiguous relationship with the league, came forward prior to sentencing, described him as a neighbour, and testified to his good behaviour on all other occasions over the fifteen or sixteen years he had known him.[53]

The results added up to a bewildering mixture of 'true bills' against Tenant Leaguers and presentments against the authorities, particularly Curtis. In the midst of the deliberations, Whelan had commented that the grand jury included 'not a few gentlemen reputed as entertaining extreme views concerning the operations of the [Tenant] League'; he also noted the 'variety of sentiment.'[54] He was correct to draw attention to the wide range of opinion. Most conspicuously, Benjamin Davies, a former Liberal assemblyman known to harbour radical views, was foreman. But other grand jurors included George W. DeBlois, the prominent land agent, Charles Wright, both land agent and proprietor, and William Swabey, the justice of the peace who had led the three expeditions of October and November that had provided military assistance for the sheriff. Men like DeBlois and Wright were obvious partisans of the proprietors, and Davies was at the opposite end of the spectrum. Moreover, Davies was not alone. Henry J. Callbeck, another grand juror, would be elected in 1867 to the House of Assembly, where he would be identified with a minority within the Liberal caucus who were strong sympathizers with the Tenant League tradition. In 1868 in the assembly he stated without equivocation that although he had never been a member of the league, he had been and remained a sympathizer. Robert Mutch of Lot 50 was associated with the league at its beginning, and would join with James B. Gay in a testimonial on behalf of a pro–Tenant League candidate at the election of 1867.[55]

With Davies, DeBlois, Wright, Callbeck, Mutch, and Swabey on the grand jury, the debate over the various charges must have been intense, and there may well have been trade-offs. Perhaps those who sympathized with the Tenant League would only agree to indict the accused rioters of Lots 22 and 65 on condition that some action also be taken against Curtis. Probably the 'no bill' in the case of Ross would have galled some opponents of the league as much as anything that would occur at this remarkable term of the Supreme Court.

On 16 January, before the Supreme Court, Joseph Hensley, a lawyer for the accused in the case of Gorveatt *et al.*, moved to challenge the array of the grand jury and to quash the indictment found against the men of Lot 65. Hensley objected to the fact that the grand jury included the Cumberlands' land agent, Wright, whose pressing of the estate's claims for rent resulted in the confrontations that in turn led to the bill of indictment on which the grand jury had to pass judgment. The lawyers on both sides argued the issue on 17 January: Attorney General Palmer, Solicitor General T. Heath Haviland Jr, and John Longworth for the crown, and Hensley, Malcolm MacLeod, Dennis Reddin, and Charles Palmer for the defence. The court decided speedily. On 18 January, Judge Peters quashed the indictment against Gorveatt *et al.* because Wright was a member of the grand jury.

After brief summary of the Hensley motion and the facts on which it was based, Peters made a strong statement on the need for undoubted impartiality on the part of all grand jurors. In stirring language, he went to the core of the issue:

A finding of twenty-four impartial jurors [twelve grand jurors and twelve petit jurors] is required by our law to convict one accused of a criminal offence. ...

The great object of the institution of the grand jury is to prevent persons being even called on to answer for alleged crimes without reasonable ground for accusation. It has been described by great jurists as the grand bulwark of civil liberty – their proceedings are conducted in secret, so that an accused or suspected person may not, without reasonable proof of guilt, suffer the mortification of a public trial. If individuals, who were actors on either side in transactions which form any material part of the subject matter of an accusation, could sit as jurors in deciding whether an accused should be subject to a public trial or not, the principal object of the institution might, in many cases, be defeated. ...

Could Colonel [B.H.] Cumberland, whose rights it was the alleged object of the conspiracy to obstruct, have been considered free from ... reasonable suspicion of undue prejudice ... [?] The counsel for the Crown at first contended that he would, but on examining the indictment they, very properly, abandoned that idea, but then they contend that though Colonel Cumberland might be challenged, his agent was not, necessarily, open to the same objection. But we think it impossible, under the circumstances of this case, to hold that an objection, good as to the principal, is not equally good against the agent. Mr. Wright was the agent for the collection of the very rents which it was the object of the alleged conspiracy to prevent being recovered; he, in fact, instituted the legal proceedings against the tenants from whom they were due; and though he might have no pecuniary interest in their recovery, yet he might, very reasonably, be presumed to labor under what was aptly described by Mr. McLeod as that ruffling and irritation of the mind naturally felt by one who had been engaged in conflict with those suspected of a combination against him, or who have by predetermined violence or intimidation prevented, or attempted to obstruct, the due course of proceedings which he (though on another's behalf) may have instituted against them.

The motion is, in substance, a plea that Mr. Wright was incompetent; and, as the facts, in our opinion, sustain the objection, the rule for quashing the indictment must be made absolute.[56]

Thus Peters, the same person who directed Curtis to take down the names of the Tenant League rioters on Lot 22 and be sure he could identify them, and who then called upon Hodgson, as administrator, to use the troops at his command to assist the sheriff before it was too late, struck down an indictment against another set of rioters in a parallel case. He was indeed the former land agent for the Cunard estate, but when, as a judge, he was forced to choose between letting the accused Tenant League rioters of Lot 65 escape indictment and letting stand an indictment that might taint the reputation of justice, he chose the former without apparent hesitation.

It was a remarkable moment in a career that was outstanding in several areas. For those who had watched Peters closely, the decision would not have been entirely surprising. In politically sensitive lawsuits involving controversial journalists like Whelan, W.H. Pope, and Duncan Maclean, Pope's predecessor as editor of the *Islander*, Peters had already shown that he was not disposed to fall back on easy answers and that his rulings could not be predicted on the basis of his supposed political predilections.[57] He was a towering figure in the public life of Prince Edward Island during the British colonial era, and the case of 'The Queen *v.* Gorveatt *et al.*' displayed his talents to good advantage.

For most Tenant Leaguers facing Supreme Court proceedings in January of 1866, the decision in the Gorveatt case was the pivotal event. After Peters quashed that indictment, Attorney General Palmer stated that he would enter a *Nolle prosequi* – a stay of proceedings – in the case of the accused rioters from Lot 22, 'The Queen *v.* James Devine *et al.*' In that instance, a parallel objection to the presence of Laurence Sulivan's land agent, DeBlois, would invalidate the indictment just as the presence of the Cumberlands' agent, Wright, had nullified the Lot 65 indictment. Although in both cases the crown could have undertaken prosecution at a later date by taking bills of indictment to a new grand jury, this never happened. The crown allowed the cases to die quietly. On 26 June 1866 the defendants entered into recognizances to appear at the Hilary Term in 1867; but they were never called upon.[58]

In circumstances like these, not proceeding with the prosecution is a non-event that those in authority are unlikely to explain. Hensley, whose motion had carried the argument with Peters, had these comments when speaking in the assembly in 1867: 'the matter was allowed to drop. It was thought, I suppose, that the agitation had run its course; or perhaps there was some object to serve in not pursuing the cases further – probably an election was coming on.'[59] His remarks reflect sound judgment. As the government assessed the situation over the next several months, there would have been persuasive reasons not to proceed. Convictions would be difficult to obtain and acquittals would do no good for the general cause of law and order. Moreover, it is likely that the government was satisfied to let the prosecutions end without further controversy, since the Tenant League as a threat to public order was dead. On the very eve of the court term, Whelan had observed that one factor in favour of the accused was that 'the excitement which prevailed at the time of the disturbances last summer, and for some considerable time after, has now wholly subsided.'[60] Given this altered atmosphere, deterrence did not need to play a major role in the thinking of the prosecution, since the challenge to the rule of law was over.

But there remained three indicted Tenant Leaguers to be dealt with. On the day after Peters quashed the Gorveatt indictment, a petit jury found Doucette guilty of assaulting constables at Wheatley River following the testimony of ten witnesses, eight for the crown and two for the defence. On the following day, Saturday the 20th of January, the court dealt with the charges arising out of the Curtisdale affray: indicted for assault and rescue, Dickieson, Doucette, and a third defendant, Peter Gallant, were found guilty of common assault but acquitted of 'rescue.'[61] Apart from Gallant's conviction, almost nothing is known of him; the Islander described him as 'a man to follow others rather than to lead in any violent affray.'[62] The court set 24 January for sentencing.

Peters, rigorous in his regard for proper, untainted legal procedure, proved draconian when it came to punishing Tenant Leaguers who had been convicted. He sentenced Dickieson to eighteen months in jail and a £50 fine. Doucette and Gallant were to be incarcerated one year and fined £20 each for the Curtisdale incident; clearly, if Peters believed Gallant to be less culpable than Doucette, his sentencing did not reflect that belief. Doucette was given an additional year in jail for his role in the battle at his house. Each was to remain in custody until his fine was paid. The sentences Peters gave the three Tenant Leaguers were undeniably severe when compared with others for comparable offences at the same time. Two other persons sentenced on the same day were Francis McKenna, convicted of assaulting Constable Bernard McKenna, and James McCullough, convicted of assaulting a bailiff in the execution of his office. According to the *Islander*, when sentencing McKenna, Peters 'declared these crimes were becoming too common; the officers of the law must be protected.'[63] He sentenced McKenna to three months in prison, and McCullough to four months.[64]

Peters acknowledged the severity of his sentences for the Tenant League prisoners on 24 January. When addressing Dickieson, he stated that 'in awarding punishment against offenders, I am bound to consider all the circumstances detailed in evidence,' and then made a remarkable declaration: 'as the attempt to rescue, which led to this assault, was fully proved, *the limited nature of the finding can make no material difference in the sentence* which it is now my painful duty to pass upon you.'[65] In an eloquent address on the importance of the rule of law, he assigned much weight to the degree of premeditation, the stalking of Curtis's party before the Curtisdale affray, and the use of tin trumpets to summon assistance. On the basis of his own words, Peters could be interpreted as foiling the petit jury by ignoring its limited finding and sentencing the three as though they had been convicted of the greater offence. Indeed Davies, the foreman of the grand jury, would state in the assembly in 1867 that this is what Peters had done.[66]

It is clear in fact that the Supreme Court regarded the Tenant League cases as special. In each of the two cases, involving Dickieson and others, and involving Doucette alone, it had a 'record of judgement' prepared. Written in legal language and signed by the prothonotary, these highly unusual documents detailed the offences, the not guilty pleas, the convictions, and the sentences, and named all the grand and petit jurors involved, as well as the judges, the attorney general, and the sheriff, in addition to all who had been directly connected to the actual offences. In the case of the Curtisdale incident, it placed the distraint and rescue of James Proctor's goods at the centre of the narrative – again emphasizing the importance of the rescue in the mind of the court, although there had been no conviction for that offence.[67]

Criticism of the sentencing in the Curtisdale cases, presented sometimes in coded ways, continued for months and even years.[68] By the end of January 1866, the *Herald* of Edward Reilly was calling for a petition to

Lieutenant Governor Dundas for release of the three Curtisdale prisoners and remission of their fines. Its bitter rival, the *Examiner*, endorsed the idea a few days later. In effect, Peters's severe sentences had made mercy for the three into a public cause. The *Patriot* of 24 March announced that a second and differently worded petition advocating compassion for the Curtisdale prisoners was being signed, since the first contained an adverse reflection on Peters.[69] The unpopularity of the sentences he handed down suggested an additional reason for the government not to resume their cases against the accused rioters of Lots 22 and 65. It must have been clear that if convictions resulted, harsh sentences were possible and that these could do great political damage to the Conservative government, which would soon have to face the electorate.

Over a period of several months the issue of clemency for the Curtisdale rioters became the main focus of public expressions of sympathy for the Tenant League cause. On 24 April a deputation waited on the lieutenant governor with a 5,275-name petition pleading for release of the three prisoners. On 1 August 1866 they were freed and their fines were remitted. A significant figure in the campaign was the Reverend George-Antoine Bellecourt, a French-Canadian Roman Catholic priest who numbered Doucette and Gallant among his parishioners at Rustico. In his correspondence he claimed much of the credit for the release, but there is no strong reason to believe that his influence was decisive.[70] It is likelier that the local government simply wished to quell continuing public discussion of the harsh sentencing.

In fact, the release did not silence some outspoken adherents of the Tenant League. Stewart wrote to the *Islander* shortly after the liberation of the prisoners; he was defiant and still describing himself as a Tenant Leaguer. In the same issue of the *Islander*, there was an editorial cautioning Stewart, but the writer conceded that 'many others' might share his opinion.[71] On 31 August Dickieson himself had a letter in the same newspaper, in which he reiterated points Stewart had made, and drew attention to the fact that the leaders of the league continued to meet every month in Charlottetown, and yet the government never attempted to prosecute them.[72] These letters were mainly significant because they illustrated that despite the demise of the league as an organization, some of the best-known Tenant Leaguers were absolutely unrepentant about their actions and their connection with it.

There were other manifestations of lingering sentiments and resentments concerning the Tenant League. One of the most peculiar was a Queens County grand jury presentment at the Trinity Term of the

Supreme Court in 1866. On 30 June the grand jury expressed its dis-approval over the careless and improper taking of oaths. At issue was a petition against the renewal of a licence to retail spirits which John Binns of New Glasgow, Lot 23 held. The petition was purportedly signed by forty resident householders, and authenticated under oath by three persons: James Laird, the former justice of the peace fired for his connection with the Tenant League, Andrew Dickieson, Charles's older brother, and Angus Gregor. On 24 April the deputy colonial secretary had written to Binns informing him that his licence could not be renewed because he had that day received a certificate and affidavit signed by two-thirds of the inhabitants of the school district requesting its non-renewal.[73] Under the most recent statute regulating the sale of spirits, passed in the previous year, such a certificate with the signatures of that proportion of the male inhabitants entitled to vote for the local school trustees could prevent renewal of a licence. The provision was part of a trend, which had commenced in 1860, to give householders more direct control over the granting of such licences within their dis-tricts.[74] Binns must have complained and been directed to take his griev-ance to the grand jury – hence the grand jury presentment. As it appeared that Laird, Dickieson, and Gregor 'did not of their own knowl-edge know all the signatures to have been properly obtained, the Grand Jury wish *thus strongly* to express their disapprobation of Oaths being carelessly and improperly taken.'[75]

The Supreme Court ordered the presentment published and also took the most unusual action of directing that a copy be forwarded to the Executive Council.[76] In the presentment there is no mention of a link with the Tenant League, but the circumstances suggest a connection, and the response of the judges – both Hodgson and Peters appear to have been on the bench – indicates that they believed there was one. Binns had probably crossed the league in some way, and this was the means certain activists chose for taking revenge; there may also have been an element of using the Tenant League issue to victimize a local commercial competitor, for according to the Farmers' Directory for 1864, Binns and Laird were the two 'general dealers' in New Glasgow.[77] On 17 July the Executive Council considered both a petition by Binns to have his licence restored and the counter-petition from Laird and oth-ers, and they asked the attorney general for a report. Two weeks later, Attorney General Palmer gave his opinion that the proper course for Binns was to apply to the Supreme Court for a mandamus to the Island's colonial secretary, directing him to issue the licence.[78] A manda-

mus – literally, 'We command' – was a writ issued to compel performance of a public duty in the absence of the availability of any other
effective redress.[79] Binns's enemies within his community had succeeded in tying him up in red tape, in addition to depriving him of his
licence for several months.

At the same term of the Supreme Court some league-related cases
remained to be dealt with. On 23 January the court had scheduled jury
trials for Curtis regarding his alleged assault against Elizabeth Devine,
and for Curtis and his co-accused, Swan, for the accusation of assault
and false imprisonment of three persons. Both cases were 'settled with
the consent of the attorney General.' No further details of substance survive, and these cases could have been dropped because of apologies, or
because the complainants had been prevailed upon not to proceed, or
for other reasons. There may even have been some linkage between
these cases and the crown's willingness not to resume prosecution of the
accused rioters from Lots 22 and 65; the cases against Curtis and Swan
were disposed of the same day that the Tenant Leaguers in the Gorveatt
and James Devine cases were bound over for another six months.[80]

Although there are unanswered questions, it is clear that distaste for
Curtis and the way he fulfilled his role had been involved in the decision of the grand jury to indict him. In 1868 the former grand juror Callbeck would recall that

we made a presentment against the Deputy Sheriff for stopping men on the
highway, searching their pockets, presenting his pistol at them, and holding
them prisoner for a length of time, without any reason for doing so. The most
obscene language had been used by the Deputy Sheriff and his officers in serving writs upon the people – language that would provoke the most loyal and
sober men to violate the law.

In contrast, Callbeck said, 'I believe, if Mr. [Thomas W.] Dodd [the High
Sheriff] had travelled through the district [2nd Queens, consisting of
Lots 23, 24, 31, 32, and 65] of which I am the representative, even when
the Tenant League excitement was at its greatest height, he would have
been unmolested.' Regardless of the words Curtis and his assistants
used, this has the ring of hyperbole, as does Callbeck's assertion that
'The principal cause of that [Tenant League] excitement, was the insulting language of the Deputy Sheriff and his officers.'[81] There is nothing in
the fairly extensive documentation to suggest that Dodd himself
believed that if he had served the writs personally, the opposition

would have melted away. Certainly there had been resistance when he and nine assistants captured Doucette at Wheatley River. It is difficult to avoid the conclusion that, although Curtis may have been abrasive and sometimes lacked judgment, he was being used as a scapegoat when such assertions were made. On 6 May 1866 he would be reappointed deputy sheriff of Queens County.[82] The most that could be fairly laid at his door was that he had aggravated a tense situation, for there would have been agrarian disturbances in Prince Edward Island in 1865 with or without him.

Withdrawal of the Troops

On 7 September 1866 Lieutenant Governor Dundas responded to an inquiry from London as to whether troops – since 22 April 1866 a third detachment, two companies of the 4th Regiment of the King's Own Royals, formerly in Malta[83] –were really still required. His answer was affirmative. He conceded that since his return to the Island in December of 1865 there had been no attempt at resistance to the law, but he stated that this was 'entirely owing to' the presence of soldiers. He believed the Tenant League was still a force to be reckoned with, although he recognized that it might have changed its tactics 'for the present.' Given that a general election was close, and that the league's 'object, of course, is a political one,' he asserted that it would be dangerous to withdraw the troops, for it was 'more than probable' that the leaguers would renew their agitation. If a general election were held without the presence of soldiers, he feared that the leaders of the organization might 'attempt to obtain popularity by an agrarian excitement, which it might again require the presence of Troops to control.' His notion that the election could be the catalyst for the renewal of disturbance was important at a practical level, because the election was likely to occur in winter, 'a season during which it would be next to impossible to send Troops to this Island.'[84] Dundas certainly seemed to take seriously statements of intent such as those made three days earlier at a meeting of the Central Committee: that the leaguers would be taking political action, and that Ross would be resuming publication of his pro-league newspaper.

In London there was some exasperation when the governor's despatch arrived at the Colonial Office. No doubt the annoyance bubbled up in part because he had worded his despatch in a way that made it appear politically unwise for Downing Street to withdraw the troops before the election; and, under the circumstances, that meant leaving them on the Island over another winter. Arthur Blackwood stated that the detachment should not be allowed to remain there 'an hour longer,' since 'The occasion for its services has passed away.'[85] T.F. Elliot leaned towards withdrawal of the troops, yet acknowledged the possibility of 'a sort of petty civil war' that would require rapid redeployment. But in their dealings with the Island government, Elliot and others at the Colonial Office, including the Earl of Carnarvon, the colonial secretary, were thinking tactically in terms of treasury considerations, as well as pondering imperial high policy. The apparent desire to have the troops remain meant that the threat of withdrawing them was a handy lever – to be 'avowedly used as a *screw*,' in Elliot's words – by which to exert pressure on the Island government for payment of additional financial charges associated with the presence of the soldiers.[86] In the end, the British authorities allowed them to remain a second winter. The disputes over who should pay for what continued for more than a year after the departure of the troops.[87]

The soldiers were withdrawn on 27 June 1867. In the spring of that year Major General Charles Hastings Doyle had recommended that they be returned to Halifax 'to work on the Defences.' On 8 April, when requesting withdrawal, he stated that their services 'have only been required in one or two instances.'[88] This was a remarkable restatement of the facts since 21 November 1865, when he had advocated a special allowance for officers on grounds that 'The Requisitions for the Service of the Military are very frequent,' with the result that they were 'being constantly employed' to assist the civil power.[89] From Doyle's perspective, clearly, by 1867 the time for their presence had passed. But he did state that he might be obliged to have them leave Halifax during the winter of 1867–8 for lack of barracks, and that in such a situation he would be happy to have them return to Victoria Barracks in Charlottetown before the close of navigation. The rationale, though, would be military convenience, and not the prevention of resistance to the sheriff.[90]

By the time the troops left, there had been a change of government in Charlottetown, and although Tenant Leaguers played a role in the elections for both the Legislative Council in December of 1866 and the House of Assembly in February of 1867, there had been no disorder, and no resumption of organized rent resistance. In the end, one can only speculate whether there would have been renewed disturbance in the countryside if the soldiers had not remained over the winter of 1866–7, but the weight of evidence suggests that it was highly unlikely. An underlying reason for the lack of tenant militancy was a perception in the colony that the land question was going to be solved by means other than refusal to pay rent – a dramatic contrast in mood from the intense frustration that spawned the Tenant League in 1864. Then there had been a highly focused sense of futility over the prospect of ever getting freehold tenure by conventional means.

Much had changed since the spring of 1864. In the summer of 1866 the government of James C. Pope purchased the Cunard estate, the largest on the Island, consisting of 212,885 acres,[91] from Edward and the heirs of Sir Samuel; and following the death of Laurence Sulivan in February of 1866 there had been considerable speculation about the likelihood of purchasing his estate, the second largest owned by an absentee. With the surrender of the Cunards and deaths of both the influential Sulivan and his powerful political protector Lord Palmerston, the long-entrenched *status quo* no longer had the same economic and political weight behind it, and new possibilities were opening up. On 30 April 1866 the Liberal *Examiner* expressed the optimism which was blossoming: 'When these estates shall have fallen into the hands of the Government, the leasehold tenure in Prince Edward Island will be well-nigh abolished – at all events, its colossal strength will be broken.'[92] With approximately 280,000 acres,[93] or 20 per cent of the Island, involved, Whelan was scarcely exaggerating the significance of the hoped-for purchases. A few days later, the Conservative *Islander* cautioned that, notwithstanding popular conjecture, the intentions of Charlotte Antonia Sulivan, who had inherited her father's property on the Island, were unknown. But even if the government was able to purchase only the Cunard estate, in the words of the *Islander*, 'the keystone of the tenant system would be removed; and hereafter stone after stone would come down under the stroke of the Government hammer until the debris alone would remain.'[94]

As matters turned out, Charlotte Sulivan would decide not to sell. But in addition to the Cunard purchase and the hope of a Sulivan purchase, there had been several other positive developments since the middle of 1864, most regarding the Montgomery properties. In 1864 and 1865 there had been two estate sales through the agency of the league, from Haythorne and the Reverend James F. Montgomery, although with differing degrees of transparency. During the same two years the Island government had used the Land Purchase Act of 1853 to acquire some of the remaining land owned by members of the Montgomery family, George, Robert, and William in these instances, for resale to occupiers.

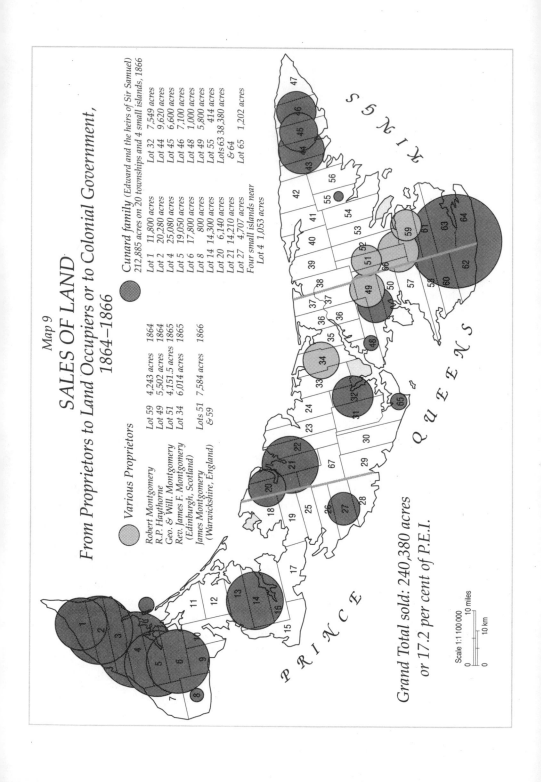

Map 9

SALES OF LAND

From Proprietors to Land Occupiers or to Colonial Government, 1864–1866

Various Proprietors

Robert Montgomery	Lot 59	4,243 acres	1864
R.P. Haythorne	Lot 49	5,502 acres	1864
Geo. & Will. Montgomery	Lot 51	4,151.5 acres	1865
Rev. James F. Montgomery (Edinburgh, Scotland)	Lot 34	6,014 acres	1865
James Montgomery (Warwickshire, England)	Lots 51 & 59	7,584 acres	1866

Cunard family (Edward and the heirs of Sir Samuel)
212,885 acres on 20 townships and 4 small islands, 1866

Lot 1	11,800 acres	Lot 32	7,549 acres
Lot 2	20,280 acres	Lot 44	9,620 acres
Lot 4	25,080 acres	Lot 45	6,600 acres
Lot 5	19,050 acres	Lot 46	7,100 acres
Lot 6	17,800 acres	Lot 48	1,000 acres
Lot 8	800 acres	Lot 49	5,800 acres
Lot 14	14,300 acres	Lot 55	414 acres
Lot 20	6,140 acres	Lots 63	38,380 acres
Lot 21	14,210 acres	& 64	
Lot 27	4,707 acres	Lot 65	1,202 acres

Four small islands near
Lot 4 1,053 acres

KINGS

QUEENS

PRINCE

Grand Total sold: 240,380 acres or 17.2 per cent of P.E.I.

Scale 1:1 100 000

0 — 10 miles

0 — 10 km

By April of 1866 it had reached an agreement, under the same legislation, to purchase 7,584 acres on Lots 51 and 59 from James Montgomery of Warwickshire.[95]

The cumulative impact of the changes, both great and small, on the overall situation had been enormous. At the 1861 census freeholders occupied only 32.6 per cent of the land mass of the Island. Between 1864 and 1866, the Cunard, Haythorne, and Montgomery transactions freed 17.2 per cent of the land on the Island from the proprietary grasp; the Sulivan purchase, if realized, would have made another 4.8 per cent of the Island's land mass eligible for freehold tenure, for a total of 22 per cent.[96] As well as the size of estates, the perception that numerous proprietors – especially when they bore names like Cunard, Montgomery, and Sulivan – were involved was important in the minds of contemporaries, for it was another indication that the pace of change was accelerating. In the face of such rapid change and expectation of further change, the general climate of opinion on the Island had shifted from near-desperation to anticipation that the end of the leasehold system was almost coming into view. These were heady times for Islanders who had watched the leasehold system survive the Escheat movement, the Liberal governments of the 1850s, the land commission of 1860, and the delegation of 1863–4.

Tenant Leaguers, supporters of Tenant League ideas, and sentimental sympathizers with the legacy of the Tenant League were present, active, and visible in public life on Prince Edward Island in 1866 and 1867. Yet there was no effective rebirth of the league, in part because of the troops, but also because, thanks in large measure to the league, there was much less need for direct action. Although the Tenant League no longer existed in any recognizable form, it was certainly arguable that its program was on its way to realization. It was indeed safe to allow the soldiers to leave in June of 1867. No public disorder or rent resistance ensued. Rent resistance had always been a means, not the objective, for the leaguers. The outcome the Tenant Leaguers had sought, namely, conversion to freehold tenure on palatable terms, was occurring at a rapid pace.

12

The Impact and Significance of the Tenant League

Redress long due – long promised, but withheld,
Each party's rule subservient to a few,
And mainly by delusion's power upheld,
Have forced the wide belief that, both untrue,
Have been by private hopes alone impelled,
And kept in action by the prize in view.
Thus trust in party seems forever gone,
And party chiefs are left, deserted and alone.
From anonymous poem in *Examiner*, 9 April 1866

Is that organization still alive?
Examiner, 6 May 1867

The Tenant League was instituted in order to harass proprietors. There never was a greater fabrication than the statement that the League was the means of causing the purchase of the Cunard Estate.
Premier and former proprietor James Colledge Pope, 15 March 1871[1]

The Tenant League as a Factor in Politics after 1865

The Tenant League disappeared as a formal organization sometime in 1866, but pro-league sentiment remained a force in Prince Edward Island life. Some of the most persuasive proofs come from elections held in 1866 and 1867. The Legislative Council provided the first evidence. It had become elective earlier in the decade, with eight-year terms, staggered elections, and a voting qualification of £100 freehold or leasehold. The chamber consisted of thirteen members, four for each county, and one for Charlottetown; on 19 December 1866 there were six seats at stake, including two in rural Queens County, where troops had been used to support the sheriff. The ruling Conservatives lost both Queens seats, although they had won them at the previous election, in February of 1863. One newspaper attributed the results to 'the stop-at-home disease'[2] among Conservative voters. John McEachern, the 'moderate Conservative' on Lot 65 who did not pay his rent until the soldiers were marching on his township, reported in recording the tally at his own poll, Nine Mile Creek, that 'many remained home.' Indeed the decline since 1863 in the number who voted at Nine Mile Creek was so dramatic – from ninety-one to fifty-one – as to require a special explanation. The Liberal vote had remained constant and the Conservative vote had shrunk to a fraction of what it had been less than four years earlier.[3]

Even more significant were the identities of the two victors in Queens, and the margins by which they won. John Balderston, who took 62.7 per cent of the votes cast in the district of 1st Queens, consisting of the western half of the county, had been dismissed as a commissioner of small debts exactly fifteen months earlier, on 19 September 1865, for his connection with the Tenant League. He had been a delegate to the founding convention of the league exactly sixteen months before that, he was unrepentant, and during the election campaign he used Tenant League rhetoric. 'The people desired to support men of independent principles – not the willing tools of the High Tory School, nor of the old Liberal party, but men who would never consent to have unfortunate Tenants driven by the bayonet, to a compliance with the wishes of grasping land-owners,' he declared on the occasion of his official nomination in Charlottetown.[4] On 10 May 1867 in the Legislative Council he would state that although 'some of them [Tenant Leaguers] went too far ... I was a member of that organization, and I have not altered my views, for I believe it was based on sound principles.'[5]

Robert P. Haythorne, the former estate owner who had disposed of his property according to the league's prescription, captured 58.9 per cent of the votes in defeating an incumbent member for 2nd Queens, in the eastern half of the county.[6] Many years later, Haythorne would acknowledge in the Legislative Council that 'I was returned ... [in 1866] in a great measure through the influence of that body [the Tenant League].'[7] Given that the league apparently ceased functioning as an organization prior to the election of December 1866, it is probable that what he really meant was that he had been elected in large part through the influence of individuals who had been active in the league formerly,

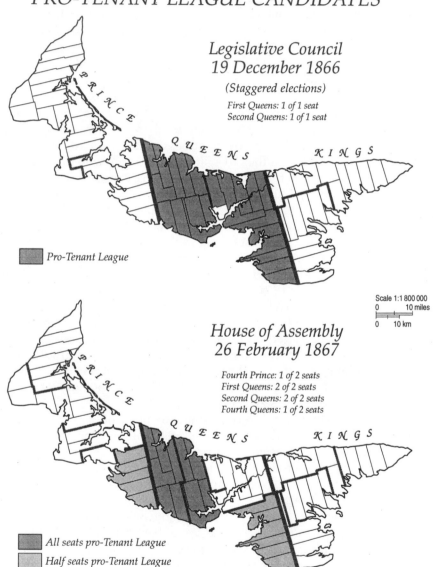

Map 10
ELECTION OF
PRO-TENANT LEAGUE CANDIDATES

Legislative Council
19 December 1866

(Staggered elections)

First Queens: 1 of 1 seat
Second Queens: 1 of 1 seat

Pro-Tenant League

Scale 1:1 800 000
0   10 miles
0   10 km

House of Assembly
26 February 1867

Fourth Prince: 1 of 2 seats
First Queens: 2 of 2 seats
Second Queens: 2 of 2 seats
Fourth Queens: 1 of 2 seats

All seats pro-Tenant League
Half seats pro-Tenant League

and who held to league ideas still. Certainly, known Tenant League activists had played a public role in bringing forward both winning candidates in Queens: at the official nominations, Robert Stewart, who had been a founding member of the league's Central Committee, was one of the two who nominated Haythorne, and James Laird, who had been dismissed as a justice of the peace because of his activities as a league organizer, was one of the two nominating Balderston.[8]

Close examination of the results of the general election for the House of Assembly on 26 February 1867 suggests that events surrounding the history of the league played a deciding role in the outcome. Six newly elected assemblymen, five of them from rural Queens, the focal point of league support during its heyday, were believed to be sympathetic to the movement. During their political careers, all would make statements consistent with support for the league, and none of them ever disavowed sympathy with it, although there was a tendency from the beginning to concede, as did Balderston, that some Tenant League actions had been mistaken.[9] Early in the 1867 session, one of the new assemblymen, William S. McNeill, a former organizer for the league, who, like Balderston, had attended the founding convention, would refer to 'the members put in by the Tenant League'; in the last speech reported in the session he stated explicitly that 'He was quite willing to be considered as a sympathizer with the Tenant League, and that his name should go down to posterity as such.'[10] Each of the six replaced a Conservative, and these reversals were sufficient to determine which party won the election. Without them, the Liberals would have lost, 17 to 13, instead of winning, 19 to 11.

A more detailed analysis, focusing on the available results for individual polls, supports the hypothesis that suppression of the Tenant League engendered bad feelings which caused a decisive shift in the assembly election of 1867. In the western half of Queens, comprising the 1st and 2nd Districts, all four successful Liberal candidates were considered to be pro–Tenant League, and they all won by wide margins. In the 1st District, Peter Sinclair and Donald Cameron took more than 94 per cent of the votes at the poll on Lot 22, the first township to which troops had been dispatched in October of 1865. In the 2nd District, the New Glasgow and Wheatley River polls, located in the home districts of the most prominent Tenant League prisoners, provided the margin of victory for McNeill and Henry J. Callbeck. Finally, Benjamin Davies, the pro-league candidate elected for the 4th District, in eastern Queens, gained his victory through overwhelming support at the polls on Lots

49 and 50, the home townships of several league leaders, including George F. Adams; troops had been sent to both in November of 1865.[11] The returns from these polls in 1st, 2nd, and 4th Queens confirm Edward Palmer's appraisal of the election in a letter to his anti-Confederate ally in Nova Scotia, Joseph Howe. The government's loss, he stated, 'proceeded more from the effects of the Tenant League, than from the confederation differences within our Conservative party.'[12] Rural Queens had delivered six of eight seats to the Conservatives in 1863, but four years later only one Tory survived.

The results from the elections for both houses of the legislature in the aftermath of the Tenant League disturbances reinforce the conclusion that Queens, the most prosperous, the most Protestant, and at recent elections the most Tory of the three Island counties, had suddenly become hostile territory for the Conservatives, and that their repressive measures against the league explained the change. It is worth noting, in addition, that all five townships in 4th Prince, where Alexander Laird Jr replaced James C. Pope, bordered on western Queens; Laird, a younger brother of the James Laird who had been dismissed as a justice of the peace, was believed to be a league sympathizer. Another factor in the outcome may have been Pope's behaviour as a landlord. The constituency included Lot 27, approximately 7,500 acres of which belonged to Pope. In 1868 Lieutenant Governor George Dundas reported to London that during the preceding five years Pope's average *actual* rental had been exceeding his *nominal* rental. This meant that he had been collecting arrears during the mid-1860s, when it was very controversial to be doing so.[13] The bitterness associated with events surrounding the Tenant League remained an ingredient in Island politics for years: in 1872, with Pope back in the legislature, McNeill, the most vocally pro-league assemblyman, would denounce his 'attempt to rule the Island by the bayonet, the handcuffs and the jail.'[14] Later in the same session, Pope called McNeill 'the Communist General' and, in an explicit reference to the Paris Commune, described him as 'a man whom [sic] he believed might be guilty of committing crimes similar to those perpetrated by the communists of France.'[15]

One of the most striking facts about the several pro–Tenant League MPPs elected in 1866 and 1867 was the extent to which, as individuals, they had been quite peripheral to the formal leadership of the organization. None of them was nearly as well known in connection with, or as identified with, the league as Adams or Alexander McNeill. Adams seems to have been nowhere to be found during the campaign, and indeed he may have left the Island by this time. McNeill had been more or less silenced, but he was not entirely neutralized, for at a public meeting in his own district of Mount Mellick he, along with Stewart, nominated Haythorne.[16] Stewart was one prominent Tenant League spokesman who did contest the election for the assembly, but he was defeated in 4th Queens, although he carried the polls on Lots 49 and 50 by overwhelming margins.[17] Evidently, even the voters in a constituency willing to elect the radical Davies preferred not to be represented in the legislature by someone so blatantly identified with the leadership of the league; the defeat of former sheriff Thomas W. Dodd in Charlottetown may also indicate a desire not to place in the assembly someone whose presence would be almost certain to become a focus for acrimonious controversy surrounding the league. The election of W.S. McNeill, Balderston, Davies, and the others did not indicate a desire to revive the Tenant League in a new form, or even an attempt to pass legislation based on its program, but it was a statement of sympathy with what the league had stood for and had accomplished.

Another notable characteristic of some of the pro–Tenant League legislators elected in 1866–7 was a personal history of support for the Conservatives prior to the emergence of the league. According to William Henry Pope's *Islander*, both Cameron and Laird had previously been Tories; Balderston would state during his first session as a legislative councillor that he had assisted in placing the previous government in office; and Callbeck would recount that although he had not been particularly active in politics, he had formerly voted Conservative.[18] This was consistent with Edward Whelan's early analysis of the Tenant League as appealing to disgruntled Conservatives who felt betrayed by the failure of the commission of 1860 to resolve the land question, and by the limitations of the Fifteen Years Purchase Bill. There was also a double-edged quality to the entry of at least one new assemblyman, which was reminiscent of the league's suspicion of professional politicians: at a meeting on the formerly disturbed township of Lot 65, Callbeck promised that, if elected, he would resign his seat should he and his constituents 'at any time differ on important questions.'[19] Like the Tenant League itself, he apparently believed that he represented the rank and file, not a party establishment, and he was willing to go to unusual lengths to reassure the electors on that point.

On 5 March 1867 the government led by James Pope resigned. A week later, when the Liberals met to form a new administration under George Coles, pro–Tenant League members of their caucus strongly opposed restoration of Whelan as queen's printer. He had been the leading reform journalist for more than two decades, and when the Liberals had been in office in the 1850s he had received the queen's printership, a major patronage post, as a matter of course. But his outspoken opposition to the league had alienated many Liberal voters, who were determined that he should not reap post-election benefits. John Ross had recently revived his newspaper, and he apparently had support among some Liberals. After party deliberations which were protracted, Coles was able to form a government on the 13th, and take office on the 14th. He assumed the Island's colonial secretaryship, which he had held in the 1850s, and since he had accepted an office of emolument under the crown, he had to face a by-election. Whelan was named queen's printer, no doubt in part because, as one anonymous correspondent to the *Islander* put it, if he had been denied, the Liberals were 'well aware of his ability to sever their ranks and drive them into dismay.'[20] Ross, perhaps vexed at not being made queen's printer, put his name forward in opposition, a gesture which delayed Coles's return to the legislature; Coles crushed him, 851–25, a ratio of thirty-four votes to one.[21]

Whelan was not so fortunate in his by-election. Over the previous twenty-one years he had compiled a formidable electoral record, winning every election. In 1863 he had taken his constituency by a margin of eight to one. But at the general election in February of 1867 he experienced a reduction in support which suggested that he was not invincible: with more than nine hundred voting, he led the poll in 2nd Kings by 112.[22] The contestants for the two seats were three Liberals, and therefore conventional partisanship was not a consideration, at least in a straightforward or obvious way, although it is possible that once the small number of Conservative electors in the constituency realized that his defeat was possible, they may have voted strategically at the by-election in order to rid themselves of their long-time nemesis.

But the vital struggle in 2nd Kings was for the votes of Liberal electors, and Whelan was no longer their automatic choice. Having been proved vulnerable in February, he lost the April by-election to Edward Reilly, the editor of the *Herald*, a former employee, a Liberal, and a bitter rival for political leadership within the Irish Roman Catholic population of the Island. In terms of issues, there were several which contributed to the switch in local Liberal preferences – the league, Confederation, Fenianism, and a strained relationship between Whelan and the Catholic clergy – and in each case Reilly represented an alternative posture. It is difficult to determine which question was decisive in Whelan's defeat.

With a majority in the constituency being Catholics, he was convinced that undue clerical influence was the key, whereas W.H. Pope's *Islander* believed that the major cause was his opposition to the league.[23] Whatever the relative weight of the causes, his rejection by people who had been his enthusiastic supporters for two decades was a devastating blow. He died before the year was out, at age forty-three.[24]

The government was clearly divided between pro– and anti–Tenant Leaguers. Within a few days of the opening of the session, Attorney General Joseph Hensley was distancing himself in the assembly from remarks Davies, his fellow-executive councillor, had made concerning the land question and the sentencing of the Tenant League prisoners. He then promptly proceeded to defend Davies's personal history, in effect drawing a distinction between his words and his actions, virtually excusing the former by reference to the latter.[25] It must have seemed confusing to many observers. In the words of land agent and surveyor Henry Jones Cundall, 'the leaguers are thorns in the side of the liberals, yet they cannot afford to dispense with them.'[26] It was a fair analysis. Although Coles had been an opponent of the league, he appears to have recognized a need to reach out to the leaguers. On the Executive Council of nine there were four members – Callbeck, Davies, Haythorne, and Laird – who clearly had been elected with the support of Tenant League sympathizers, and a fifth, the veteran Liberal politician W.W. Lord, who would state in the legislature more than once, without apology, that he had contributed money to the league.[27]

On 16 April, Charles Dickieson received the minor office of commissioner of highways for Queens County road district number 3; it involved disbursing £117 11s. 6d. for road work on Lots 23 and 24.[28] Less than a year earlier he had been in prison because of the Curtisdale affray, and indeed he would still have been there had the lieutenant governor not remitted the remainder of his sentence. When the legislature opened two days later, members of the assembly appointed Robert Gordon, a former justice of the peace who had been fired for his association with the Tenant League, assistant reporter to the house over the objection of Opposition Leader T. Heath Haviland Jr, who attempted to have his name deleted from the relevant resolution.[29] On 30 April, the government named Benjamin Balderston Jr, a teacher who had been secretary of his local branch of the league, on Lot 31, to the major office of registrar of deeds and keeper of plans, although evidence of his active role in the league was part of the public record, in an appendix to the Assembly *Journal* for 1866.[30] He was also a brother of John Balderston, the new legislative councillor and one of the founders of the league. Several other lesser-known league activists received minor appointments in the first few months the Liberals were back in office. Whelan's newspaper, the *Examiner*, was moved to ask the following rhetorical question about the league: 'is that organization still alive?'[31]

The Dickieson and Balderston appointments particularly embarrassed the new government, and in fact they could not be sustained. On 6 May, Haviland attempted to bring a motion on the two before the assembly, but his effort was rebuffed by the speaker, who ruled that adequate notice had not been given; perhaps ironically, the speaker was Joseph Wightman, whom the Tory Executive Council had investigated in 1865 for reported association with the Tenant League.[32] Despite the lack of a resolution, there was a rancorous exchange between Haviland, who compared the influence of the league on the Liberal government to that of the Jacobin Club on the government of France, and Coles, who retorted by reminding the Conservatives that they had continued Balderston's pay as schoolmaster despite his public identification with the league.[33]

The appointments incensed Lieutenant Governor Dundas, and according to a despatch he sent to the Colonial Office, he 'cancelled' Dickieson's appointment, and 'was unable to approve' the nomination of Balderston.[34] The *Examiner* reported that Balderston had withdrawn his claim, and in fact there is support for that interpretation in the Executive Council Minutes.[35] But Dundas related in another, confidential despatch that although the executive councillors had unanimously agreed to the cancellation in the case of Dickieson, with respect to Balderston they had divided after a long discussion, six to three, with the majority favouring the appointment, and Premier Coles in the minority. Only his own adamant refusal to give his approval prevented the choice of Balderston being finalized, the governor wrote.[36]

Dundas did not have the last word. On the basis of a notice in the *Royal Gazette*, it appears that by 7 June Andrew Dickieson was occupying the post of highways commissioner for Queens road district number 3. He was the brother of Charles who had been involved in the attempt to cancel the licence of John Binns to retail spirits in New Glasgow, Lot 23 in 1866, and this substitute appointment would represent a compromise between the two factions of Liberals who had divided over the Tenant League.[37] As for Balderston, in 1868 he would be named reporter to the assembly – which Haviland would find 'extraordinary.'[38] This position, and that of Gordon, as assistant reporter, would not require the approval of the lieutenant governor, as they were in the gift of the majority in the assembly.

These bruising encounters demonstrated that the Tenant League, or at least memories, loyalties, and enmities associated with it, remained a potent factor in Prince Edward Island politics. If political power includes the capacity to reward and punish, the sympathizers with the league had a measure of it. In the opinion of Frederick Brecken, a Conservative assemblyman for Charlottetown, the appointment of Balderston as reporter to the assembly in 1868 indicated that 'a section of the Government wanted some public recognition of the services of at least one prominent member of the Tenant League.'[39] More broadly, the league had a divisive impact on the Liberal party comparable to the effect of the Confederation issue on the Conservatives, creating a schism between the radicals and the 'Old Liberals.'[40]

Impact on Proprietary Policies and Morale

It is possible to argue that although the Tenant League became a major political nuisance for the Liberal party of Prince Edward Island, this had not been its purpose, and that unless it made a difference in the outcome of the land question, or at least the timing of the outcome, the league would have to be classified as a failure. Tenant Leaguers present at the founding convention of 19 May 1864 would certainly have agreed about the content of the crucial question: how was the Island, with respect to land tenure, different because of the league? Or, to use an old metaphor, did it turn the wheel of history?

The argument of this study is that the Tenant League was decisive in ensuring the end of the leasehold system on Prince Edward Island. But it has not received this recognition from historians. Most who have commented on the termination of leasehold tenure have simply noted that the promise of a compulsory land purchase act was included in the terms of Confederation in 1873; they have not made an explicit link with the Tenant League. For example, although in writing on Confederation Francis W.P. Bolger has stated that settlement of the land question was the greatest benefit the Island received from union with Canada, he does not link its resolution to the Tenant League. Aside from the league's contributions to an atmosphere of disorder and to the defeat of the Conservative government in 1867, the only significance he attributes to it is that the financial costs associated with the military presence brought in to quell it were a factor in relations between the anti-Confederate local government and the pro-Confederate imperial government, giving the British another means of applying pressure and indicating displeasure. In effect, Bolger presents the league solely as a sidelight to the story of Confederation, and makes no attempt to estimate its contribution to settlement of the land tenure problem, prior to the commitment, in the terms of union, to legislated abolition. By implication, Bolger dismisses it as inconsequential with respect to the land question, and therefore futile in terms of its own *raison d'être*.[41]

Scholars who have commented on the impact of the Tenant League on the land question, however briefly, have tended to be negative in their assessments. In 1933 Daniel Cobb Harvey, the most eminent Maritime historian of his generation, wrote in an article on Prince Edward Island and Confederation that the activities of the league 'had no other effect than to bring back the military to the island for two years, at a cost of some £10,000.'[42] This was, of course, an opinion offered in passing during a brief study of another topic. There is no suggestion of careful examination of the Tenant League or indeed of the land question in the 1860s. The judgment is nonetheless categorical.

Another distinguished scholar who made a statement on the subject is Andrew Hill Clark, in his *Three Centuries and the Island: A Historical Geography of Settlement and Agriculture in Prince Edward Island, Canada*, published in 1959. Clark's volume stretches from the first European contacts to the 1950s, and he did not pretend to deal with the Tenant League in any depth, but he did remark that it 'may have done more harm than good to the cause of land reform.' His reason for this tentative judgment is apparently that 'The proprietors successfully tagged it with the label "rebellious."' He did not elaborate on the counter-productive impact of the league and, later in the same paragraph, he conceded that its activities led to more sales of proprietary estates 'and an increasing realization by both the Imperial Government and the proprietors that some solution to the problem, satisfactory to the tenants, had to be found.'[43]

If this assessment by Clark of what the Tenant League had accomplished – causing both proprietors and the Colonial Office to lose heart – is accepted as correct, it is difficult to understand how the league could be judged, on balance, to have hurt the cause of land reform, and to have been, by implication, a failure. Indeed, after the crisis surrounding the league had subsided, hostile contemporaries tended to argue that it had, in a general way, disgraced the Island because of its defiance of the law, but did not claim that it had retarded settlement of the land question.[44] Clark's interpretation would appear to be the product either of an assumption that resistance to the rule of law is always self-defeating, or of a confusion of the fate of the league as an organization with the cause of land reform, or possibly both. Had he inquired systematically into the impact of the league in concrete terms, he would almost certainly have concluded that its efforts were not injurious to the cause it espoused.

It is necessary to examine the record in some detail to establish precisely the contribution of the Tenant League to the resolution of the land question. Portions of the story have already been recounted in this study, and need only to be summarized. The league had played a crucial part in two unprecedented mass purchases of land by tenants directly from their proprietors, Haythorne in 1864 and the Reverend James F. Montgomery in 1865. Although varying degrees of camouflage were used, these transactions fitted the program of the league, and leaguers were involved. It is impossible to separate the two purchases from the emergence of the league, and it is highly unlikely they would have occurred without the league; nothing like them had happened before or would happen again. Secondly, at the start of 1865 the Cunard estate announced a reduction in rents on Lots 63 and 64, an area of tenurial insecurity and large arrears of rent where the league had first taken hold. This was a major benefit for the tenants involved, and the timing was not a mere coincidence, for the landlord had had more than one-quarter of a century in which to reduce their rents if that had been his intent. In fact, he did so only after he received a prompt from the league. Thirdly, a few months later Cunard and Laurence Sulivan, two absentee proprietors who owned, between them, one-fifth of the colony, informed the public that they would grant no more new leases, and that all future agreements with settlers would be for the sale of land. This alteration in policy was momentous, virtually a recognition that the leasehold system of Prince Edward Island had no future. All these changes had occurred within a year of the formation of the league, and they grew out of the atmosphere of increasing confrontation of which it was the product, embodiment, and stimulus.

Yet the greatest achievement which can be attributed to the league came in the summer of 1866. The resistance of the landlords had hardened for a period, and some were particularly assertive regarding arrears. But after the events of the summer and autumn of 1865, proprietors apparently began to take stock. Collecting rent in Prince Edward Island was a complex matter, requiring attention to the politics of the Island, as well as lobbying in London, and the political climate was unfriendly, perhaps more so than ever before. On 1 July 1866 the Cunard family holdings, which included land on twenty townships, were sold to the local government for resale to the tenantry. Including the tenants of Edward Cunard, the number of leaseholders who would become freeholders through this transaction probably approximated 1,350.[45] The sale of the huge Cunard estate, comprising more than 15 per cent of the land mass of the Island, represented a decisive turning-point in the struggle over leasehold tenure. For a period it was widely believed that the property of Sulivan, who died in February of 1866, would also be sold. During the session of the legislature that year there was much optimistic talk about 'breaking the neck' of the leasehold system through purchasing both the Cunard and the Sulivan estates. The anticipation was no longer of gradual change; cataclysm was in the air. Finally Sulivan's daughter, Charlotte Antonia, visited the colony in 1867 and made it clear that she intended to remain a proprietor.[46]

Despite the disappointment regarding the Sulivan estate, a vital transformation had occurred. The terms of the debate would never again be the same. It was as though many of the supporters and beneficiaries of the old order knew that its days were numbered. There was at least one significant estate purchase by the Island government each year afterwards through 1871. The vendors tended to be resident proprietors with political connections. They included James Pope, Haviland, and the Palmer family.[47] On 22 February 1871 Haviland, then the Island's colonial secretary, declared in the Legislative Council that

I am not any longer a proprietor. I considered it for the interest of the country, as well as my own, to get rid of the small quantity of land I held; ... When proprietors are obliged to resort to the rigors of the law to obtain their rents, a bad feeling will of course arise. In fact it is altogether contrary to the spirit of the people to remain any longer under the proprietary system. I do not think any party would have the boldness, at the present time, to rise up in the Legislature, or in any public place and say that it should be perpetuated. The hand writing is on the wall and it must go down.[48]

Several weeks later in the assembly, Pope, once again premier, declared that 'there is not a proprietor at present in the House'[49] – quite a change from 1856, when Coles had asserted that one-third of assemblymen were either landlords or land agents.[50] Die-hard proprietors remained in legal possession of their lands, but they had little or no political support on the Island. There were virtually no defenders of leasehold tenure left in the legislature, and when unpopular landlords were criticized, no one rose to explain away their actions.[51] With the purchase of the Cunard estate, the deaths of Laurence Sulivan and Viscount Palmerston, and then the purchases of estates owned by leading local Tories, the proprietors appeared more isolated than ever before when under attack.

The most important estate purchase the Island government ever made and the key indicator for the future in many minds was the Cunard transaction. In appraising the role of the league in Island history, much hangs on how that sale is interpreted. In 1871 James Pope put his unequivocal opinion on the record: 'There never was a greater fabrication than the statement that the [Tenant] League was the means of causing the purchase of the Cunard Estate.'[52] It is understandable why he would take this position. The Cunard purchase was the major achievement of his first term as premier, 1865–7, and it was natural that he would want to receive political credit for it; the fact that this had not happened probably sharpened his views on the subject. Secondly, he made no apologies for his government's actions in suppressing the league and therefore, as a corollary, it seemed, he had no intention of attributing to it beneficial measures.[53] Indeed, one could add that he had particular reasons not to acknowledge the league as a positive force in Island history: it had played a central role in driving his first government from office, and for years in the assembly there remained a deep animosity between him and pro-league assemblymen. He clearly would have liked to drive them from public life, and he was not willing to participate in a posthumous rehabilitation of the league's reputation, especially not while some of its former supporters, not at all penitent, were still a factor in Island politics, and particularly since they were on the other side of the house. As Robert Hodgson put it in a letter to Sir John A. Macdonald in 1873, Pope was 'what Dr. Johnson terms "a good hater" and of a very unforgiving disposition.'[54]

The Cunard purchase was so important both in itself and in precipitating what followed over the ensuing several years that the judgment of one who had political ends in mind and a grudge to satisfy cannot be accepted as definitive. Furthermore, not everyone who had opposed the league in its heyday agreed with Pope. Whelan, who had spent much time and energy decrying its efforts as futile, conceded in October of 1867 that 'the pressure of the Tenant League' had been a factor in the Cunard purchase.[55] The difficulty with attempting to be categorical on this subject is that no known document states exactly why the transaction took place. But there is disinterested scholarly testimony. According to Phyllis R. Blakeley, the biographer of Sir Samuel in the *Dictionary of Canadian Biography*, 'the rents had become difficult to collect, and cash was needed to pay the legacies left to his daughters'[56] – a reference to the directive in the will that £20,000 be paid to each of his six female offspring.

Given that Cunard had decided in 1865 to grant no more new leases, the executors were simply taking his pessimistic analysis of the prospects of leasehold tenure on the Island to its logical conclusion. In fact, the decision to sell is consistent with the trend of development in his policy concerning the estate over the preceding several years. This had been driven by the spectre of rent resistance and social disorder. In the early 1860s, the forces of defiance were more or less formless, yet they had sufficed to make him compromise on the rights of property. With the foundation of the Tenant League in 1864, those forces took clear and menacing shape. What had seemed to amount to a nuisance, albeit an embarrassing one, had suddenly a strength and a vitality that could be checked only by the use of soldiers. The emergence of the league, with its widespread support, must have confirmed fears that tenants in the colony would never accept landlordism as permanent. Hence the Cunard family, treading the path of retreat the patriarch had blazed, decided to sell the Island property in order to assist in raising the cash necessary to honour the provisions of his will.

Sir Samuel had left many other assets to which the executors could have turned, but the land on Prince Edward Island had become a questionable blessing, and with such formidable allies as Sulivan and Palmerston no longer part of the political establishment in London, the time had come to salvage what could be saved, and to abandon the notion of transplanting a neo-feudal system of land tenure in the New World. The Cunard decision to sell should be understood as a surrender, an acknowledgment that the sentiment in favour of freehold could no longer be resisted; and the cutting edge of that current of opinion in the crucial years of the middle 1860s had been the Tenant League. If there had been no such organization in the 1860s, or if its early supporters had opted for the policies of the Liberals, that is, reliance on parliamentary means, then there is no reason to believe that the Prince Edward Island property would have been singled out among the assets of Sir Samuel and sold when it was.

In the eyes of contemporaries, including other proprietors, the Cunard purchase was a pivotal moment in the long history of the land question.[57] Once the Cunards had washed their hands of the system, then the politically astute landlords residing in Prince Edward Island – Haviland, the Palmers, Pope – began to make their bargains before facing new, compulsory legislation. The Tenant League had played a determining role in bringing the leasehold system to its severely weakened position as the 1860s ended. Perhaps reflecting the frustration of someone who had campaigned for decades for resolution of the land question the orderly, constitutional way, through legislation, Whelan wrote during the summer of 1867 that 'It is not improbable that if the purchase of the Cunard Estates had been completed in 1865, instead of 1866, there would have been no Tenant League "riots."'[58] But no agreement to purchase had been reached in 1865; the 'riots' had occurred; and only then was there an agreement to purchase. The sequence was more than coincidence. The time had come to sell. But it had come because of rent resistance by the tenantry, in the form of the Tenant League.

The 1867 election was the last time the land question was a significant divisive factor in Prince Edward Island politics. The necessity of resolving it had become part of the political consensus, and the major portion of credit belonged to the Tenant Leaguers, for their agitation and resistance had helped to persuade the Tories that the cost of maintaining and defending the system was too great both politically and in terms of the negative impact of disorders and repression. Under the terms of union with Canada in 1873, negotiated in part by a Liberal government led by Haythorne and in part by a Conservative government led by Pope, the final demise of the leasehold system was arranged. But the critical years in undermining it were the middle 1860s, when the league flourished, and the league itself was the key. In 1874 in the Legislative Council Haythorne would give some revealing testimony on this point. He recalled that 'In 1868 the [Liberal] Government [of which he was a member] was in communication with the Proprietors and some of them were a little more reasonable than they are now. They had the fear of the tenant league before their eyes at that time.'[59]

The Tenant League in Perspective

On Friday, 19 October 1866, John Francis Maguire, a leading Irish Member of Parliament at Westminster and the founding proprietor of the Cork *Examiner*, the most important Irish nationalist and Roman Catholic newspaper outside Dublin, arrived in Charlottetown. His purpose was to gather material for a book on the Irish in North America, and on the following day he travelled close to sixty miles in the course of visiting two 'exclusively' Irish settlements in rural Queens County; one was Monaghan on Lot 36, for which Deputy Sheriff James Curtis had set out on 10 March 1865. Maguire took a particular interest in Catholic institutions and, as a layman who had written a book which was regarded as the standard contemporary English-language defence of the papacy, on Sunday he very probably attended church. On the evening of Monday the 22nd at the North American Hotel in Charlottetown there was a banquet in his honour which continued until 11 p.m.; he then went directly to the boat, which sailed away that night.[60]

Although on the Island for only four days, Maguire formed some strong impressions, and he related them in the second chapter of his book, *The Irish in America*, published in 1868. He was apparently surprised to find that the politics of the colony revolved around the land question, and that the organization which had been pressing the issue recently was called the Tenant League. This was a name with reverberations in his personal history, for he had been the last leader of the parliamentary group of the defunct Irish Tenant League. Historian E.D. Steele, the authority on the Irish 'tenant right' movement of the late 1860s, describes him as being by this time 'a comparative moderate in agrarian politics.'[61] Maguire noted that the demands of the Island Tenant League went far beyond the Irish organization's advocacy of fixity of tenure, that is, continued possession of the land by the tenant who paid his rent and observed the covenants in his lease. The Tenant Leaguers of Prince Edward Island, of course, demanded an end to the payment of rent, and the right to purchase the land they worked. Maguire stated that, given the nature of the demands, he was further surprised to learn that the rents were as low as they were, since he had assumed that for such a radical program to take root the rents must be onerous. He reported in an air of wonderment that there was a consensus on both sides in local politics that the leasehold system must be wound down. In his words, the Fifteen Years Purchase Bill which the Island Conservatives had passed in 1864 'would be regarded in the British House of Commons as a measure of sweeping confiscation worthy of the French Revolution, or the days of Jack Cade.' Maguire clearly believed that there were lessons in the Island experience for British 'statesmen who recoil with dismay from the least invasion of the "rights of property."'[62] He would be consulted extensively by Prime Minister William Ewart Gladstone when the latter was framing his Irish land reform legislation of 1870.[63]

Maguire's account regarding the land question on the Island was more or less confined to a description of what he had found, some of it presented in the form of dialogue, letting the reader draw the obvious conclusions. During the autumn of the year previous to his visit, another writer from the United Kingdom had visited the Island, and had been drawn to the land question also. He was Charles Mackay, a Scottish journalist, lecturer, song-writer, and poet who published 'A Week in Prince Edward Island' in the *Fortnightly Review* of London in June of 1866. At the time of his visit, he was based in New York, where he had been covering the American Civil War as special correspondent for *The Times* of London. By his own report, his intent had been to make a general tour of the Island and to study the working of representative institutions in the smallest of the British North American colonies. He found that the land question dominated conversation on the Island when he was there, and he gave a fairly substantial account of the Tenant League and its issues. He stayed for eight days at Ardgowan, the home of W.H. Pope, about a mile and a half outside of Charlottetown, and his perspective reflected that fact; indeed, he did not venture outside 'the immediate neighbourhood' of the capital on account of the wet weather and consequent state of the roads, which were 'ankle-deep in brick-coloured and very adhesive slush.'[64] Aside from the matter of bias against the league, the most significantly misleading weakness in Mackay's version is his failure to mention the fact that some landlords refused to sell under any conditions; he treated purchase as an available option which Tenant Leaguers failed to pursue, whereas Maguire recognized that the league's 'object ... was not so much to abolish the payment of rent, as to compel the proprietors to sell their estates on fair terms.'[65]

Mackay wanted to see Islanders punished for their defiance of the law, but he was not sure what would suit the circumstances of the case best: revocation of responsible government, annexation to Nova Scotia, or simply handing the Island government a steep bill for military costs. Perhaps in part because his views fitted so well with those of Pope, the latter's newspaper, the *Islander*, reprinted his article on 3 August 1866.[66] Mackay did recognize, though, that the land commissioners of 1860 had pleased no one with their attempt at compromise, and that there was a North American factor which was a powerful force. He informed his readers that

It should ... be taken into account in England, that except in great cities, where rent is paid for houses, rent is all but unknown in America. Land is so cheap, that any hard-working thrifty man can speedily obtain as much of it as he can cultivate. To this must be added the feeling, widely spread in democratic America and the equally democratic British colonies – that rent is an aristocratic institution, unsuited to the New World, where every man ought to till his own ground, and sit 'under the shadow of his own fig-tree.'

Prince Edward Islanders lived within that North American context, where the payment of rent for the land one worked seemed unnatural and degrading, and consciousness of the hegemony of freehold tenure everywhere else nearby had been a major factor in resistance to the leasehold system in the 1860s. In Mackay's words, 'tenants all over the island made up their minds half a century ago that ... the exaction of rent was an act of oppression which they were bound to resist.'[67]

There was, to coin a term, a 'freeholder ideology' whose potency could not be denied. Sir Andrew Macphail, the Canadian man of letters, who was born on the former Selkirk estate in the year the Tenant League was founded, stated succinctly the central article of faith in this ideology: 'A man who lives on his own land and owes no man anything develops all the dignity inherent in his nature.'[68] The term freeholder, rather than 'yeoman,' is significant because Islanders developed their perspective in specific opposition to the leasehold system.[69] They believed that leasehold tenure acted as a stimulus to out-migration, and as an obstacle to agricultural improvement. In fact, the 1861 census had revealed that on Lots 19, 34, and 49, three of the best-situated townships on the Island, net population loss had begun; on each, a majority of land occupiers were tenants.[70] Furthermore, during the decade of the 1870s, with the winding down of the leasehold system, the rate of land clearing would almost double.[71] It is impossible to disentangle cause and effect with precision in the latter instance: to determine whether the increased rate of improvement could be attributed to the decline of leasehold tenure as a system of land-holding, to the freeholder ideology referred to above, to other causes, or to some combination. But it is probable that the ideology which had been so important in building support for the Tenant League also contributed to optimism about the future once the league had done its work. This is an example of the power of ideas in history.

There are a number of frameworks within which the Tenant League as an historical phenomenon has to be understood. Most obviously, these are the Prince Edward Island setting itself, the broader British North American context, and the past and contemporaneous agrarian issues in the British Isles, from which the vast majority of the settlers or their parents had come. Both Prince Edward Island and the parallels in the British Isles are undoubtedly important for an appreciation of the significance of the league.

But undue focus on the precedents in the United Kingdom could be misleading, and intense awareness of Irish history certainly led astray a major actor in this story, Whelan, on one key point. He opposed the Tenant League almost from the start. For some, his motives were suspect, since he was the chief Liberal journalist in the colony, and his constituency was slipping away; furthermore, over the years the young radical had become more conservative, and more appreciative of the advantages of the liberal constitutionalism embedded in the British political tradition. In part, he argued, the league did not deserve to win. He was firmly attached to constitutional methods, and in his view the Tenant Leaguers had not exhausted all the constitutional avenues open to them; thus their rejection of conventional politics was untenable and their resistance to the law was immoral. Yet his overweening objection to the Tenant League, which he emphasized over and over again, perhaps because he thought this was the argument which would have most impact with militants, was that it could not succeed. Their defiance of the law would provoke repression which would ruin many followers of a leadership which, he was convinced, was either naive or opportunistic or incompetent – or two or all three of these. Many poor people might lose everything in the course of accomplishing nothing. Therefore, he believed, it had been 'one of those situations in which a middle course is not admissable [sic]. .... The plain duty of every lover of his country, ... was not only to hold himself aloof from any direct complicity with the men of the Tenant Union, but also actively to oppose the spread of their principles.'[72]

Whelan reckoned without taking sufficient account of an achievement to which he had contributed mightily in the 1840s and 1850s: responsible government. With a local executive council responsible to assemblymen elected on a franchise which was virtually universal for adult males, the authorities would tread as lightly as they possibly could in their use of the troops. They would have to account to the electors for their behaviour, and these included the tenantry and their allies. This was not Ireland without Home Rule, and with Irish parliamentarians always greatly outnumbered at Westminster. Responsible government made an enormous difference, one which Maguire hinted at when he referred to the Island as having 'a system of government based upon popular suffrage and amenable to popular control.'[73] Prince Edward

Islanders had their own House of Assembly which determined who sat in government offices in Charlottetown, and those government members knew they were electorally answerable to the mass of Islanders. The Tenant League ruined few tenants but did undermine landlordism.

In the end, the most violent single event was the attack on Deputy Sheriff Curtis at Curtisdale in July of 1865.[74] He sustained a broken arm, but was back on duty within weeks. That Islanders came so close to serious bloodshed and yet did not slip over the precipice is a tribute to their good sense and also a testimony to their respect for the same traditions and institutions as Whelan. They were not out to make a political revolution. They wanted freehold tenure, like everyone else in North America. Mackay referred in the *Fortnightly Review* to Tenant Leaguers congregating at the sound of a tin trumpet 'sometimes ... in armed gangs of 100 or 200 at a time.'[75] Technically, or literally, this was correct, but there was a ritualistic quality to these appearances and to the display of arms. No one on either side during the Tenant League disturbances appears at any time to have fired a weapon with the intent of hitting someone.

There was a sense of limits to be observed. Historians have documented similar ingredients in instances of riot or protest elsewhere. Scott W. See has noted a ritualistic aspect of rioting in colonial New Brunswick, where groups who were heavily armed exercised restraint in the actual use of their weapons, limiting themselves to the least lethal tools in their arsenals, such as rocks, sticks, and their fists, and simply displaying revolvers, cannons, pitchforks, and 'treenails' (hard wooden pins for securing timber).[76] In Eric Richards's work on the history of protest in Highland Scotland, he quotes an estate factor as writing in 1853 that 'I am persuaded the people think they may do anything they please short of destroying life.'[77]

Likewise, in Prince Edward Island, dropping everything and fashioning an armed demonstration in the path of a sheriff was making an important point about the degree of anger in rural districts and the loss of legitimacy of leasehold tenure. Throwing a stone, a stick, or a piece of earth, or even swinging a stick at him served to add force to the point. There was no need to shoot the sheriff, and moreover such a deed would almost certainly provoke serious repression and alienate public opinion in the colony and elsewhere. Self-restraint was essential to the effectiveness of league tactics. The contrast with what happened in eastern New York in 1845 is significant: anti-rent activists shot to death a deputy sheriff, with the result that public opinion was shocked, and their movement divided and weakened.[78]

The issues in Prince Edward Island resembled more closely those of the Irish land question than any other obviously parallel situation. But as Maguire had noted, the attitudes were different. He did not emphasize the fact, but the circumstances of the Island tenantry were also quite dissimilar to those of the Irish tenantry. In the Old World, the tenant typically rented a fully developed farm, ready to be worked. If the tenancy was not renewed and he had to leave, that might be unsatisfactory, but it was radically different from the plight of many Prince Edward Island settlers. The vast majority of farms in the colony had been carved out of the wilderness within living memory. In many instances in the 1860s the farms were at least partially the creations of the resident farmers or their parents or grandparents, and this gave a visceral quality to the issue that can hardly be overstated.[79] In their readiness to resort to illegal methods and violence, the farmers of Prince Edward Island fell somewhere between the behaviour of the Highland Scots of the Clearance period, and that of Irish tenants in the nineteenth century. Given the depth of feeling, the restraint was remarkable, yet there were examples of the staples of 'agrarian outrage' in Ireland: incendiarism, the threatening letter, and animal maiming.

In addition to the Old World context which informed Tenant Leaguers, it must be remembered that the league arose, flourished, and declined at precisely the same time as the Confederation movement was dominating the political agenda in British North America. There was a tangential relationship between the two phenomena, for what each accomplished or did not accomplish had an impact on the other. The failure of the Confederation conferences to address the land question would have appeared to Prince Edward Island tenants to be one more piece of evidence that they would have to take matters into their own hands, and that no one was going to solve the problem of leasehold tenure for them – not land commissioners from the other Maritime colonies, not the Colonial Office, not local politicians, and now, not Canadian politicians.[80] They would have to do it themselves, and the Tenant League was their chosen vehicle. That was how matters stood in 1864 and 1865; but once the Cunard estate had been purchased in 1866, and once some other proprietors deserted the system over the succeeding several years, it was clear that leasehold tenure in the colony was crumbling, and that it was doing so at that time in large measure because of the impact the league, with its calculated defiance of the law and of the political establishment, had had on the foundations of its support.

The momentum had shifted against the leasehold system permanently, and there was a perception that its demise was inevitable. The non-partisan consensus and the sense of accomplishment that developed on this issue had the effect of removing the land question as the possible source of a decisive lure for entry into Confederation. Those uncommitted to union of the colonies seemed to believe that they could resolve the problem on their own, sooner or later, without the Dominion of Canada. Thus the lack of sympathy of the delegates from other colonies in 1864 left the field open for the Tenant League to play the crucial role in breaking the strength of the leasehold system, and the success of the league in doing precisely that meant that pro-Confederates on the Island would have to find other means to draw the colony towards Canada. In the end, this turned out to be a railway whose construction costs would cause a financial crisis which made entry into Confederation seem necessary.[81]

The Tenant League should also be considered within the context of popular movements in Canadian history as a whole. To begin with, both the Escheat movement of the 1830s and 1840s and the Tenant League of the 1860s deserve recognition as the first manifestations of a tradition of organized agrarian protest and revolt which represented, in numerical and political terms, the most important constituency of progressive forces in first British North America and then the Dominion of Canada until at least the First World War. The farmers of Prince Edward Island were the forerunners of the many agrarian movements which flourished in other regions late in the nineteenth century and in the early part of the twentieth. The leaguers' intense suspicion of party politicians, whom they excluded from their founding convention, is an obvious feature they share with later movements. To adopt the unfriendly phrase of the lieutenant governor, 'men of obscure position' – ordinary people – were coming to the fore.[82]

As well as being, with the Escheat movement, first within the tradition of organized agrarian revolt in Canada, the Tenant League was, much more than Escheat, remarkably successful in advancing its cause. Indeed, within Canadian history it is difficult to name another extra-parliamentary movement for radical change, willing to resort to civil disobedience, which was as successful as the league; and within North American history as a whole only the African-American desegregation movement of the 1950s and 1960s comes to mind.[83] In Canada this is a tradition which includes the Winnipeg General Strike of 1919, a demonstration of solidarity among the majority of the labouring population in that city which is unique in North American labour history for its combination of comprehensiveness and duration. Yet, despite the powerful display of support for a cause, the degree of its success is a matter on which there is a wide range of opinion among historians, much depending upon how long a temporal perspective one adopts.

Like the Winnipeg General Strike, the Tenant League had the support of an extraordinary proportion of the producing class in the society of which it was a part. In a population numbering between eighty and ninety thousand, even its enemies apparently accepted that eleven thousand was a realistic estimate of its membership, a degree of support which is phenomenal. This breadth of backing explains the intelligence system which assisted in warning the leaguers when the sheriff was on his way; and after the soldiers arrived, the depth of the league's support enabled it to work successfully to induce desertion, in the hope of provoking the Halifax authorities to withdraw the men. The real strength of the league was in that grassroots support, but there must also have been extensive coordination in order for it to spirit out of the colony as many soldiers as it did without being detected. The organizational capacity of the leadership would have been crucial in arranging the escapes of deserting soldiers from the colony; yet it seems probable that the leaders may have disappointed the local militants by their apparent invisibility once the troops went into action in support of the sheriff in October and November of 1865.

The history of the Tenant League has further significance within broader Canadian history in terms of resort to military aid for the civil power. The use of military assistance was a serious matter in this case, probably the most important, in terms of the challenge, between the Rebellions of the 1830s in the Canadas and the disturbances in the West after Confederation. This was not insurrection, to be sure, but it was defiance of the civil authorities on basic matters concerning the functioning of society and property relations under the British flag. Unlike situations where the military was used to deal with labour conflicts at Lachine in Lower Canada, election-related violence in Newfoundland, the Gavazzi riots in Quebec City and Montreal, or disorderly lumberers in the Miramichi region of New Brunswick, this was not an episode provoked by the special circumstances of violent quarrelling over construction jobs, by a specific electoral contest, by an inflammatory public speaker, or by liquor. Since rent and arrears of rent were legally collectible debts, a landlord faced with non-payment could obtain the assistance of the law. Supporting those who failed to pay could lead, logically, to involvement in resisting the sheriff and his assistants. The eleven thousand cited above had taken a pledge to refuse payment of rent or to support those who did so, as the case demanded, and by the summer and autumn of 1865 hundreds were proving ready to act on that pledge whenever they heard the blast of a tin trumpet. In the words of historian Charles Townshend regarding a parallel situation in Ireland, 'It was the derogation of law, the unenforceability of the royal writ, which constituted the gage which the government could not avoid picking up. The direct personal peril to land agents and process-servers meant, in abstract political terms, the breakdown of law and order.'[84] The challenge to civil authority on Prince Edward Island in 1865 was persistent, systematic, and fundamental, and it appeared to be escalating in audacity as well. The state could not ignore it.

Yet the league did not resort to insurrection. Despite its willingness to flout conventional values through such measures as promoting desertion from the military, there is no evidence to suggest that it ever thought of an armed uprising as a possibility. Here it parts company with the Rebellion of 1837 in Upper Canada, an occurrence also rooted in widely shared popular grievances, particularly concerning land, which is much more familiar to students of Canadian history. The rebellion was attended by tragic costs for many people, and in the end those costs may be its greatest significance. They are a reminder that, as well as a degree of personal courage and some dramatic interest, it represents blundering miscalculation and foolhardiness, so much so that it might never have happened if the leadership had had a more realistic understanding of what was involved.[85]

In sophistication and success, without the price of much suffering, the Tenant League compares favourably with the Upper Canadian Rebellion. Those who mapped its strategy had a sense of what was possible, and what had to be avoided in order to contain the sacrifices by individuals within acceptable limits. In 1986 the grandson of the Tenant Leaguer who received the greatest legal penalty for actions associated with the league would remember that his grandfather had never rued the personal price he had paid; and he had remained on his farm, to be succeeded there by three generations of descendants who would work the same land over the 125 years following his arrest.[86] In the case of the Tenant League, the point is not the disproportionate costs in terms of what was accomplished, but *vice versa*.

To date, the Tenant League has not received even token mention in the scholarly histories of Canadian popular movements, which focus overwhelmingly on central and western Canada.[87] Those interested in such phenomena who examine the development, success, tactical boldness, and discipline of the Tenant Leaguers will find that few movements of the common people in Canadian history succeeded so quickly and impressively against entrenched economic interests and political establishments. The Tenant Union of Prince Edward Island was more than a curiosity of social history, a unique product of the peculiar leasehold system of land tenure bequeathed to the colony by the land lottery of 1767. It successfully broke the mould which a great imperial power had imposed, and changed the direction of the history of the society of which it was a part.

Appendix 1:

Tenurial Insecurity Index

This is an index of the tenurial insecurity per land occupier on particular townships. It represents an attempt to measure insecurity attributable to tenurial status as indicated in the 1861 census, and does not take into account the factor of rent arrears, concerning which there are no broad-based data available. The following values have been assigned:

Freeholder – 0
Tenant with lease or agreement = 1
   Additional value for tenant with 'short lease':
   100–anything less than 999 years = 0.5 (1.5)
   50–99 years = 0.75 (1.75)
   30–49 years = 1.25 (2.25)
   under 30 years = 1.75 (2.75)
Tenant at will (verbal agreement) = 3
Squatter = 5

The Index is arrived at by calculating the total Tenurial Insecurity values for a township and dividing by the number of occupiers of land.

Appendix 2:

Significant Proprietors of Prince Edward Island, by Township, *ca.* 1864

Prince County

Lot 1: Edward Cunard, Palmer family
Lot 2: Sir Samuel Cunard
Lot 3: predominantly freehold tenure and squatting
Lot 4: Edward Cunard
Lot 5: Edward Cunard
Lot 6: Edward Cunard
Lot 7: Robert Bruce Stewart
Lot 8: predominantly freehold tenure
Lot 9: Laurence Sulivan
Lot 10: Robert Bruce Stewart
Lot 11: predominantly freehold tenure, formerly Walsh estate
Lot 12: Robert Bruce Stewart
Lot 13: James Yeo Sr
Lot 14: Sir Samuel Cunard
Lot 15: predominantly freehold tenure (escheated lands)
Lot 16: Laurence Sulivan
Lot 17: predominantly freehold tenure
Lot 18: several small proprietors
Lot 19: Andrew T. Todd
Lot 25: predominantly freehold tenure
Lot 26: Holland and Thompson families
Lot 27: Sir Samuel Cunard, James C. Pope, Robert Bruce Stewart
Lot 28: Holland and Irving families

Western Queens

Lot 20: Sir Samuel Cunard, Cundall family
Lot 21: Sir Samuel Cunard
Lot 22: Laurence Sulivan
Lot 23: Rennie family, Daniel Hodgson
Lot 24: Winsloe family
Lot 29: Viscount Melville, Lady Georgiana Fane
Lot 30: Robert Bruce Stewart
Lot 31: Douse family
Lot 32: Sir Samuel Cunard
Lot 65: Colonel B.H. Cumberland, Wright family, Sir Samuel Cunard
Lot 67: Lady Louisa Wood

Eastern Queens

Lot 33: Winsloe family
Lot 34: Montgomery family
Lot 35: McDonald family
Lot 36: McDonald family
Lot 37: John Roach Bourke and others
Lot 48: several small proprietors
Lot 49: Sir Samuel Cunard, Robert Poore Haythorne
Lot 50: Lady Louisa Wood, Maria Matilda Fanning
Lots 57, 58, 60, 62: predominantly freehold tenure, formerly Selkirk estate

Kings County

Lots 38, 39, 41, 42: predominantly freehold tenure, formerly Worrell estate
Lot 40: various
Lot 43: T.H. Haviland Sr, Townshend family, but predominantly freehold
        tenure
Lot 44: Sir Samuel Cunard
Lot 45: Sir Samuel Cunard, but predominantly freehold tenure and squatting
Lot 46: Sir Samuel Cunard, but predominantly freehold tenure and squatting
Lot 47: Robert Bruce Stewart, but predominantly freehold tenure
Lot 51: Montgomery family
Lot 52: predominantly freehold tenure and squatting
Lot 53: Viscount Melville, Lady Georgiana Fane, but predominantly squatting
Lot 54: predominantly freehold tenure and squatting

Lot 55: predominantly freehold tenure (escheated lands)
Lot 56: T.H. Haviland Sr
Lot 59: Montgomery family
Lot 61: Laurence Sulivan
Lot 63: Sir Samuel Cunard
Lot 64: Sir Samuel Cunard
Lot 66: predominantly squatting

Appendix 3:

The Fifteen Years Purchase Bill and the Purchase of the Estate of the Reverend James F. Montgomery

The relevance of the Fifteen Years Purchase Bill is moot. Unlike his relatives who owned the other two-thirds of Lot 34, the Reverend James F. Montgomery had not been one of the twelve consenting proprietors enumerated in the attached 'Schedule (A).' In fact, on 28 March 1865 in the Legislative Council Edward Palmer (coincidentally, one of the twelve himself) had presented a petition 'Of divers Tenants settled on the south part of Township No. 34, the property of the Rev. James F. Montgomery, praying that an Act may be passed to enable the tenantry of this Island to avail themselves of the provisions of the Fifteen Years' Purchase Bill of the last Session.'[1] One incidental function of the legislation, therefore, seems to have been to set a benchmark price for land to which it did *not* apply in law: a minimum in the mind of Montgomery, an acceptable figure in the minds of his tenants.

In any event, an adjusted figure for the savings which Montgomery's terms offered the tenants over the terms of the Fifteen Years Purchase Bill would take into account the arrears that accumulated after 1 May 1858 (and therefore would be collectible under the terms of the Fifteen Years Purchase Bill) and also the expenses. According to Montgomery, fifteen years' rent alone totalled £1,146 more than the purchase price he was willing to accept. There is no known figure for the post–1 May 1858 arrears on the estate, but it is possible to make informed estimates. The total arrears owing in 1861 had been £961 2s. 1½d., and by 1864 the total had grown to £1,253 6s. 9d.[2] This means an increase of £292 4s. 7½d. in three years, for an annual rate of increase of approximately £97. But in 1875 R.P. Haythorne, who was privy to the negotiations, estimated the arrears due at time of purchase to have been £1,200, which would mean a more modest annual rate of increase of £48 between 1861 and 1866. Assuming these figures to provide the minimum and maximum rates of increase in arrears, the accumulation between 1858 and 1866 would be between £384 (8 × 48) and £776 (8 × 97). The expenses of the transaction can be estimated at £150 7s., based on Haythorne's report of a difference of 6d. per acre between the rates the tenant paid and the proprietor received.[3] In other words, the arrears Montgomery forgave far offset the expenses he passed on to the purchasing tenants, and his terms represented a saving to the tenants (as compared with the terms of the Fifteen Years Purchase Bill) of at least £1,380,[4] and possibly as much as £1,772.[5]

Notes

Introduction

1 For the locations of these incidents, see map 3.
2 David Maldwyn Ellis, *Landlords and Farmers in the Hudson-Mohawk Region 1790–1850* (New York, 1967 ed. [orig. publ. 1946]), 3–4, 14, 225–67, especially 238, 242, 248–9; Henry Christman, *Tin Horns and Calico: A Decisive Episode in the Emergence of Democracy* (Cornwallville, NY, 1975 ed. [orig. publ. 1945]), 24–6, 28, 41, 74, 81, 90, 96, 111, 132. Both these books are quite partisan: Christman adopts a simplistic Good vs. Evil approach, always giving the benefit of the doubt to the anti-renters; Ellis maintains a more dispassionate tone, and provides footnote references, but insists that many leaseholders should be considered freeholders because of the liberality of the terms of

their leases, in spite of the fact that the landlords retained the right to distrain the tenants' property for non-fulfilment of any of the terms (227–8). For Ellis's critique of Christman's approach, see 226n1. The anti-rent movement in New York received coverage in the newspaper press of Prince Edward Island, where it was often referred to as 'the patroon war,' after the term for landlords of Dutch origin; see, for example, *Royal Gazette*, 28 Jan. 1840.

3 Precision about national origins is impossible because the census recorded birthplaces, not ethnicity, and by mid-century the vast majority of Islanders were native-born. Throughout the remainder of this study, the term 'Scots' when used in a Prince Edward Island context will refer to Islanders of Scottish ancestry regardless of birthplace, 'Irish' to those of Irish ancestry, etc., unless the passage makes it clear that the intent is to refer to persons in the United Kingdom.

4 Andrew Hill Clark, *Three Centuries and the Island: A Historical Geography of Settlement and Agriculture in Prince Edward Island, Canada* (Toronto, 1959), 91

5 See James Hunter, *The Making of the Crofting Community* (Edinburgh, 1976); Willie Orr, *Deer Forests, Landlords and Crofters: The West Highlands in Victorian and Edwardian Times* (Edinburgh, 1982); T.M. Devine, *The Great Highland Famine: Hunger, Emigration and the Scottish Highlands in the Nineteenth Century* (Edinburgh, 1988); T.M. Devine, *Clanship to Crofters' War: The Social Transformation of the Scottish Highlands* (Manchester, 1994); Charles W.J. Withers, *Gaelic Scotland: The Transformation of a Culture Region* (London, 1988); and the work of Eric Richards: 'How Tame Were the Highlanders during the Clearances?,' *Scottish Studies*, 17/1 (1973), 35–50; 'Patterns of Highland Discontent, 1790–1860,' in Roland Quinault and John Stevenson, eds., *Popular Protest and Public Order: Six Studies in British History 1790–1920* (London, 1974), 75–114; *A History of the Highland Clearances*, vol. 1: *Agrarian Transformation and the Evictions 1746–1886* (London, 1982); *A History of the Highland Clearances*, vol 2: *Emigration, Protest, Reasons* (London, 1985).

6 The literature on the Irish land question is substantial, but see particularly Michael J. Winstanley, *Ireland and the Land Question 1800–1922* (London, 1984); Samuel Clark, 'Landlord Domination in Nineteenth-Century Ireland,' UNESCO *Yearbook on Peace and Conflict Studies 1986* (Paris, 1988), 5–29; and the work of W.E. Vaughan: 'Landlord and Tenant Relations in Ireland between the Famine and the Land War, 1850–78,' in L.M. Cullen and T.C. Smout, eds., *Comparative Aspects of Scottish and Irish Economic and Social History 1600–1900* (Edinburgh, [1977]), 216–26; *Landlords and Tenants in Ireland 1848–1904* ([Dublin], 1984); and *Landlords and Tenants in Mid-Victorian Ireland* (Oxford, 1994).

7 On the origin of the Ulster custom of tenant right, see W.H. Crawford, 'Land-

lord-Tenant Relations in Ulster 1609–1820,' *Irish Economic and Social History*, 2 (1975), 8–12.

8 Vaughan, *Landlords and Tenants in Ireland 1848–1904*, 17

9 In this study, the terms 'Old World' and 'New World' are used exclusively as metaphors regarding the relationship, including contrasts in sensibility, between Europe collectively, including the British Isles, and settler societies in North America, and do not involve any assumption about aboriginal societies in North America.

10 See David Weale, 'The Gloomy Forest,' *Island Magazine*, no. 13 (Spring–Summer 1983), 9, 13.

11 Among recent one- and two-volume general histories of Canada, only one mentions the Tenant League: see Margaret Conrad, Alvin Finkel, and Cornelius Jaenen, *History of the Canadian Peoples*, vol. 1: *Beginnings to 1867* (Toronto, 1993), 591; Finkel and Conrad, with Veronica Strong-Boag, *History of the Canadian Peoples*, vol. 2: *1867 to the Present* (Toronto, 1993), 39.

12 As historian John S. Moir has pointed out: 'the "rebellion of 1837" in Upper Canada *started* as a peaceful show of force like the Tenant League action ... but ... the "rebels" in that tragic-comic uprising lacked the degree of sophisticated organization that was developed in PEI' (Moir to author, 15 Sept. 1994). See Colin Read and Ronald J. Stagg, 'Introduction' to *The Rebellion of 1837 in Upper Canada: A Collection of Documents* (Toronto, 1985); Colin Read, *The Rebellion of 1837 in Upper Canada*, Canadian Historical Association Booklet no. 46 (Ottawa, 1988).

1: Social and Political Background

1 Letterbooks of Palmerston, Add. MSS. 48579, British Library, London (original emphasis). I am grateful to Professor David R. Raynor of the University of Ottawa and Professor Arthur N. Sheps of Scarborough College, University of Toronto, for their assistance in providing references to the Palmerston Letterbooks.

2 In Ian Ross Robertson, ed., *The Prince Edward Island Land Commission of 1860* (Fredericton, 1988), 195

3 MS 62, Palmerston Papers, LB/1, folio 99, University of Southampton Library, Southampton, England. I am grateful to Dr C.M. Woolgar, Archivist and Head of Special Collections, at the University of Southampton Library, Southampton, England for providing a transcription of this passage.

4 The census listed three categories of occupiers of land (below, *i–iii*) who were not freeholders. In the text I am including as 'tenants' the following: those classified as holding land (*i*) 'Under Lease or agreement for Lease' (the latter

meaning a *written* agreement to lease), and those holding (*ii*) 'By verbal agreement' (also known popularly as 'tenants at will'); and I am designating as 'squatters' those listed in the census as (*iii*) 'Occupants being neither free-holders nor tenants.' Of these three census categories, the first was by far the largest: 5,357 of 6,813, or 78.6 per cent. Calculations are based on corrected totals of the data in PEI, House of Assembly, *Journal*, 1862, app. A.

5 Calculations based on PEI, Assembly, *Journal*, 1862, app. A, and 1872, app. I. The 1881 census, following legal abolition as a consequence of the Island's entry into the Dominion of Canada in 1873, does provide statistics on the proportion of landownership by numbers of persons. By then, more than 93 per cent of occupiers of land were owners; Andrew Hill Clark, *Three Centuries and the Island: A Historical Geography of Settlement and Agriculture in Prince Edward Island, Canada* (Toronto, 1959), 132.

6 For a critical survey of the published historical literature on the land question see Ian Ross Robertson, 'The Maritime Colonies, 1784 to Confederation,' in M. Brook Taylor, ed., *Canadian History: A Reader's Guide*, vol. 1: *Beginnings to Confederation* (Toronto, 1994), 269–73.

7 Calculation based on Clark, *Three Centuries and the Island*, 261

8 J.M. Bumsted, 'Sir James Montgomery and Prince Edward Island, 1767–1803,' *Acadiensis*, 7/2 (Spring 1978), 76–102

9 Frank MacKinnon, *The Government of Prince Edward Island* (Toronto, 1951), 12

10 See Harry Baglole, 'Walter Patterson,' *DCB*, vol. 4 (Toronto, 1979), 608–11; and these works by J.M. Bumsted: 'The Origins of the Land Question on Prince Edward Island, 1767–1805,' *Acadiensis*, 11/1 (Autumn 1981), 50–2; 'The Patterson Regime and the Impact of the American Revolution on the Island of St. John, 1775–1786,' *Acadiensis*, 13/1 (Autumn 1983), 58–67; *Land, Settlement, and Politics on Eighteenth-Century Prince Edward Island* (Kingston and Montreal, 1987), 83–97, 195.

11 Seymour to Joseph Sidney Dealey, 2 June 1840, Seymour of Ragley Papers, CR114A/508/4, Letterbooks, WCRO. Also see H.T. Holman, 'James Bardin Palmer,' *DCB*, vol. 6 (Toronto, 1987), 565–9; power of attorney, Seymour to Palmer, 4 April 1820, in RG 16, Land registry records, Conveyance registers, liber 27: ff. 70–4, PAPEI. The registrations of the transactions that led to Palmer's dismissal are spread over libers 29 and 32.

12 J.M. Bumsted, 'Captain John MacDonald and the Island,' *Island Magazine*, no. 6 (Spring–Summer 1979), 15–20; F.L. Pigot, 'John MacDonald of Glenaladale,' *DCB*, vol. 5 (Toronto, 1983), 515; Ian Ross Robertson, 'Donald McDonald,' *DCB*, vol. 8 (Toronto, 1985), 530–3; Ian Ross Robertson, '[Rev.] John McDonald,' *DCB*, vol. 10 (Toronto, 1972), 460–2. The spelling of the family's surname appears to have changed over the generations; for individuals I have

accepted the spellings they most frequently used, and for the family as a group I have used the 'Mc' form.

13 For a mapping of the Worrell estate, see Clark, *Three Centuries and the Island*, 97, fig. 46.

14 Worrell to Lord John Russell, 17 Dec. 1840, CO 226/60, 553, microfilm, PAC. See M. Brook Taylor, 'Charles Worrell,' DCB, vol. 8, 953–5. The £2,000 estimate comes from Seymour of Ragley Papers, CR114A/380, Tour Journal of Seymour, 1 Sept. 1840, WCRO; Seymour's source was probably Worrell's kinsman and major-domo John Edward Worrell Alleyne.

15 Deborah Stewart, 'Robert Bruce Stewart and the Land Question,' *Island Magazine*, no. 21 (Spring–Summer 1987), 3–11

16 On the situation of squatters, see Clark, *Three Centuries and the Island*, 99 and 242n19; Robertson, ed., *The Prince Edward Island Land Commission of 1860*, xii–xiv; Ian Ross Robertson, 'Reform, Literacy, and the Lease: The Prince Edward Island Free Education Act of 1852,' *Acadiensis*, 20/1 (Autumn 1990), 60–3.

17 PEI, Assembly, *Journal*, 1853, app. E

18 Calculation based on PEI, Assembly, *Journal*, 1862, app. A

19 PEI, Assembly, *Journal*, 1863, app. A

20 For evidence on this point, see the testimony of Joseph Wightman in J.D. Gordon and D. Laird, eds., *Abstract of the Proceedings before the Land Commissioners' Court, Held During the Summer of 1860, to Inquire into the Differences Relative to the Rights of Landowners and Tenants in Prince Edward Island* (Charlottetown, 1862) (hereafter, Gordon and Laird, eds., *Abstract ...*), Proceedings, 148; page references to this volume will indicate whether the page occurs in the Proceedings or the Report, since the two are not numbered consecutively. Robertson, ed., *The Prince Edward Island Land Commission of 1860* is an abridgment of Gordon and Laird, eds., *Abstract ...* , and because of its greater accessibility will be cited when possible.

21 Diary of Horatio Mann, entry for 9 Sept. 1840, in Marilyn Bell, ed., 'Mr. Mann's Island: The Journal of an Absentee Proprietor, 1840,' *Island Magazine*, no. 33 (Spring–Summer 1993), 21

22 The most detailed study of this movement is K. Rusty Bittermann, 'Escheat!: Rural Protest on Prince Edward Island, 1832–1842,' PHD thesis, University of New Brunswick, 1991.

23 Harry Baglole, comp. and ed., *The Land Question: A Study Kit of Primary Documents* (Charlottetown, 1975), section A, document M. Bittermann has devoted chapter 5 of 'Escheat!: Rural Protest on Prince Edward Island, 1832–1842' to the organized lobbying of the proprietors in London, but he seems unaware of the Palmerston-Sulivan linkage, which was exceptionally close.

Harry Baglole was the first historian to write about the lobbying activity by Prince Edward Island landlords, and his paper, 'A Reassessment of the Role of Absentee Proprietors in Prince Edward Island History' (Memorial University of Newfoundland, Dec. 1970), although unpublished, has long been available to researchers at PAPEI; some of the insights in this suggestive work are incorporated into the 'Guide to Land Question Study Kit' included in *The Land Question: ... Primary Documents*, and into the annotations to the documents.

24 Palmerston to Labouchere, 19 Dec. 1855, Letterbooks of Palmerston, Add. MSS. 48579, British Library, London

25 Rogers, Memorandum, 24 Oct. 1865, CO 226/101, 690

26 For a mapping of the Cunard estate, see Clark, *Three Centuries and the Island*, 97, fig. 46.

27 Buckner, 'The Colonial Office and British North America, 1801–50,' *DCB*, vol. 8, xxxii

28 Murdoch to Elliot, 22 Dec. 1866, CO 226/102, 528

29 Buckner, 'The Colonial Office and British North America, 1801–50,' xxxiii

30 See P.A. Buckner, 'Charles Douglass Smith,' *DCB*, vol. 8, 826. The most penetrating account of Smith's experiment with escheat is in M. Brook Taylor, 'William Johnston,' *DCB*, vol. 6, 359–60.

31 Harry Baglole, 'William Cooper,' *DCB*, vol. 9 (Toronto, 1976), 156

32 Harry Baglole, 'William Cooper of Sailor's Hope,' *Island Magazine*, no. 7 (Fall–Winter 1979), 10

33 Eric Richards, 'How Tame Were the Highlanders during the Clearances?,' *Scottish Studies*, 17/1 (1973), 41. Although Cooper fits that pattern, the circumstances and eventual developments on Prince Edward Island in the 1830s had been dramatically different. There had been no 'fundamental lack of a rallying ideology' for the common people, with the result that resistance had developed into a politicized movement – unlike the Scottish cases Richards has studied.

34 The two terms were used interchangeably, but 'Reform' seemed to be the favoured name in the 1840s, and it was gradually supplanted by 'Liberal.'

35 Baglole, 'William Cooper,' 158

36 In terms of practicability, it would have been exceptionally difficult to devise an equitable procedure for carrying out a 'partial' escheat. Taking, for example, the return of quitrents as of 1801, reproduced in Baglole, comp. and ed., *The Land Question*, section A, as document E, one finds that some of the best colonizing landlords, like John MacDonald of Glenaladale, had failed almost completely in the payment of quitrents, whereas a family like the Sulivans, with an apparently non-existent record as colonizers, had done much better

than average in paying quitrents. The situation, tangled at the start of the nineteenth century, only became more complex with time, for it would be necessary to consider the interests of those landlords who had acquired their lands by purchase.

37 Ian Ross Robertson, 'Edward Whelan,' DCB, vol. 9, 828–9

38 See Ian Ross Robertson, 'George Coles,' DCB, vol. 10, 182–5; Ian Ross Robertson, 'The 1850s: Maturity and Reform,' in Phillip A. Buckner and John G. Reid, eds., *The Atlantic Region to Confederation: A History* (Toronto and Fredericton, 1994), 343–6. For the link of educational reform with the land question see Robertson, 'Reform, Literacy, and the Lease.'

39 The *actual* rental could be much less than the nominal rental from the landlord's rent roll, and therefore under the proposed legislation a landlord could be taxed on money he did not receive. If a landlord was receiving much less than the nominal rental, then the effective rate of taxation could be much more than 5 per cent. But by the same token, the actual rental could be greater if the landlord was able to collect arrears. An example of a landlord whose actual rental exceeded his nominal rental over the years 1863–8 is cited in Ian Ross Robertson, 'Political Realignment in Pre-Confederation Prince Edward Island, 1863–1870,' *Acadiensis*, 15/1 (Autumn 1985), 42n28.

40 For the texts for the two disallowed acts, namely, the Rent Roll Act and the Tenants Compensation Act, see *British Parliamentary Papers: Correspondence and Papers Relating to Canada 1854–58*, vol. 21 (Shannon, Ireland, 1970), 443–51; for the text of the 'One-Ninth Act' which was passed in 1854 and received royal allowance in 1855, see PEI, *Statutes*, 17 Vic., c. 6. The objection to the earlier version is given in PEI, Assembly, *Journal*, 1854, app. C.

41 Ian Ross Robertson, 'Religion, Politics, and Education in Prince Edward Island, from 1856 to 1877,' MA thesis, McGill University, 1968

42 See Cunard and five others to Duke of Newcastle, 13 Feb. 1860, in PEI, Assembly, *Journal*, 1860, app. V.

43 Robertson, ed., *The Prince Edward Island Land Commission of 1860*, 169, 171; Stewart's claim that the witness had in fact attorned was presented in Gordon and Laird, eds., *Abstract ...* , Proceedings, 311.

44 Gordon and Laird, eds., *Abstract ...* , Proceedings, 243

45 For a list of memorials and statements received, see CO 226/95, 183–9. According to the commissioners, they sent a circular to all proprietors 'asking them to file their Abstract of Title and Rent Roll.' Draft interim report, n.p., initialled by the three commissioners, dated 1 Oct. 1860, Joseph Howe Papers, vol. 65, PAC.

46 For background on Wightman and the commissioners' conception of his task, see address by Howe on 20 Aug. 1861, reprinted in C. Birch Bagster, *The*

Progress and Prospects of Prince Edward Island, written during the leisure of a visit in 1861. A sketch intended to supply information upon which enquiring emigrants may rely, and actual settlers adopt as the basis of a wider knowledge of their beautiful island home (Charlottetown, 1861), appendix, xxvii–xxviii, or in *Examiner*, 26 Aug. 1861. The nature and secrecy of Wightman's mission had led Whelan's *Examiner* to label him 'the spy' (15 July).

47 Draft interim report, n.p., 1 Oct. 1860, Howe Papers, vol. 65, PAC

48 Robertson, ed., *The Prince Edward Island Land Commission of 1860*, 198. In their draft interim report of 1 Oct. 1860, the commissioners had indicated that they intended to fix a purchase price, but by the time they prepared their final report they had concluded that such was impossible. See draft interim report, n.p., Howe Papers, vol. 65, PAC; Robertson, ed., *The Prince Edward Island Land Commission of 1860*, 198–9.

49 See Phyllis R. Blakeley, 'Sir Samuel Cunard,' DCB, vol. 9, 183–4.

50 Palmerston to Newcastle, 30 June 1863, MS 62, Palmerston Papers, LB/1, folio 99, University of Southampton Library, Southampton, England. From Palmerston's phrasing, it appears that he was familiar with at least excerpts from the report of the commissioners.

51 Newcastle to George Dundas, 9 Aug. 1862, in PEI, Assembly, *Journal*, 1863, app. C; Law Officers (William Atherton and Roundell Palmer) to Newcastle, 9 June 1863, in PEI, Assembly, *Journal*, 1864, app. W

52 *Statutes of Prince Edward Island*, 27 Vic., c. 2. Its official name was *An Act for settling differences between Landlord and Tenant, and to enable tenants on certain townships to purchase the fee simple of their farms.*

53 PEI, Legislative Council, *Debates and Proceedings*, 1864, 84

54 The five were Sir Samuel and Edward Cunard, Sir Graham and James Montgomery, and Laurence Sulivan. The sixth, the Earl of Selkirk, had sold his estate to the local government in the meantime.

55 The seven residents were John Roach Bourke, William Cundall, T.H. Haviland Sr, Daniel Hodgson, John Archibald McDonald, and Edward and Henry Palmer. For a listing of significant proprietors on Prince Edward Island townships *ca.* 1864, see appendix 2.

56 Cunard to Rogers, 5 Dec. 1863, and the draft legislation entitled *A Bill for settling differences between Landlord and Tenant, and to enable Tenants on certain Townships to purchase the reversion of their Farms.*

57 See Robertson, ed., *The Prince Edward Island Land Commission of 1860*, 198–9.

58 In 1865 the Island government passed an ancillary statute, 28 Vic., c. 5, titled *An Act to assist leaseholders in the purchase of the fee-simple of their Farms.* Known popularly as 'the Loan Bill,' it enabled a tenant to borrow up to 50 per cent of the purchase price from the commissioner of public lands, who

would in turn pay the entire amount to the proprietor, providing it did not exceed 16s. 8d. currency per acre, which was the equivalent of 15 years' purchase @ 1s. sterling per acre. The Loan Bill was both more inclusive and more exclusive than the Fifteen Years Purchase Bill. A loan was available to any tenants whose terms of purchase fit the requirements, not simply to those whose landlords were specified in 27 Vic., c. 2; but no allowance was made for arrears, an exclusion which would eliminate many tenants from eligibility. By 1874, 6,876¾ acres had been purchased under the act; see PEI, Assembly, *Journal*, 1875, app. E.

59 From *The Tenant's Song*, dated 22 March 1864, in *Ross's Weekly*, 7 April 1864. 'The cut-stone house so grand' refers to the Colonial Building, where the legislature sat; it was constructed of stone from Nova Scotia.

60 See Cunard to Rogers, 5 Dec. 1863, in PEI, Assembly, *Journal*, 1864, app. F; Cunard to Edward Cardwell, 6 July 1864, CO 226/100, 542.

61 Dundas, 'Statistics respecting Land Tenure,' 20 May 1868, CO 226/104, 311

62 This is a reference to the Land Purchase Act of 1853 and the two large estates which had been purchased under its authority. Most of the Worrell estate, purchased by a Liberal government in the 1850s, lay in 2nd Kings, the district Whelan represented. Most of the Selkirk estate, purchased by the Conservatives in 1860, lay in 4th Queens, which had elected only Conservatives since its creation, in 1864 its representatives were Premier Gray and Colonial Secretary Pope.

63 This refers to the 'fishery reserves' question, one of the more complex issues between landlords and tenants.

2: A Tenantry in Ferment

1 *Examiner*, 14 March 1864

2 The Tenurial Insecurity Index used for map 2 and explained in appendix 1 measures this type of insecurity.

3 Duke of Newcastle to George Dundas, 16 June 1860, PEI, House of Assembly, *Journal*, 1861, app. E. By the time the unanimous report was released officially, that factor had been reduced from 'double' to 'additional'; see Newcastle to Dundas, 7 Feb. 1862, in PEI, Assembly, *Journal*, 1862, app. O.

4 Entry for 11 June 1861, Joseph Howe Papers, vol. 48, 123, PAC

5 Whelan to Howe, 14 April 1860, Howe Papers, vol. 3, 11–18, PAC

6 See, as examples, the statement of grievances by a committee from Lot 23, in Ian Ross Robertson, ed., *The Prince Edward Island Land Commission of 1860* (Fredericton, 1988), 43; memorial of Lot 22 tenants, 48; testimony of Lot 34 delegates, 52; testimony of James Howatt, 94.

7 Eric W. Sager and Lewis R. Fischer, 'Patterns of Investment in the Shipping Industries of Atlantic Canada, 1820–1900,' *Acadiensis*, 9/1 (Autumn 1979), 28, table 3; 'visible' means excluding ship sales and revenues from shipowning.

8 [Anonymous], 'Historical Sketch of the Province of Prince Edward Island,' in *Illustrated Historical Atlas of the Province of Prince Edward Island* (Belleville, Ont., 1972 [orig. publ. 1880]), 8; John Ross, 'The Land Question of Prince Edward Island. – Continued,' *Prince Edward Island Magazine*, 1/9 (Nov. 1899), 324–5; Lorne C. Callbeck, *The Cradle of Confederation: A Brief History of Prince Edward Island from Its Discovery in 1534 to the Present Time* (Fredericton, 1964), 177; Wayne E. MacKinnon, *The Life of the Party: A History of the Liberal Party in Prince Edward Island* (Summerside, 1973), 39–40; Peter McGuigan, 'The Lot 61 Irish: Settlement and Stabilization,' *Abegweit Review*, 6/1 (Spring 1988), 42; Peter McGuigan, 'Tenants and Troopers: The Hazel Grove Road, 1865–68,' *Island Magazine*, no. 32 (Fall–Winter 1992), 22. In his 1988 article, McGuigan has confused the 'small' meeting in Sturgeon in December 1863 to which Ross referred in 1899 with the 'large' one on the adjacent Lot 63 which his newspaper, *Ross's Weekly*, reported on 28 Jan. 1864 as having been held 'lately'; consequently, McGuigan has written as though the evidence on the Lot 63 meeting related to the Lot 61 meeting.

9 PEI, House of Assembly, *Debates and Proceedings*, 1866, 11

10 Letter dated 17 May 1866, in *Examiner*, 11 June 1866

11 PEI, Assembly, *Debates*, 1875, 157

12 *Examiner*, 14 Dec. 1863 (original emphasis). See Ian Ross Robertson, 'Edward Whelan,' *DCB*, vol. 9 (Toronto, 1976), 828–32.

13 Alan Rayburn, *Geographical Names of Prince Edward Island* (Ottawa, 1973), 116

14 Letter of Bourke to 'Mr. Editor,' dated 16 Dec. 1863, in *Examiner*, 21 Dec. 1863. Despite his threat, it appears from a reading of the Minutes of the Supreme Court for Queens County that no charges resulted from the disturbance on Lot 29.

15 For an account of the incident, see Dundas to Newcastle, 28 Nov., 8 Dec. 1859, CO 226/91, 196–8, 201–2, microfilm, PAC. Also see Supreme Court Minute Book, Kings County, 13, 14, 15 March 1860, RG 6, PAPEI.

16 *Examiner*, 21 Dec. 1863 (original emphasis)

17 *Islander*, 18 Dec. 1863

18 Many years later, Ross would be one of those authors to state that the Tenant League had originated in December 1863, but in southern Kings County.

19 *Semi-Weekly Advertiser*, 25 Dec. 1863. This newspaper may have commented on the incident earlier, but two of the three numbers between Whelan's first comment and Christmas Day are missing.

20 Letter of 'A Tenant,' dated 15 Jan. 1864, in *Examiner*, 25 Jan. 1864

21 See Elinor Vass, 'George Wastie DeBlois,' DCB, vol. 11 (Toronto, 1982), 241–2.
22 This assumption was recommended by George Wightman, the researcher for the land commission of 1860; see PEI, Assembly, *Journal*, 1863, app. A.
23 These calculations are based on data supplied by the Cunard estate, which are in PEI, Assembly, *Journal*, 1863, app. A.
24 As cited in PEI, Assembly, *Journal*, 1863, app. A
25 PEI, Assembly, *Journal*, 1862, app. A
26 PEI, Assembly, *Journal*, 1863, app. A
27 Calculations based on the 1861 census in PEI, Assembly, *Journal*, 1862, app. A
28 McGuigan has provided a useful and vivid portrait of the situation in 'The Lot 61 Irish: Settlement and Stabilization'; in his judgment, 'The only people [on the township] worse off than the Irish were the recently arrived Skyemen of St. Mary's Road West.' (41) At the census of 1861 only 10.4 per cent of Lot 61 inhabitants gave Ireland as their birthplace, but since 70.5 per cent were Prince Edward Island–born, family names and religious affiliations are better indicators of Irish ethnicity. The evidence in McGuigan's article and the fact that 51.4 per cent were Roman Catholics, with the overwhelming majority of those probably being of Irish extraction, suggest that the Irish were the largest ethnic group on the township.
29 Letter of 'A Tenant,' dated 15 Jan. 1864, in *Examiner*, 25 Jan. 1864
30 Eric Richards, *A History of the Highland Clearances*, vol. 2: *Emigration, Protest, Reasons* (London, 1985), 334. Also see 312–17, 324–7, 332–3, 341–2; and other works by Richards: *The Leviathan of Wealth: The Sutherland Fortune in the Industrial Revolution* (London and Toronto, 1973), 213–14, 251–2; 'Patterns of Highland Discontent, 1790–1860,' in Roland Quinault and John Stevenson, eds., *Popular Protest and Public Order: Six Studies in British History 1790–1920* (London, 1974), 90–2, 97, 101–2, 104, 106–7, table 4 on 96; *A History of the Highland Clearances*, vol. 1: *Agrarian Transformation and the Evictions 1746–1886* (London, 1982), 249, 373, 425, 442, 446, 461–3, 466–8, 489–90; plus, W. Hamish Fraser, 'Patterns of Protest,' in T.M. Devine and Rosalind Mitchison, eds., *People and Society in Scotland I: 1760–1830* (Edinburgh, 1988), 271; Charles W.J. Withers, *Gaelic Scotland: The Transformation of a Culture Region* (London, 1988), 368; T.M. Devine, *Clanship to Crofters' War: The Social Transformation of the Scottish Highlands* (Manchester, 1994), 210.
31 Historian Rusty Bittermann has attributed a prominent role to women in incidents of physical resistance and destruction of property associated with the Escheat movement. But his work must be used with caution. In one instance in which he claims there was an accusation that a woman had participated in animal maiming, a source he cites indicates that the 'female' was in fact a male. This was an Acadian with the given name 'Fidele,' and among

Prince Edward Island Acadians 'Fidele' has been a name of males, not females, a point confirmed by several authorities. These include the Acadian researcher and writer J. Wilmer Blanchard, who describes it as being 'as far as I know, ... strictly a male name,' and Acadian folklorist and historian Georges Arsenault, who writes that 'il est toujours masculin à l'Île comme ailleurs en Acadie.' Blanchard to author, 5 Dec. 1994, and Arsenault to author, 2 May 1995. See Bittermann, 'Women and the Escheat Movement: The Politics of Everyday Life on Prince Edward Island,' in Janet Guildford and Suzanne Morton, eds., *Separate Spheres: Women's Worlds in the Nineteenth-Century Maritimes* (Fredericton, 1994), 23–38, particularly 32n32, but also *Royal Gazette*, 21 Jan. 1834.

32 For other reports, see James Curtis to John Morris, 14 March 1865, and Bernard McKenna to the Administrator in Council, 4 Sept. 1865, both in Accession 2514/10, PAPEI; Henry Jones Cundall Diaries, vol. 5, entry for 27 Sept. 1865, PEIM.

33 PEI, Assembly, *Debates*, 1866, 32

34 For an account of the rise of Orangeism as a movement and as an issue in the colony, see Ian Ross Robertson, 'Party Politics and Religious Controversialism in Prince Edward Island from 1860 to 1863,' *Acadiensis*, 7/2 (Spring 1978), 46–58.

35 For a mapping of the Selkirk estate, see Andrew Hill Clark, *Three Centuries and the Island: A Historical Geography of Settlement and Agriculture in Prince Edward Island, Canada* (Toronto, 1959), 97, fig. 46.

36 *Ross's Weekly*, 28 Jan. 1864. For a reference to Ross's Orangeism, see *Monitor*, 29 Oct. 1863.

37 *Examiner*, 1 Feb. 1864.

38 Letter of 'James,' Georgetown, dated 19 Feb. 1864, in *Protestant*, 27 Feb. 1864; advertisements in *Royal Gazette*, 6, 13 Jan. 1864. Thwarting sheriff's sales had been an Escheat tactic, and Kings had been the county where the Escheat movement had been strongest.

39 The correspondence appeared in *Islander*, 26 Feb. 1864.

40 *Islander*, 25 March 1864

41 *Ross's Weekly*, 31 March 1864; calculations based on PEI, Assembly, *Journal*, 1862, app. A.

42 See *Examiner*, 7, 14, 28 March 1864.

43 See letter of Coles 'To the Tenantry of Prince Edward Island,' dated 5 March 1864, in *Examiner*, 7 March 1864; paraphrase of remarks by Kelly in *Examiner*, 28 March 1864; Coles in PEI, Assembly, *Debates* 1864, 11.

44 Calculation based on PEI, Assembly, *Journal*, 1862, app. A. Also see the following by Ian Ross Robertson: 'Highlanders, Irishmen, and the Land Ques-

tion in Nineteenth-Century Prince Edward Island,' in L.M. Cullen and T.C. Smout, eds., *Comparative Aspects of Scottish and Irish Economic and Social History 1600–1900* (Edinburgh, [1977]), 232–3; '[Rev.] John McDonald,' DCB, vol. 10 (Toronto, 1972), 460–2; 'Donald McDonald,' DCB, vol. 8 (Toronto, 1985), 530–3.

45 Richards, *A History of the Highland Clearances*, vol. 2, 309; Vaughan, *Sin, Sheep and Scotsmen: John George Adair and the Derryveagh Evictions, 1861* (Belfast, 1983), 33–4

46 *Examiner*, 14 March 1864

47 *Islander*, 25 March 1864

48 As paraphrased in *Examiner*, 28 March 1864

49 For the sake of simplicity I am adapting the boundaries of the electoral districts, placing First and Second Queens in 'western Queens County' and Third and Fourth Queens in 'eastern Queens County.' The line of division, therefore, has Lots 24 and 32 on one side and Lot 33 on the other. An alternative would be to divide the townships as closely as possible along a north-south line running through Charlottetown. The only resulting change in classification would be to place Lot 33 in 'western' rather than 'eastern' Queens.

50 Calculations based on 1861 census in PEI, Assembly, *Journal*, 1862, app. A

51 PEI, Assembly, *Journal*, 1862, app. A

52 Calculations based on PEI, Assembly, *Journal*, 1862, app. A

53 Robertson, ed., *The Prince Edward Island Land Commission of 1860*, 121

54 See for example the comments of George Beer in PEI, Legislative Council, *Debates and Proceedings*, 1867, 35.

55 Donald E. Jordan Jr, *Land and Popular Politics in Ireland: County Mayo from the Plantation to the Land War* (Cambridge, 1994), 153–4

56 *Examiner*, 9 May 1864

57 *Islander*, 22 April 1864

58 *Islander*, 29 April 1864 (original emphasis)

3: A New Tenant Organization

1 *Ross's Weekly*, 26 May 1864

2 For a description of the hotel, which contained a dining room capable of accommodating one hundred guests, see *Examiner*, 8 Aug. 1864.

3 Unless otherwise indicated, the source for information on the convention is *Ross's Weekly*, 26 May 1864.

4 See letter of Archibald McNeill, dated 3 June 1864, in *Protestant*, 11 June 1864; letter of John Bassett, dated 8 June 1864, in *Ross's Weekly*, 16 June 1864.

5 *Ross's Weekly*, 26 May 1864. One may safely dismiss the extraordinary state-

ment on 17 May 1867, almost exactly three years later, by Benjamin Davies, a long-time radical on the land question, that 'That pledge, though a few attempted to get it passed, was never recognized by the League' (PEI, House of Assembly, *Debates and Proceedings*, 1867, 145). There is no record of anyone else making this claim.

6 John Balderston in PEI, Legislative Council, *Debates and Proceedings*, 1870, 125

7 F.S.L. Lyons, *Ireland since the Famine* (London, 1973 ed.), 121; also see 115–16, 122; Michael J. Winstanley, *Ireland and the Land Question 1800–1922* (London, 1984), 27; Paul Bew, *Land and the National Question in Ireland 1858–82* (Dublin, 1978), 34

8 T.W. Moody, 'Fenianism, Home Rule and the Land War (1850–91),' in T.W. Moody and F.X. Martin, eds., *The Course of Irish History* (Cork, 1967), 278

9 CO 226/100, 228–9, microfilm, PAC

10 Letter of Archibald McNeill, dated 3 June 1864, in *Protestant*, 11 June 1864

11 Ewen MacNeill, 'McNeill of Barra on P.E. Island: Ancestors and Descendants of Alexander McNeill' (privately circulated, 1982), unpaginated, copy at PAPEI; Ian Ross Robertson, ed., *The Prince Edward Island Land Commission of 1860* (Fredericton, 1988), 123. In interviews on 26 July and 1 Sept. 1995 Ewen MacNeill stated that Alexander had left the ring to another grandson (his half-brother), long since deceased, who lived in Regina.

12 *Ross's Weekly*, 9 June, 7 July 1864

13 William M. Glen, 'The Making of a Gentleman? The Curious Life of W.W. Irving,' *Island Magazine*, no. 30 (Fall–Winter 1991), 6

14 Dundas to Arthur Blackwood, 6 June 1864, private, CO 226/100, 232

15 For parallel linked phenomena in Ireland, see Samuel Clark, *Social Origins of the Irish Land War* (Princeton, 1979), 262–7 with respect to the emergence of 'farmers' allies,' and Donald E. Jordan Jr, *Land and Popular Politics in Ireland: County Mayo from the Plantation to the Land War* (Cambridge, 1994), 156–7, 233 regarding the loss of legitimacy of debts *to landlords* among Irish tenants.

16 *Ross's Weekly*, 26 May 1864 (original emphasis). If Adams's address was published in pamphlet form, no copy is known to survive.

17 *Ross's Weekly*, 26 May 1864

18 *Examiner*, 28 Jan. 1880; *Islander*, 8 Nov. 1867. I owe these references to genealogist George F. Sanborn Jr, 12 Aug. 1981.

19 Dundas to Blackwood, 6 June 1864, private, CO 226/100, 232

20 Henry Jones Cundall Diaries, vol. 2, entry for 4 Nov. 1858, PEIM

21 See Francis L. Haszard and A. Bannerman Warburton, eds., *Reports of Cases Determined in the Supreme Court, Court of Chancery, and Vice Admiralty Court of Prince Edward Island ... [1850–1874]*, vol. 1 (Charlottetown, 1885), 209–18, 228–31, 242–4. Convenient access to a set of nineteenth-century court reports was

provided through the generosity of the late John P. Nicholson, chief justice of the Prince Edward Island Supreme Court.

22 Testimony of John Ramsay in P.S. Macgowan, ed., *Report of Proceedings before the Commissioners Appointed under the Provisions of 'The Land Purchase Act, 1875'* (Charlottetown, 1876), 112–13

23 Sanborn to author, 23 Feb. 1995; it was Mr Sanborn who drew the Adams-Ramsay connection to my attention. Information on the exact cause of the suit between Adams and his two in-laws is lacking. His obituary mentions leaving a grieving wife and child, which suggests that there was no permanent marriage breakdown.

24 *Ross's Weekly*, 8 Sept. 1864

25 *Examiner*, 15 Feb. 1864; the letter, dated 30 Jan. 1864, appeared in *Protestant*, 5 March 1864.

26 Dundas to Blackwood, 6 June 1864, private, CO 226/100, 230; also see E. Mac-Neill, 'McNeill of Barra on P.E. Island.'

27 *Ross's Weekly*, 7 July 1864

28 *Ross's Weekly*, 9 June 1864

29 Letter of Bassett in *Ross's Weekly*, 7 July 1864

30 *Examiner*, 23 May 1864

31 *Examiner*, 30 May 1864

32 *Examiner*, 23 Oct. 1860

33 *Examiner*, 30 May 1864

34 Dundas to Blackwood, 6 June 1864, private, CO 226/100, 231–2

35 Letter of Bassett, dated 19 June 1864, in *Ross's Weekly*, 7 July 1864

36 For an example, see Ian Ross Robertson, 'Religion, Politics, and Education in Prince Edward Island, from 1856 to 1877,' MA thesis, McGill University, 1968, 127–8.

37 *Ross's Weekly*, 7 July 1864

38 *Ross's Weekly*, 9 June 1864

39 Clark, *Social Origins of the Irish Land War*, 3, 248, 311; Bew, *Land and the National Question in Ireland 1858–82*, 124, 221–2; W.E. Vaughan, *Landlords and Tenants in Mid-Victorian Ireland* (Oxford, 1994), 178–9; Jordan, *Land and Popular Politics in Ireland*, 284–93, 295; R.F. Foster, *Modern Ireland, 1600–1972* (Harmondsworth, 1989 ed.), 406; Lyons, *Ireland since the Famine*, 168; Joyce Marlow, *Captain Boycott and the Irish* (London, 1973), 101–4, 134–42, 145, 159–60, 198–200, 214–19; James S. Donnelly Jr, *The Land and the People of Nineteenth-Century Cork: The Rural Economy and the Land Question* (London and Boston, 1975), 272, 327–30, 380; Charles Townshend, *Political Violence in Ireland: Government and Resistance since 1848* (Oxford, 1983), 173–4. As Clark and Vaughan point out, such measures did not originate with this case. Contem-

porary terms for variants included 'social excommunication' and 'exclusive dealing.'

40 *Ross's Weekly*, 9 June 1864

41 Letter of 'Observer,' dated 8 Nov. 1864, in *Ross's Weekly*, 10 Nov. 1864

42 Clark, *Social Origins of the Irish Land War*, 312

43 *Ross's Weekly*, 9 June 1864

44 Calculations based on PEI, Assembly, *Journal*, 1862, app. A

45 See draft offer, dated 3 July 1864, in *Ross's Weekly*, 4 Aug. 1864.

46 *Ross's Weekly*, 4 Aug. 1864

47 Dundas, 'Statistics respecting Land Tenure,' 20 May 1868, CO 226/104, 309; also see Dundas to Duke of Buckingham and Chandos, 20 May 1868, confidential, CO 226/104, 283. There is no indication of the significance of the purchase, or the fact that it occurred when it did, in Andrew Robb, 'Robert Poore Haythorne,' *DCB*, vol. 12 (Toronto, 1990), 418–19; moreover, the author has confused the location of Haythorne's farm on Lot 34, 'Marshfield,' which he *leased* from the Rev. James F. Montgomery, with the estate he *owned* on Lot 49.

48 See report of meeting at Mount Mellick in *Ross's Weekly*, 7 July 1864.

49 Haythorne to the Editor of *Islander*, 25 April 1865, in *Islander*, 28 April 1865. Also see testimony by Haythorne in Macgowan, ed., *Proceedings of Land Commissioners of 1875–76*, 49.

50 See, for example, *Examiner*, 19 June 1865

51 PEI, Assembly, *Debates*, 1866, 13

52 Dundas, 'Statistics respecting Land Tenure,' 20 May 1868, CO 226/104, 309

53 Minute dated 30 Aug. 1864, CO 226/100, 292. Elliot was not alone in his feelings of superiority, for Dundas had written on 8 May that 'The common people here are very ignorant and very stupid.' Dundas to Blackwood, 8 May 1864, private, CO 226/100, 477.

54 CO 226/100, 228

55 See, for example, Donnelly, *The Land and the People of Nineteenth-Century Cork*, 6, 200, 249–52, where the concept has primarily an economic significance, linked to verifiable improvements in living standards (housing, diet, education). The theory is associated with political scientist James C. Davies, 'Toward a Theory of Revolution,' *American Sociological Review*, 27/1 (Feb. 1962), 5–19, who builds upon passages in Marx and de Tocqueville. Davies expressly disclaims any suggestion that his theory is universally applicable: 'No claim is made that all rebellions follow the pattern, but just that the ones here presented do' (8).

56 See Jim Hornby, *Black Islanders: Prince Edward Island's Historical Black Community* (Charlottetown, 1991). One former slave, a native of Africa who had become a freeholder through purchase of a farm, had lived until 1845 (30–1).

57 J.M. Bumsted, 'Lord Selkirk of Prince Edward Island,' *Island Magazine*, no. 5 (Fall–Winter 1978), 4–5; H.T. Holman, 'William Douse,' *DCB*, vol. 9 (Toronto, 1976), 223

58 Calculations based on the 1861 census as published in PEI, Assembly, *Journal*, 1862, app. A. The only townships on the Island with comparable rates of tenancy were in north-western Prince County, where Edward Cunard, a son of Sir Samuel, was proprietor: Lots 4, 5, and 6. In Kings the highest percentage was 83.3, on Sulivan's Lot 61. The other three townships where Sulivan owned land also featured high rates of tenancy: 81.8 per cent on Lot 9, 83.1 on Lot 16, and 88.9 on Lot 22.

59 PEI, Assembly, *Journal*, 1860, 165

60 Dundas to Blackwood, 6 June 1864, private, CO 226/100, 232; PEI, Assembly, *Journal*, 1863, app. A

61 Calculations based on PEI, Assembly, *Journal*, 1862, app. A. A 'significant' number of leaseholders on a township is taken to be at least 50 per cent of the average number of holders of written leases or agreements to lease listed in the 1861 census, that is, at least forty.

4: The Gathering Conflict

1 CO 226/100, 612, microfilm, PAC

2 Accession 2514/10, PAPEI

3 The first number of a new pro-Liberal, pro-tenant newspaper, the *Herald*, incorrectly reported disallowance of the bill, and had to retract a week later; see *Herald*, 12, 19 Oct. 1864. Could the initial error have been an attention-seeking device? The *Herald* had reprinted speculation by *Ross's Weekly*, not a paragon of reliability, as fact. Edward Reilly, editor and publisher of the *Herald*, although young, was already a veteran Island newspaper man, and would be aware of Ross's frailties. See Ian Ross Robertson, 'Edward Reilly,' *DCB*, vol. 10 (Toronto, 1972), 612–13.

4 PEI, *Statutes*, 27 Vic., c. 2, Preamble

5 Henry Jones Cundall to W.C.A. Williams, 29 June 1864, Henry Jones Cundall Letterbook, PAPEI

6 Calculation based on various tabular statements in PEI, Assembly, *Journal*, 1875, app. E

7 The advertisement was dated 27 October.

8 There are many parallels with the approach of Irish and Scottish landlords to the collection of rent. Some land agents in Ireland routinely prepared notices to quit as part of their strategy for extracting rent, and the limited research which has been done on the use of summonses of removal by Scottish estates

suggests similar practices. See Samuel Clark, *Social Origins of the Irish Land War* (Princeton, 1979), 168–9; T.M. Devine, *The Great Highland Famine: Hunger, Emigration and the Scottish Highlands in the Nineteenth Century* (Edinburgh, 1988), 174–5. Irish historian W.E. Vaughan estimates a ratio of five threats of eviction to each actual eviction for the 1848–80 period, although he states that once a landlord had a decree from the courts ordering the sheriff to put him in possession of his lands, 'practically nothing could stop [him] from evicting his tenants, if he were determined to do so.' *Landlords and Tenants in Ireland 1848–1904* ([Dublin], 24; *Sin, Sheep and Scotsmen: John George Adair and the Derryveagh Evictions, 1861* (Belfast, 1983), 24.

9 *Islander*, 4, 11, 29 Nov. 1864

10 Clark, *Social Origins of the Irish Land War*, 165

11 Samuel Clark, 'Landlord Domination in Nineteenth-Century Ireland,' *UNESCO Yearbook on Peace and Conflict Studies 1986* (Paris, 1988), 12–14; also see Clark, *Social Origins of the Irish Land War*, 164–71, 179–81, 184.

12 Clark, *Social Origins of the Irish Land War*, 235; see 344n for examples of withdrawal.

13 *Osborn's Concise Law Dictionary*, 6th edition (London, 1976), 193

14 PEI, Assembly, *Journal*, 1854, app. C, James Wilson (financial secretary to the Treasury) to the Board of Trade, 3 Nov. 1853; PEI, *Statutes*, 17 Vic., c. 6

15 For a certified copy of the petition together with a list of the signers, see CO 226/100, 613–14.

16 The exchange rate of currency to sterling was 2:3, or 50 per cent.

17 Wood and Fanning to Cardwell, 31 Oct. 1864, CO 226/100, 612

18 *Ross's Weekly*, 15 Dec. 1864

19 *Ross's Weekly*, 22 Dec. 1864

20 *Ross's Weekly*, 15 Dec. 1864

21 Gregory S. Kealey and Bryan D. Palmer, *Dreaming of What Might Be: The Knights of Labor in Ontario, 1880–1900* (Toronto, 1987 ed. [orig. publ. 1982]), 65; also see 90–1. They defend their emphasis on a volatile, as opposed to a static, count, on 62–4.

22 On 11 April 1866 John Hamilton Gray, who had been premier until 22 Dec. 1864, denied that Dundas had been responsible, and stated that 'the boon was altogether of a private nature,' without being any more precise; PEI, House of Assembly, *Debates and Proceedings*, 1866, 25. The *Islander*, 15 June 1866 asserted that it was Gray who made the crucial approach to Cunard.

23 Term drawn from Clark, 'Landlord Domination in Nineteenth-Century Ireland,' 14, who in turn is citing sociological theorists of bureaucratic management.

24 James Corley and eight others to DeBlois, 4 Jan. 1865, in *Protestant*, 11 Feb. 1865

25 DeBlois to Corley and eight others, dated 4 Feb. 1865; DeBlois to the Editor of the *Protestant*, dated 9 Feb. 1865, in *Protestant*, 11 Feb. 1865

26 Robert Stewart to Benjamin DesBrisay, undated; B. DesBrisay to Robert Stewart, 9 Dec. 1864, in *Ross's Weekly*, 29 Dec. 1864

27 When alluding to this incident, the local press habitually referred to Lot 35, although Fort Augustus and the Monaghan Settlement are in fact on Lot 36. Curtis had, however, to pass through Lot 35 on his way to Lot 36.

28 *Examiner*, 7 March 1864; CO 226/100, 229; *Herald*, 11 Jan. 1865

29 Ian Ross Robertson, '[Rev.] John McDonald,' *DCB*, vol. 10, 460–2. The poet Milton Acorn, upon reading the preceding biographical sketch, composed the poem *John Dhru Macintosh Stands Up in Church*, published in Acorn, *The Island Means Minago* (Toronto, 1975), 37.

30 Curtis could not discover the given name of this son.

31 As Robert Gordon, a student at Scarborough College, University of Toronto, pointed out on 9 Feb. 1988, there was a note of belligerence in the demand by Curtis, a law enforcement officer, to know whether Callaghan's tavern was 'licensed.' Under PEI, *Statutes*, 23 Vic., c. 12 s. 1, licensees were required to take an oath to 'entertain such proper guests as may offer' and to enter into a £15 bond to adhere to regulations and conditions surrounding licenses; PEI, *Statutes*, 19 Vic., c. 2 s. 4 provided that a licensee who 'shall refuse to accommodate travellers' was subject to a fine of 40s. per offence; and s. 14 of the latter statute allowed the grand jury or justices of the peace to remove a licence upon complaint of irregular or improper behaviour.

32 Curtis to John Morris, 14 March 1865, Accession 2514/10, PAPEI

33 *Islander*, 23 June 1865 (original emphasis). Elsewhere in British North America there are documented incidents of the use of horns to warn of the arrival of the authorities and, sometimes, to summon those prepared to resist. See Michael S. Cross, '"The Laws Are Like Cobwebs": Popular Resistance to Authority in Mid-Nineteenth Century British North America,' in Peter B. Waite, Sandra Oxner, and Thomas Barnes, eds., *Law in a Colonial Society: The Nova Scotia Experience* (Toronto, 1984), 103, 117–18.

34 Curtis to Morris, 14 March 1865, Accession 2514/10, PAPEI

35 PEI, Assembly, *Debates*, 1866, 32

36 Morris to J.C. Pope, 15 March 1865, Accession 2514/10, PAPEI

37 David Webber, *A Thousand Young Men: The Colonial Volunteer Militia of Prince Edward Island 1775–1874* (Charlottetown, 1990), 101; on the theoretical distinctions between militia and Volunteers, which did not always apply in Prince Edward Island, see 42, 119–20.

38 Morris to J.C. Pope, 15 March 1865, Accession 2514/10, PAPEI

39 *Examiner*, 11 April 1859

40 See Ian Ross Robertson, 'William Henry Pope,' *DCB*, vol. 10, 593–9; 'James Colledge Pope,' *DCB*, vol. 11 (Toronto, 1982), 699–705; 'Edward Palmer,' *DCB*, vol. 11, 664–70.

41 PEI, Assembly, *Debates*, 1861, 110 (original emphasis). Issues surrounding the possible uses of the Volunteers and who precisely was being armed had caused acrimonious debate in the House of Assembly in 1861; see Ian Ross Robertson, 'Religion, Politics, and Education in Prince Edward Island, from 1856 to 1877,' MA thesis, McGill University, 1968, 82–90.

42 *Ross's Weekly*, 15 Dec. 1864; *Islander*, 13, 20, 27 Jan., 10, 17, 24 Feb., 17 March 1865; *Lady Wood & an' v. Thos Beers*, affidavit of debt, PEI, Supreme Court Case Papers, 1865, PAPEI

43 *Herald*, 25 Jan. 1865; *Protestant*, 31 Dec. 1864; *Islander*, 20 Jan. 1865

44 *Ross's Weekly*, 6 Oct. 1864

45 *Herald*, 11 Jan. 1865

46 PEI, Assembly, *Debates*, 1866, 33; also see James Duncan in PEI, Assembly, *Debates*, 1866, 11

47 PEI, Assembly, *Debates*, 1868, 42

48 See *Examiner*, 27 May 1867; Coles in PEI, Assembly, *Debates*, 1866, 28

49 PEI, Assembly, *Debates*, 1872 (2nd session), 38

50 For a denial of partisanship, see the Address of Grand Master David Kaye on 9 Feb. 1865, GOL, Fourth Annual Report (1865), 6, in private possession; for evidence of partisanship, see Address of Grand Master J.B. Cooper, GOL, Sixth Annual Report (1867), 15; Report of Grand Secretary Thomas J. Leeming, GOL, Sixth Annual Report, 24.

51 Address of Kaye, GOL, Fourth Annual Report, 6

52 Report of Committee on Provincial Grand Master's Address, GOL, Fourth Annual Report, 10–11; Report of Committee on Correspondence, 13; Report of Grand Secretary Leeming, 7; Report of the Committee on the Grand Secretary's Report, 15

53 Sixth resolution, GOL, Fourth Annual Report, 16

54 Address of Kaye, GOL, Fourth Annual Report, 6; report of Leeming, GOL, Fourth Annual Report, 6–8; report of Grand Secretary Leeming, GOL, Fifth Annual Report (1866), 5; letter of Kaye, dated 5 Feb. 1866, GOL, Fifth Annual Report, 16; report of Grand Secretary Leeming, GOL, Sixth Annual Report, 24

55 Clark, *Social Origins of the Irish Land War*, 80–2, 255, 282, 291, 307

56 Ian Ross Robertson, 'Party Politics and Religious Controversialism in Prince Edward Island from 1860 to 1863,' *Acadiensis*, 7/2 (Spring 1978), 46–58. Significantly, the most violent incident involving religious and ethnic tensions

in Prince Edward Island, the Belfast Riot of 1847, occurred more than a decade before the rapid growth of the local Orange movement.

57 See Ian Ross Robertson, 'George Dundas,' DCB, vol. 10, 264–5; 'Sir Robert Hodgson,' DCB, vol. 10, 352–3; 'James Horsfield Peters,' DCB, vol. 12 (Toronto, 1990), 838–42.

58 PEI, Executive Council Minutes, 16 March 1865. The actual letter sent to Morris added something to the order recorded in the Executive Council Minutes: he was to call out the *posse* '*in all cases* where resistance of this kind is offered' (emphasis added) (Charles DesBrisay [Clerk of the Executive Council] to Morris, 17 March 1865, in *Islander*, 14 April 1865).

59 *The Concise Oxford Dictionary*, 7th edition (1982), 800

60 *Patriot*, 20 Jan. 1866, as reprinted in *Examiner*, 22 Jan. 1866 (original emphasis)

5: Popular Defiance and Samuel Fletcher

1 Accession 2514/10, PAPEI

2 Sidney and Beatrice Webb, *The Parish and the County* (London, 1963 ed. [orig. publ. 1906]), 306–7

3 *Islander*, 24 March 1865

4 Frederick Brecken in PEI, House of Assembly, *Debates and Proceedings*, 1867, 7. Davies, who was back in the assembly by 1867, was present, and did not deny Brecken's statement.

5 *Islander*, 20 Jan. 1865

6 John Morris to J.C. Pope, 18 March 1865, Accession 2514/10, PAPEI; *Islander*, 24 March 1865; Robert Hodgson to Edward Cardwell, 2 Aug. 1865, in PEI, Assembly, *Journal*, 1866, app. G

7 Date of birth from tombstone of Fletcher, courtesy of genealogist James K. Raywalt, 26 Nov. 1989; CO 226/100, 338, 614, microfilm, PAC; 'Louisa A. Lady Wood & other *v.* Samuel Fletcher,' Affidavit of Debt, 4 March 1865, PEI, SCCP, 1865, PAPEI

8 T. DesBrisay to Morris, 18 March 1865; Morris to J.C. Pope, 18 March 1865, Accession 2514/10, PAPEI; George Sinclair in PEI, Assembly, *Debates*, 1866, 36; *Islander*, 14 March 1865; *Examiner*, 10 April 1865

9 Morris to J.C. Pope, 18 March 1865, Accession 2514/10, PAPEI

10 T. DesBrisay to Morris, 18 March 1865, Accession 2514/10, PAPEI

11 S. and B. Webb, *The Parish and the County*, 306–7

12 T. DesBrisay to Morris, 18 March 1865, Accession 2514/10, PAPEI

13 Morris to J.C. Pope, 18 March 1865, Accession 2514/10, PAPEI

14 George Dundas to Duke of Newcastle, 28 Nov. 1859, CO 226/91, 197–8; *Protestant*, 15 April 1865

15 *Examiner,* 11 April 1859; also see Ian Ross Robertson, 'William Henry Pope,' *DCB,* vol. 10 (Toronto, 1972), 593–9; 'Party Politics and Religious Controversialism in Prince Edward Island from 1860 to 1863,' *Acadiensis,* 7/2 (Spring 1978), 29–59, especially 36, 53–4. Pope was appointed colonial secretary despite not being elected, in accordance with the Palmer government's policy of 'non-departmentalism,' by which members of the legislature were not to hold offices of emolument. See D.C. Harvey, 'Dishing the Reformers,' *Transactions* of the Royal Society of Canada, 3rd ser., vol. 25, sect. 2 (1931), 37–44; Ian Ross Robertson, 'Edward Palmer,' *DCB,* vol. 11 (Toronto, 1982), 666.

16 *Islander,* 27 Jan. 1865 (original emphasis)

17 *Islander,* 27 Jan. 1865

18 Max Beloff, *Public Order and Popular Disturbances 1660–1714* (London, 1963 ed. [orig. publ. 1938]), 133, 139, 140. Also see 153; S. and B. Webb, *The Parish and the County,* 488–9n4; F.C. Mather, *Public Order in the Age of the Chartists* (Manchester, 1959), 47, 80–1.

19 *Islander,* 17 March 1865; see Ian Ross Robertson, 'Political Realignment in Pre-Confederation Prince Edward Island, 1863–1870,' *Acadiensis,* 15/1 (Autumn 1985), 35–58, especially 48.

20 *Islander,* 24 March 1865 (emphasis added)

21 PEI, Executive Council Minutes, 21 March 1865. The Minutes, although correctly conveying the difference in opinion between the Executive Council and the sheriff, contain the statement that Morris had declared he would 'cheerfully' obey the instructions to call out the *posse* if his advice were rejected. The adverb is a gratuitous addition that is inconsistent with the tone of Morris's letter, and although perhaps it may seem a small matter, this editorial work is of a piece with the 'damage control' operation the government mounted after the calling out of the *posse comitatus.* The sheriff would be made to appear the initiator.

22 *Royal Gazette,* 22 March 1865 (emphasis added)

23 See Ian Ross Robertson, 'Religion, Politics, and Education in Prince Edward Island, from 1856 to 1877,' MA thesis, McGill University, 1968, 92–3

24 Dundas to Cardwell, 25 Aug. 1865, confidential, CO 226/101, 666

25 PEI, Assembly, *Debates,* 1865, 36

26 Charles DesBrisay to Morris, 23 March 1865, in *Islander,* 14 April 1865

27 *Islander,* 24 March 1865

28 This of course is a reference to John Morris, and for some reason Le Page chose not to use his surname; in another stanza, he rendered Curtis's surname as 'C——.'

29 John Le Page, *The Island Minstrel: A Collection of Some of the Poetical Writings of*

John Le Page, vol. 2 (Charlottetown, 1867), 199; see Thomas B. Vincent, 'John LePage,' *DCB*, vol. 11, 511–12

30 George D. Atkinson (private secretary to Dundas) to J. [?] Reddin *et al.*, 5 April 1865, Government House Letterbooks, PAPEI

31 *Herald*, 12 April 1865

32 Interview with Maurice Pope, Ottawa, 18 May 1968

33 *Islander*, 24 March 1865

34 No two accounts of the *posse comitatus* incident are identical. Even the number of marchers is uncertain, with 150 being more or less a median of the estimates; but in most instances the doubtful elements concern non-essentials. For more detailed commentary on the sources, see Ian Ross Robertson, 'The *Posse Comitatus* Incident of 1865,' *Island Magazine*, no. 24 (Fall–Winter 1988), 10.

35 *Herald*, 12 April 1865

36 Letter of 'One of the Posse,' n.d., in *Examiner*, 17 April 1865

37 *Herald*, 12 April 1865

38 See Brecken in PEI, Assembly, *Debates*, 1867, 118; John Ross, 'Tenant League Results,' *Prince Edward Island Magazine*, 2/1 (March 1900), 23; 'Celt,' 'Chips,' *Prince Edward Island Magazine*, 2/2 (April 1900), 52; 'Rambler,' 'The Tenant League Articles,' *Prince Edward Island Magazine*, 2/5 (July 1900), 162.

39 PEI, Executive Council Minutes, 18 April 1865; also see Atkinson to Wright, 13 April 1865, Government House Letterbooks, PAPEI; John Ross, *Reminiscences from the Life of John Ross, Formerly Publisher of 'Ross's Weekly,' the Avowed Champion of the Tenant Union Organization, also Advocate of the Prince Edward Island Railway, and Now Manufacturer and Proprietor of the Celebrated 'Magic Healer' Salve* (Charlottetown, 1892), 34; Ross, 'Tenant League Results,' *The Prince Edward Island Magazine*, 2/1 (March 1900), 23. The PEI Executive Council Minutes, 4 March 1867 reveal that Curtis made a claim for his services in calling out the *posse* almost two years after the event. This was the final meeting the Tory Executive Council held before leaving office after a defeat at the polls which involved election of several new assemblymen known to be sympathetic to the Tenant League, and Curtis may have believed it to be his last opportunity to collect; otherwise his timing might seem inexplicable. But members refused the claim.

40 Ross, 'Tenant League Results,' *Prince Edward Island Magazine*, 2/1 (March 1900), 22

41 Henry Jones Cundall Diaries, vol. 5, entry for 7 April 1865, PEIM

42 Anonymous first-person account published as part of the editorial columns in *Herald*, 12 April 1865

43 See, for example, *Herald*, 24 May 1865; Joseph Hensley in PEI, Assembly, *Debates*, 1866, 27.

44 *Herald,* 26 April 1865

45 A second anonymous correspondent in the *Examiner,* 'One of the Possy [*sic*],' writing ostensibly 'To the Editor of the *Islander,*' raised this question in a letter dated 15 April 1865; see *Examiner,* 17 April 1865. The writer makes a sarcastic play on some of the phrasing in an editorial in the *Islander* of 14 April, and this may account for its non-appearance in the *Islander.*

46 R. Hodgson to Cardwell, 2 Aug. 1865, in PEI, Assembly, *Journal,* 1866, app. G

47 *Examiner,* 10 April 1865

48 *Examiner,* 10 April 1865

49 *Herald,* 12 April 1865

50 *Protestant,* 15 April 1865

51 See Le Page, *The Island Minstrel,* vol. 2, 196–201

52 This referred to Adams's employment as a postman.

53 Ross, *Reminiscences* ..., 32. The shorter version of Le Page's poem had other significant deletions and minor changes, all unacknowledged and unexplained; see Ross, 'Tenant League Results,' *Prince Edward Island Magazine,* 1/12 (Feb. 1900), 434–5. Dr G. Edward MacDonald, curator of history at the Prince Edward Island Museum and Heritage Foundation, and a poet himself, alerted me to the fact that Ross's rendition of Le Page's poem in the *Prince Edward Island Magazine* is a shortened version.

54 *Herald,* 26 July 1865. I owe this reference to Dr G. Edward MacDonald.

55 *Examiner,* 10 April 1865

56 Ross, 'Tenant League Results,' *Prince Edward Island Magazine,* 2/1 (March 1900), 24

57 *Islander,* 14 April 1865

58 Le Page, *The Island Minstrel,* vol. 2, 199

59 Morris to J.C. Pope, 18 March 1865, Accession 2514/10, PAPEI. The five published in *Islander,* 14 April 1865 are: Curtis to Morris, 14 March 1865; Morris to J.C. Pope, 15 March 1865; C. DesBrisay to Morris, 17 March 1865; T. DesBrisay to Morris, 18 March 1865; C. DesBrisay to Morris, 22 March 1865.

60 See, for example, Solicitor General and Executive Councillor T. Heath Haviland Jr in PEI, Assembly, *Debates,* 1866, 23, who hewed to the line that sole responsibility lay with the sheriff.

61 *Herald,* 12 April 1865

62 *Royal Gazette,* 24 May 1865

63 PEI, Assembly, *Debates,* 1866, 27; see G. Edward MacDonald, 'Joseph Hensley,' DCB, vol. 12 (Toronto, 1990), 425–7

64 Lorne C. Callbeck, *The Cradle of Confederation: A Brief History of Prince Edward Island from Its Discovery in 1534 to the Present Time* (Fredericton, 1964), 177–80;

George E. Hart, *The Story of Old Abegweit: A Sketch of Prince Edward Island History* ([Charlottetown?], [1935?]), 55; Errol Sharpe, *A People's History of Prince Edward Island* (Toronto, 1976), 107–8; Reg Phelan, 'P.E.I. Land Struggle,' *Cooper Review*, no. 3 (1988), 30–1; Douglas Baldwin, *Land of the Red Soil: A Popular History of Prince Edward Island* (Charlottetown, 1990), 90–1. In his text for grade 6 students, Baldwin also focuses on the two incidents; see *Abegweit: Land of the Red Soil* (Charlottetown, 1985), 194–5.

65 A.L. Morrison, *My Island Pictures: The Story of Prince Edward Island* (Charlottetown, 1980), painting # 19. In 1892 and again in 1900 Ross referred to 'Fort Fletcher'; see *Reminiscences ...* , 36; 'Tenant League Results,' *Prince Edward Island Magazine*, 2/1 (March 1900), 24.

66 *Islander*, 29 Sept. 1865

67 'A Liberal,' to 'Mr. Editor,' n.d., in *Examiner*, 23 Oct. 1865

68 PEI, Supreme Court Minutes, 8 May 1865, PAPEI

69 'The Queen *v.* Samuel Fletcher,' Attachment, signed by Daniel Hodgson, 22 May 1865, PEI, SCCP, 1865, PAPEI; the sheriff's answer is written on the attachment.

70 PEI, Executive Council Minutes, 2 Sept. 1865; the actual letter from DesBrisay to J.C. Pope, 2 Sept. 1865, survives in Accession 2514/10, PAPEI.

71 PEI, Executive Council Minutes, 12 Dec. 1865

72 PEI, Executive Council Minutes, 16 Jan., 17 April 1866

73 Fletcher, receipt for sale of farm, 31 Dec. 1866, Land Title Documents, Leases and Related Documents, Lease No. 31, Lot 50, PAPEI. This file also contains Fletcher's lease from Fanning and Wood, dated 10 Oct. 1853, which leases the land to Fletcher for a term of 999 years commencing 1 Jan. 1853, with rent to commence 1 Jan. 1854; one should not automatically infer that Fletcher was a newcomer to the farm in 1853.

74 *Herald*, 15 Jan. 1868

75 Information provided by James K. Raywalt of Washington, DC: telephone interview, 26 Nov. 1989, and Raywalt to author, 8 July 1990. Mr Raywalt kindly responded to my statement in the 'Sources' note at the end of my article 'The *Posse Comitatus* Incident of 1865,' *Island Magazine*, no. 24 (Fall–Winter 1988), 10, that I 'would welcome information concerning the fate of Samuel Fletcher after 1866.' In the body of the article I had concluded by stating that 'it is uncertain, on the basis of the available documentary record, whether Sam Fletcher sought a new life elsewhere off the Island, or whether he led a quiet life somewhere on the Island, perhaps assuming a new identity.' Mr Raywalt has cleared up that mystery.

76 *Islander*, 14 April 1865

77 Information from genealogist Robert C. Anderson, 12 Aug. 1981

78 Cooper was notably quiet during the Tenant League disorders. He had been his usual vocal self during the hearings of the land commission of 1860.

6: Landlords on the Retreat and under Attack

1 'The Queen at the prosecution of Bernard McKenna *v.* Francis McKenna, Hugh McKenna, and Patrick McKenna,' Indictment, marked 'True Bill,' January 1866, PEI, SCCP, PAPEI
2 *Protestant*, 13 May 1865
3 The news of Cunard's death was apparently published for the first time in the local press by the *Islander*, 12 May 1865.
4 PEI, *Statutes*, 27 Vic., c. 2, s. 6
5 PEI, Legislative Council, *Debates and Proceedings*, 1864, 84
6 Pope to T.F. Elliot, 13 Jan. 1864, in PEI, House of Assembly, *Journal*, 1864, app. F
7 'Direct,' because the history of the Tenant League demonstrates that freeholders – often former leaseholders themselves – could be quite willing to become deeply involved in the struggles over landlord-tenant relations.
8 Information on the Reverend James F. Montgomery comes from testimony by his first cousin, also named James F. Montgomery, before the land commission of 1875–6: see P.S. Macgowan, ed., *Report of Proceedings before the Commissioners Appointed under the Provisions of 'The Land Purchase Act, 1875'* (Charlottetown, 1876), 345.
9 Concerning the relevance of the Fifteen Years Purchase Bill and an adjusted figure for the projected saving to the tenantry when arrears and expenses are taken into account, see appendix 3.
10 Montgomery to James Robertson *et al.* (a committee of five), dated 13 April, in *Islander*, 19 May 1865. The acreage comes from PEI, Executive Council Minutes, 9 Jan. 1868, app.; the number of acres mentioned in the initial story in *Islander*, 19 May 1865 was 6,300.
11 PEI, Assembly, *Journal*, 1865, 43
12 *Islander*, 19 May 1865
13 Macgowan, ed., *Proceedings of Land Commissioners of 1875–76*, 280, 345; also see 346.
14 See, for example, *Examiner*, 25 June 1866.
15 Montgomery to T. Heath Haviland Jr, 15 Sept. 1865, in *Islander*, 13 Oct. 1865
16 Haviland to the Editor of the *Islander*, dated 12 Oct. 1865, in *Islander*, 13 Oct. 1865
17 *Islander*, 25 March 1864
18 Another Alexander Robertson, of Lot 30, was a member of the Central Committee.

19 See 'General Order,' dated 30 Aug. 1865, over the name of P.D. Stewart, AG (adjutant general of the militia), in *Royal Gazette*, 30 Aug. 1865.

20 Montgomery to Haviland, 15 Sept. 1865, in *Islander*, 13 Oct. 1865

21 See Kenneth Henderson in PEI, Legislative Council, *Debates*, 1866, 35; Dundas (in Pitlochry, Perthshire) to Edward Cardwell, 25 Aug. 1865, confidential, CO 226/101, 664–7, microfilm, PAC.

22 *Examiner*, 25 June 1866. In 1875 R.P. Haythorne stated that the 13s. 9d. figure included the costs of the transaction and that the proprietor received approximately 13s. 3d. per acre, a figure that fits perfectly with the acreage of 6,014 cited in PEI, Executive Council Minutes, 9 Jan. 1868, app.; see Macgowan, ed., *Proceedings of Land Commissioners of 1875–76*, 255.

23 Robert Robertson to Charles Palmer, 19 June 1866, in *Examiner*, 25 June 1866; regarding Robertson, see PEI, Executive Council Minutes, 4 April 1865.

24 Legislative Council, *Debates*, 1866, 35

25 Haythorne to the Editor of *Islander*, dated 20 Dec. 1865, in *Islander*, 22 Dec. 1865; see, for example, the comments on the estate in PEI, Executive Council Minutes, 9 Jan. 1868, app.

26 Macgowan, ed., *Proceedings of Land Commissioners of 1875–76*, 336

27 Macgowan, ed., *Proceedings of Land Commissioners of 1875–76*, 49

28 *Herald*, 17 May 1865. The *Herald* copied the report of the Tenant Union meeting from *Ross's Weekly*, 11 May 1865, which does not survive.

29 *Examiner*, 22 May, 19 June 1865. In hearings before the land commission of 1875–6, Bourke stated that 'I sold at the time of the Tenant League for 12 shillings,' but this appears to refer to purchases by isolated individuals. See Macgowan, ed., *Proceedings of Land Commissioners of 1875–76*, 662. Bourke's letter to the Editor of the *Examiner*, dated 25 Sept. 1865, in *Examiner*, 2 Oct. 1865 supports this interpretation.

30 See *Islander*, 1 Sept. 1865, including the letter of Bourke, dated 26 Aug. 1865, 'To the Editor of the *Islander*.' *Ross's Weekly*, 24 Aug. 1865, had, in an issue which does not survive (although the news item is reprinted in *Examiner*, 4 Sept. 1865), treated the Lot 29 'purchase' as an accomplished fact; Bourke was clearly vexed with the newspaper for publishing 'such unblushing untruths.' For additional background to this story, see *Ross's Weekly*, 4 Aug. 1864.

31 *Herald*, 28 June 1865 (original emphasis)

32 *Herald*, 28 June 1865

33 *Examiner*, 4 Sept. 1865

34 Dundas to Duke of Buckingham and Chandos, 20 May 1868, confidential, CO 226/104, 286, 309

35 See Dundas to Buckingham, 20 May 1868, confidential, CO 226/104, 309;

C. Palmer to Haythorne *et al*, 12 May 1865, in *Islander*, 19 May 1865; R. Robertson to C. Palmer, 19 June 1866, in *Examiner*, 25 June 1866; and C. Palmer 'To the Tenants of that part of Township No. 34, lately the Property of the Rev. James F. Montgomery,' 22 June 1866.

36 See Haythorne in PEI, Legislative Council, *Debates*, 1874, 181 and in Macgowan, ed., *Proceedings of Land Commissioners of 1875–76*, 48–9, 336.

37 PEI, Executive Council Minutes, 9 Jan. 1868, app.

38 *Examiner*, 5, 12 June 1865. Also see *Examiner*, 19 June 1865; *Royal Gazette*, 7 June 1865; *Islander*, 9 June 1865.

39 Hodgson to Cardwell, 2 Aug. 1865, CO 226/101, 335; also see PEI, Executive Council Minutes, 6 June 1865.

40 Dundas to Cardwell, confidential, 25 Aug. 1865, CO 226/101, 666

41 Letter of Coles to the Editor of the *Herald*, dated 26 June 1865, in *Herald*, 28 June 1865; PEI, House of Assembly, *Debates and Proceedings*, 1868, 72

42 *Examiner*, 17 April 1865, 12 June 1865; letter of Coles to the Editor of the *Herald*, dated 26 June 1865, in *Herald*, 28 June 1865; letter of 'Philos' to the Editor of the *Herald*, dated 20 June 1865, in *Herald*, 28 June 1865

43 See Ian Ross Robertson, 'Donald McDonald,' DCB, vol. 8 (Toronto, 1985), 530–3.

44 Macgowan, ed., *Proceedings of Land Commissioners of 1875–76*, 565–6

45 Letter of 'Philos' to the Editor of the *Herald*, dated 20 June 1865 from Tracadie Sandhills (now known as Blooming Point), in *Herald*, 28 June 1865

46 *Royal Gazette*, 7 June 1865, 15 Aug. 1866; *Examiner*, 12 June 1865; Irene Rogers, *Charlottetown: The Life in Its Buildings* (Charlottetown, 1983), 13–15

47 Michael J. Winstanley, *Ireland and the Land Question 1800–1922* (London, 1984), 22. W.E. Vaughan examines the phenomenon carefully (prevalence, definition, forms, causes, consequences, importance) in *Landlords and Tenants in Mid-Victorian Ireland* (Oxford, 1994), 138–76 and app. 19–23, on pp. 279–88; the latter are a series of statistical analyses of agrarian outrages. Also see T.W. Moody, *Davitt and Irish Revolution 1846–82* (Oxford, 1981), app. E, F, 565–8.

48 Richards, 'How Tame Were the Highlanders during the Clearances?,' *Scottish Studies*, 17/1 (1973), 42

49 *Islander*, 9 June 1865. W.H. Pope was not the first writer in the 1860s to make the comparison. George Wightman, researcher for the land commission of 1860, did so with respect to the impact of rent arrears on the incentive to improve one's farm: 'in fact, it is Ireland on a small scale!' Pope had quoted him on the point in his correspondence with Sir Samuel Cunard and the Colonial Office during the delegation to London in 1863–4. See PEI, Assembly, *Journal*, 1863, app. A; Pope to Duke of Newcastle, 18 Dec. 1863, in PEI, Assembly, *Journal*, 1864, app. F.

50 *Islander*, 9 June 1865. The 'shooting' reference was an allusion to the misfortunes of John Archibald's father in 1851.
51 *Islander*, 9 June 1865
52 *Islander*, 16 June 1865; the editorial passage came from the *Examiner* of 12 June.
53 *Examiner*, 19, 26 June 1865. One of the two financial contributors to the league was Francis Kelly, an assemblyman representing, with Coles, the area of the Tracadie estate; see PEI, Assembly, *Debates*, 1866, 33.
54 *Islander*, 30 June 1865
55 *Examiner*, 5 June 1865
56 *Examiner*, 12 June 1865
57 *Examiner*, 12 June 1865. The view of Reilly, Whelan's newspaper rival for the attention of Roman Catholics, was that conditions and traditions on the Tracadie estate were such that an 'agrarian outrage' like the fire of 27 May 1865 might have happened, 'Tenant Union or no Tenant Union' (*Herald*, 14 June 1865). He certainly had a point, for trouble on the estate was not limited to the period of the Tenant League's existence, but he was on much weaker ground in his repeated statements (in the 7, 14 June issues, for example) that there was doubt as to whether the fire was deliberately set.
58 See *Examiner*, 15 Aug. 1864; *Monitor*, 4 Aug. 1864; PEI, Assembly, *Journal*, 1865, app. BB.
59 See PEI, Executive Council Minutes, 16 May 1865; *Islander*, 19 May 1865.
60 *Examiner*, 29 May 1865
61 Certainty is impossible, but it appears that the author of the editorials may have been William Kennedy, about whom little else is known. See John Ross, *Reminiscences from the Life of John Ross, Formerly Publisher of 'Ross's Weekly,' the Avowed Champion of the Tenant Union Organization, also Advocate of the Prince Edward Island Railway, and Now Manufacturer and Proprietor of the Celebrated 'Magic Healer' Salve* (Charlottetown, 1892), 18; W.L. Cotton, 'The Press in Prince Edward Island,' in D.A. MacKinnon and A.B. Warburton, eds., *Past and Present of Prince Edward Island ...* (Charlottetown, [1906]), 117.
62 *Examiner*, 19 June 1865. The missing issues of *Ross's Weekly* bore the dates 15 and 22 June.
63 *Examiner*, 26 June 1865
64 *Herald*, 12 July 1865
65 *Examiner*, 19 June 1865
66 *Examiner*, 12 June 1865
67 Letter dated 17 Aug. 1865, CO 226/101, 680. She enclosed a letter from resident proprietor Robert Bruce Stewart to Cardwell, dated 'July 1865,' CO 226/101, 683–5.

68 *Examiner*, 19 June 1865
69 'Notice!' dated 10 July 1865 in *Examiner*, 10 July 1865

7: The Arrest and Bail Hearing of Charles Dickieson

1 CO 226/101, 334, microfilm, PAC
2 PEI, SCCP, 1865, PAPEI
3 *Examiner*, 31 July 1865
4 *Osborn's Concise Law Dictionary*, 6th edition (London, 1976), 146
5 *Herald*, 12 July 1865
6 *Examiner*, 10 July 1865
7 *Protestant*, 31 Dec. 1864; *Herald*, 25 Jan. 1865
8 *Islander*, 14 July 1865
9 'Abstracts of Titles,' CO 226/95, 131
10 H.T. Holman, 'William Douse,' *DCB*, vol. 9 (Toronto, 1976), 223
11 *William Cundall et al. (Executors for William Douse) v. William Large* and *v. George Clow*, PEI, SCCP, 1865, PAPEI
12 Calculations based on PEI, House of Assembly, *Journal*, 1862, app. A; also see John Balderston in PEI, Legislative Council, *Debates and Proceedings*, 1874, 131.
13 Ian Ross Robertson, ed., *The Prince Edward Island Land Commission of 1860* (Fredericton, 1988), 120
14 Cited by Robert Hodgson to Edward Cardwell, 16 Aug. 1865, reprinted in PEI, Assembly, *Journal*, 1866, app. G (original emphasis). According to T. Heath Haviland Jr in PEI, Assembly, *Debates and Proceedings*, 1867, 82, this report first appeared in *Ross's Weekly*.
15 *Islander*, 21 July 1865
16 *Examiner*, 24 July 1865. A *ca. sa.* for Large, dated 15 July, was also issued, and it was returned by Dodd with the statement that he 'is not found within my bailiwick.' See PEI, SCCP, 1865, PAPEI.
17 *Ross's Weekly*, 27 July 1865, in CO 226/101, 403, enclosure in despatch of R. Hodgson to Cardwell, 16 Aug. 1865
18 *Examiner*, 31 July 1865
19 Cundall to Clow, 24 Feb. 1868, 14 April 1869, Henry Jones Cundall Letter-book, PAPEI
20 'Tenancy rate' refers to the percentage of tenants among occupiers of land; calculation based on PEI, Assembly, *Journal*, 1862, app. A. Aside from two townships forming part of the Worrell and the Selkirk estates which the Island government had purchased from landlords (and therefore are not truly comparable to Lot 23), only Lots 20 and 28, with absolutely no 'tenants at will' or squatters, had as small totals of 'tenants at will' + squatters.

21 G.M. Story, 'William Eppes Cormack,' *DCB*, vol. 9, 159; 'CFH' [Catherine F. Horan], 'William Epps Cormack,' *Encyclopedia of Newfoundland and Labrador*, vol. 1 (St John's, 1981), 536; Alan Rayburn, *Geographical Names of Prince Edward Island* (Ottawa, 1973), 92; information on the intertwining of the Cormack and Rennie families comes from file no. 1 on William Frederick Rennie at the Newfoundland Historical Society, St John's.

22 P.S. Macgowan, ed., *Report of Proceedings before the Commissioners Appointed under the Provisions of 'The Land Purchase Act, 1875'* (Charlottetown, 1876), 479

23 'Abstracts of Titles,' CO 226/95, 134

24 Macgowan, ed., *Proceedings of Land Commissioners of 1875–76*, 477. Ball was testifying specifically with reference to the two-thirds of the estate under the administration of Robert Rennie in 1875, but he knew the entire estate well, and he stated regarding the remaining one-third that it 'is nearly as valuable as the other part in proportion' (478).

25 Robertson, ed., *The Prince Edward Island Land Commission of 1860*, 43

26 See *Ross's Weekly*, 10 Aug. 1865 in CO 226/101, 404, enclosure in despatch of R. Hodgson to Cardwell, 16 Aug. 1865.

27 Macgowan, ed., *Proceedings of Land Commissioners of 1875–76*, 477; *Islander*, 10 March 1865

28 Robertson, ed., *The Prince Edward Island Land Commission of 1860*, 43; PEI, Assembly, *Journal*, 1862, app. A

29 Based on the 1861 census in PEI, Assembly, *Journal*, 1862, app. A; at the land commission of 1875–6, the Prince Edward Island attorney general had a theory as to how this had happened; see Frederick Brecken in Macgowan, ed., *Proceedings of Land Commissioners of 1875–76*, 484; and testimony by George Smith, by Jacob Ling, and by Fermain Peno on 487, 493–5.

30 See E.J. Hodgson in Macgowan, ed., *Proceedings of Land Commissioners of 1875–76*, 475–6.

31 *Robert Rennie, Administrator, v. Charles Dickieson*, and *v. James Proctor*, PEI, SCCP, 1865, PAPEI. Peter McGuigan has published an account of the events of 18 July, and also 26 July, when Dickieson was taken before two justices of the peace in Charlottetown, in 'Tenants and Troopers: The Hazel Grove Road, 1865–68,' *Island Magazine*, no. 32 (Fall–Winter 1992), 24–5.

32 E.J. Hodgson to George Dundas, 19 July 1865, Accession 2514/10, PAPEI. One may wonder about Hodgson's motive in writing directly to the lieutenant governor on this matter; perhaps he was dissatisfied with the Executive Council and hoped, through Dundas, to prod it into more vigorous support of the sheriff.

33 Affidavit of Jonathan Collings, 19 July 1865, PEI, SCCP, 1865, PAPEI

34 Affidavit of James Curtis, 19 July 1865, Accession 2514/10, PAPEI

35 Affidavit of Curtis, 19 July 1865, Accession 2514/10, PAPEI

36 Affidavit of Curtis, 19 July 1865, Accession 2514/10, PAPEI. The term 'rescue' is used here in its legal meaning of 'forcibly taking back goods which have been distrained and are being taken to the pound.' Such action is known as 'pound-breach.' *Osborn's Concise Law Dictionary*, 6th edition, 290. McGuigan has reported that a granddaughter of Dickieson recalls that he emphatically denied having been at the New Glasgow bridge, and stated instead that he had joined the protesting crowd later. It should be noted, though, that Collings as well as Curtis gave sworn evidence that Dickieson was present at the New Glasgow bridge, and that, in any event, this is a matter of detail which has no bearing on the charges which would be laid against Dickieson as a result of the events of 18 July. See 'Tenants and Troopers,' 24.

37 Affidavit of Collings, 19 July 1865, PEI, SCCP, 1865, PAPEI

38 'The examination and evidence of James Curtis, Deputy Sheriff of Queens County, taken this 26th day of July 1865,' PEI, SCCP, 1865, PAPEI

39 Affidavit of Curtis, 19 July 1865, Accession 2514/10, PAPEI

40 Examination and evidence of Curtis, 26 July 1865, PEI, SCCP, 1865, PAPEI

41 Examination and evidence of Curtis, 26 July 1865, PEI, SCCP, 1865, PAPEI. The 'Curtis Dale Hotel' is shown on a detailed map of Prince Edward Island which D.J. Lake published in 1863: *Topographical Map of Prince Edward Island in the Gulf of St. Lawrence. From actual Surveys and the late Coast Survey of Capt. H.W. Bayfield* (Saint John, 1863); as a place name, 'Curtisdale' does not appear to have survived. See also Mrs Frank Bagnall, 'Road Houses and Taverns,' in Mary Brehaut, ed., *Historic Highlights of Prince Edward Island* (Charlottetown, 1955), 55.

42 Affidavit of Curtis, 19 July 1865, Accession 2514/10, PAPEI

43 Examination and evidence of Curtis, 26 July 1865, PEI, SCCP, 1865, PAPEI

44 Affidavit of Collings, 19 July 1865, PEI, SCCP, 1865, PAPEI

45 On the term 'longer,' defined as 'a long, unfinished wooden pole used especially for fencing,' see T.K. Pratt, ed., *Dictionary of Prince Edward Island English* (Toronto, 1988), 92–3; recorded references commence in 1822.

46 Affidavit of Curtis, 19 July 1865, Accession 2514/10, PAPEI

47 Examination and evidence of Curtis, 26 July 1865, PEI, SCCP, 1865, PAPEI

48 By 26 July 1865 Collings had identified this man as George Rackem; see examination and evidence of Jonathan Collings, 26 July 1865, PEI, SCCP, 1865, PAPEI. Rackem was never apprehended.

49 Affidavit of Collings, 19 July 1865, PEI, SCCP, 1865, PAPEI

50 *Examiner*, 24 July 1865

51 Examination and evidence of Collings, 26 July 1865; deposition of Jonathan Collings, 17 Aug. 1865, PEI, SCCP, 1865, PAPEI

52 *Examiner*, 29 Jan. 1866. It should be noted that no one disputed that there had been a violent attack on Curtis, whose arm was broken. But another contrary theory as to Dickieson's role might focus on the suggestion in the evidence that Dickieson and Collings may have known each other prior to 18 July, and that there may have been bad feeling between them. If they had indeed had a previous unpleasant encounter, this might have given the bailiff a motive to arrest Dickieson, and once he had been arrested it would have been natural for Curtis and Collings to picture their prisoner, with whom they returned after losing the distrained goods to the Tenant Leaguers, as the 'ringleader' and as the actual perpetrator of violence, rather than as simply one of the crowd. Both produced affidavits on the following day which described Dickieson as the most violent of the crowd and the 'ringleader.' Yet no one at the time – among the newspapers, not even *Ross's Weekly* – expressed doubt that Dickieson played a leading and aggressive role in the confrontation. Moreover, it is by no means certain that Collings and Dickieson had clashed personally prior to 18 July; given that Collings had served as bailiff and constable 'for a number of years past,' he may have been recognizable to Dickieson simply by virtue of that fact (affidavit of Collings, 19 July 1865, PEI, SCCP, 1865, PAPEI).

53 *Ross's Weekly*, 27 July 1865, in CO 226/101, 403, enclosure in despatch of R. Hodgson to Cardwell, 16 Aug. 1865. Two weeks later *Ross's Weekly* was taking the line that (a) Dickieson had been acting on the 'impression' that Curtis had exceeded his authority in refusing to complete arrangements with Proctor for the return of the distrained goods on the road, and (b) Curtis *et al.* 'were the more guilty party' (*Ross's Weekly*, 10 Aug. 1865, in CO 226/101, 404, enclosure in despatch of R. Hodgson to Cardwell, 16 Aug. 1865).

54 PEI, Legislative Council, *Debates*, 1870, 125

55 *Islander*, 21 July 1865

56 Dodd to W.H. Pope, 19 July 1865, Accession 2514/10. Both the *Islander*, 21 July 1865 and the *Examiner*, 24 July 1865 gave the distance as seven miles.

57 DesBrisay to J.C. Pope, 21 July 1865, Accession 2514/10, PAPEI

58 R. Hodgson to Joseph Pope, 6 Sept. 1865, Pope Scrapbook, PAPEI

59 Ian Ross Robertson, 'Sir Robert Hodgson,' *DCB*, vol. 10 (Toronto, 1972), 352–3

60 PEI, Executive Council Minutes, 21 July 1865; *Royal Gazette* 'Extra,' 22 July 1865. The decision to fire Laird as a justice of the peace does not show up in the Executive Council Minutes, but in a despatch dated 25 Sept. 1865 Hodgson informed Cardwell that Laird had been 'summarily dismissed' on the advice of Council; CO 226/101, 431. According to *Examiner*, 10 July 1865, Laird had resigned; if so, presumably the purpose of the dismissal was simply to make the point that if the severance had not been voluntary, it would

have occurred involuntarily. This was not the first controversy over the connection or apparent connection of a justice of the peace with the Tenant League, but in the case of Thomas Beers it appears that no disciplinary action had been taken.

61 John McNeill, Secretary of the Board of Education, to the Clerk of the Executive Council, 5 Oct. 1865; Extract from the Minutes of the Board of Education (printed), 27 July 1865, Accession 2514/9, PAPEI

62 Dundas to Cardwell, 25 Aug. 1865, confidential, CO 226/101, 667

63 Affidavit of Bernard McKenna, 5 Aug. 1865, Accession 2514/10, PAPEI. The incident is reminiscent of scenes which historian Michael S. Cross recounts as occurring in Lower Canada in 1850; see '"The Laws Are Like Cobwebs": Popular Resistance to Authority in Mid-Nineteenth Century British North America,' in Peter B. Waite, Sandra Oxner, and Thomas Barnes, eds., *Law in a Colonial Society: The Nova Scotia Experience* (Toronto, 1984), 103.

64 R. Hodgson to Cardwell, 2 Aug. 1865, in CO 226/101, 331. This despatch is the sole source, and Hodgson did not attest to a precise date ('On or about the 22nd day of July'); it does not appear to be the same incident as the one which occurred in Fort Augustus.

65 'Examination' is used here in its legal sense as 'interrogation of a person on oath,' the 'person(s)' here being the witnesses. *Osborn's Concise Law Dictionary*, 6th edition, 138. Witnesses did not include the defendant, since in English law generally an accused was not considered competent as a sworn witness on his own behalf through much of the nineteenth century (information on who was competent to be a witness courtesy of Allan Q. Shipley, a member of the Ontario bar, 1 Sept. 1992). McGuigan's account in 'Tenants and Troopers,' 25 implies that the events of 26 July occurred on 25 July.

66 Marianne Morrow, 'The Builder: Isaac Smith and Early Island Architecture,' *Island Magazine*, no. 18 (Fall–Winter 1985), 18. For descriptions of the jail, 'the old court house'-city hall, and other relevant parts of Queen Square, see Morrow, 'The Builder,' 17–23; Irene Rogers, *Charlottetown: The Life in Its Buildings* (Charlottetown, 1983), 212–13 (with a photograph of the 'old court house' on 4); Mary K. Cullen, 'Charlottetown Market Houses: 1813–1958,' *Island Magazine*, no. 6 (Spring–Summer 1979), 27–30. The 'old court house' was known as such because the Supreme Court had sat there until construction of the Colonial Building in the 1840s to accommodate it and the legislature.

67 *Islander*, 28 July 1865

68 *Examiner*, 31 July 1865

69 *Islander*, 28 July 1865

70 R. Hodgson to Cardwell, 2 Aug. 1865, printed in PEI, Assembly, *Journal*, 1866,

app. G. Also see Dodd to W.H. Pope, 26 July 1865, Accession 2514/10, PAPEI; *Islander*, 28 July 1865.

71 W.H. Pope to Dodd, 26 July 1865, Accession 2514/10, PAPEI

72 *Examiner*, 31 July 1865; also see R. Hodgson to Cardwell, 2 Aug. 1865, printed in PEI, Assembly, *Journal*, 1866, app. G.

73 Examination and evidence of Curtis, 26 July 1865; examination and evidence of Collings, 26 July 1865; recognizance of J. Laird and John Dickieson for appearance of Charles Dickieson at the Supreme Court, 26 July 1865; *Robert Rennie, Administrator, v. Charles Dickieson*, writ of *ca. sa.*, 17 July 1865, PEI, SCCP, 1865, PAPEI; *Examiner*, 31 July 1865. *Ross's Weekly* reported the bail as £400 rather than £200; yet the actual recognizance, which survives, has each of the two sureties committed to the extent of £100. Evidently, *Ross's Weekly* mistook the fact that they together had given bail for £200 as meaning £200 each. McGuigan, 'Tenants and Troopers,' 25 follows *Ross's Weekly*. This mistake by *Ross's Weekly* conformed to a pattern: it reported the number of special constables as fifty, that is, double the actual number. See *Ross's Weekly*, 27 July 1865, in CO 226/101, 404, enclosure in despatch of R. Hodgson to Cardwell, 16 Aug. 1865.

74 This account of the riot is based chiefly on *Islander*, 28 July 1865; *Examiner*, 31 July 1865; *Herald*, 2 Aug. 1865; R. Hodgson to Cardwell, 2 Aug. 1865, printed in PEI, Assembly, *Journal*, 1866, app. G; *Ross's Weekly*, 27 July 1865, in CO 226/101, 403–4, enclosure in despatch of R. Hodgson to Cardwell, 16 Aug. 1865.

75 *Examiner*, 31 July 1865

76 *Ross's Weekly*, 27 July 1865, in CO 226/101, 404, enclosure in despatch of R. Hodgson to Cardwell, 16 Aug. 1865

77 PEI, Assembly, *Debates*, 1866, 24

78 See Samuel Clark, *Social Origins of the Irish Land War* (Princeton, 1979), 309–11.

79 R. Hodgson to Cardwell, 2 Aug. 1865, printed in PEI, Assembly, *Journal*, 1866, app. G

80 *Examiner*, 31 July 1865

81 Morrison to E. Palmer, 27 July 1865, Colonial Secretary's Letterbooks, PAPEI

82 *Islander*, 28 July 1865

83 *Ross's Weekly*, 27 July 1865, in CO 226/101, 404, enclosure in despatch of R. Hodgson to Cardwell, 16 Aug. 1865

84 *Herald*, 2, 30 Aug. 1865

85 *Ross's Weekly*, 27 July 1865, in CO 226/101, 404, enclosure in despatch of R. Hodgson to Cardwell, 16 Aug. 1865

86 R. Hodgson to Cardwell, 2 Aug. 1865, printed in PEI, Assembly, *Journal*, 1866, app. G

87 *Examiner*, 24 July 1865

88 *Ross's Weekly*, 27 July 1865 (original emphasis), in co 226/101, 404, enclosure in despatch of R. Hodgson to Cardwell, 16 Aug. 1865

89 *Examiner*, 31 July 1865

90 Interview with Roy Dickieson of New Glasgow, PEI, 27 Aug. 1986

91 *Ross's Weekly*, 27 July 1865, in co 226/101, 404, enclosure in despatch of R. Hodgson to Cardwell, 16 Aug. 1865

92 *Examiner*, 31 July 1865

93 *Islander*, 18 Aug. 1865 (original emphasis)

94 *Islander*, 21 July 1865

95 Robertson, ed., *The Prince Edward Island Land Commission of 1860*, 45

96 PEI, Executive Council Minutes, 1 Aug. 1865; R. Hodgson to Cardwell, 2 Aug. 1865, in PEI, Assembly, *Journal*, 1866, app. G. For a spirited and comprehensive defence of the decision to call upon the military, see *Islander*, 18 Jan. 1867; from one perspective, this editorial could be regarded as a pre-election 'justifying' exercise.

97 *Examiner*, 29 Jan. 1866. C. Palmer also stated that there was 'a desperate personal enmity' between Curtis and Dickieson; this flies in the face of a statement Curtis made under oath, that he 'did not know the name of the prisoner until driving up the hill [in his wagon, after the arrest], when he said his name was Charles Dickieson'; examination and evidence of Curtis, 26 July 1865, PEI, SCCP, 1865, PAPEI.

98 *Examiner*, 24 July 1865

99 *Islander*, 26 Jan. 1866

100 Interviews with Roy Dickieson, 27 Aug. 1986, 4 Sept. 1994

101 Telephone interviews with Dr Herb Dickieson of O'Leary, PEI, 20 Nov. 1992, and Roy Dickieson, 21 April 1995; CBC television, 'Prime Time News,' 18 June 1993; interviews with Roy Dickieson and Dr Herb Dickieson, 4 Sept. 1994; Dedication to Dana Dickieson, *Prince Edward Island Holstein Directory*, n.d., p. 2. I am grateful to Harry Baglole, Director of the Institute of Island Studies, University of Prince Edward Island, for his indispensable assistance in putting me in touch with both Roy Dickieson and Dr Herb Dickieson.

8: The Beginning of Repression

1 Printed in PEI, House of Assembly, *Journal*, 1866, app. G

2 Printed in PEI, Assembly, *Journal*, 1866, app. G

3 co 226/101, 666, microfilm, PAC

4 Robert Hodgson to Sir Richard Graves MacDonnell, 1 Aug. 1865, printed in PEI, Assembly, *Journal*, 1866, app. G

5 George Dundas to Edward Cardwell, 23 March 1865, printed in PEI, Assembly, *Journal*, 1866, app. G

6 Hodgson to Cardwell, 2 Aug. 1865, printed in PEI, Assembly, *Journal*, 1866, app. G

7 Hodgson to Cardwell, 2 Aug. 1865, printed in PEI, Assembly, *Journal*, 1866, app. G

8 See John Longworth in PEI, House of Assembly, *Debates and Proceedings*, 1866, 68.

9 *Examiner*, 4 Sept. 1865; also see 14 Aug. 1865

10 P.D. Stewart to Dundas, 18 April 1866, in PEI, Assembly, *Journal*, 1866, app. Z

11 Hodgson to Cardwell, 2 Aug. 1865, printed in PEI, Assembly, *Journal*, 1866, app. G

12 Scott W. See, '"Mickeys and Demons" vs. "Bigots and Boobies": The Woodstock Riot of 1847,' *Acadiensis*, 21/1 (Autumn 1991), 122; *Riots in New Brunswick: Orange Nativism and Social Violence in the 1840s* (Toronto, 1993), 122, 127

13 Hodgson to MacDonnell, 1 Aug. 1865, printed in PEI, Assembly, *Journal*, 1866, app. G (emphasis added)

14 3 Aug. 1865, nos. 1 and 6, Garrison Orders, vol. 111, HQOP, PANS. For a list of the eight officers in Prince Edward Island as of 11 Aug., see CO 226/101, 662.

15 Colonel J.H. Francklyn to Hodgson, 4 Aug. 1865, printed in PEI, Assembly, *Journal*, 1866, app. G

16 6 June 1865, no. 1, General Orders, vol. 59, HQOP, PANS. Although contained under the General Orders of 6 June, Doyle's remarks were dated 5 June. See R.H. McDonald, 'Sir Charles Hastings Doyle,' *DCB*, vol. 11 (Toronto, 1982), 278–81.

17 Carol M. Whitfield, *Tommy Atkins: The British Soldier in Canada, 1759–1870* (Ottawa, 1981), 71; also see 59, 70; Peter Burroughs, 'Tackling Army Desertion in British North America,' *Canadian Historical Review*, 61/1 (March 1980), 41n31.

18 See report of Captain J.D.G. Tulloch, n.d. [1850], War Office 1/563, 184–6, microfilm, PAC; PEI, Assembly, *Journal*, 1852, app. S; PEI, Assembly, *Journal*, 1853, app. J; PEI, Assembly, 1854, *Journal*, app. E; PEI, Assembly, *Journal*, 1855, app. F.

19 See Paul Bew, *Land and the National Question in Ireland 1858–82* (Dublin, 1978), 121–6, 155–61, 170–4, 221–3, who argues the importance of 'rent at the point of the bayonet' as a strategy of confrontation (in comparison with boycotting or mass resistance to process-servers), and makes the interesting point that the dollars of Irish-American sympathizers were invaluable in supplying the means to finance the cumulatively enormous legal costs

entailed. Tenant Leaguers in Prince Edward Island are not known to have had off-shore financial backers with deep pockets.

20 See, for example, *Examiner*, 23 Oct. 1865.

21 Hodgson to Francklyn, 8 Aug. 1865, printed in PEI, Assembly, *Journal*, 1866, app. G. In this communication, which dealt mainly with the protocol to be observed when requesting military aid to the civil power, Hodgson did state that the 'presence' of the troops would 'prevent' defiance of the sheriff, implying that the planned role for them was passive rather than proactive.

22 *Islander*, 11 Aug. 1865

23 *Royal Gazette*, 16 Aug. 1865

24 PEI, Executive Council Minutes, 14 Aug. 1865

25 *Examiner*, 14 Aug. 1865

26 Doyle to Hodgson, 22 Aug. 1865, printed in PEI, Assembly, *Journal*, 1866, app. G

27 Hodgson to Doyle, 25 Aug. 1865, printed in PEI, Assembly, *Journal*, 1866, app. G

28 Tydd to Hodgson, 24 Aug. 1865, printed in PEI, Assembly, *Journal*, 1866, app. G

29 Hodgson to Cardwell, 16 Aug. (despatch no. 67), 30 Aug. (despatch no. 70) 1865, printed in PEI, Assembly, *Journal*, 1866, app. G

30 Hodgson to Cardwell, 13 Sept. 1865, CO 226/101, 426–7; memorandum of Arthur Blackwood, 25 Sept. 1865, CO 226/101, 427; *Islander*, 8 Sept. 1865

31 Regarding Doucette's role at Curtisdale, see depositions by James Curtis and Jonathan Collings, both dated 17 Aug. 1865, in PEI, SCCP, 1865, PAPEI.

32 *Islander*, 26 Jan. 1866

33 Dodd to W.H. Pope, 15 Aug. 1865, printed in PEI, Assembly, *Journal*, 1866, app. G

34 Dodd to W.H. Pope, 15 Aug. 1865, printed in PEI, Assembly, *Journal*, 1866, app. G. Also see deposition of James Curtis regarding events on 15 Aug. 1865, dated 17 Aug. 1865, in SCCP, 1865, PAPEI; all further references in this chapter to a deposition sworn by Curtis on 17 Aug. will signify this document.

35 Deposition of Curtis, 17 Aug. 1865, SCCP, 1865, PAPEI

36 Dodd to W.H. Pope, 15 Aug. 1865, printed in PEI, Assembly, *Journal*, 1866, app. G

37 Deposition of Curtis, 17 Aug. 1865, SCCP, 1865, PAPEI

38 Dodd to W.H. Pope, 15 Aug. 1865, printed in PEI, Assembly, *Journal*, 1866, app. G; also see deposition of Curtis, 17 Aug. 1865, SCCP, 1865, PAPEI.

39 Dodd to W.H. Pope, 15 Aug. 1865, printed in PEI, Assembly, *Journal*, 1866, app. G

40 See Rowe to Dodd, dated 2 July 1867, in *Herald*, 28 Aug. 1867. Also see letters

of Rowe, dated 25 Sept., 23 Dec. 1867, in *Herald*, 2 Oct. 1867, 15 Jan. 1868 and letter from 'A Resident of Queen's County,' dated 10 Dec. 1867, in *Herald*, 18 Dec. 1867.

41 PEI, Assembly, *Debates*, 1868, 18–19

42 Deposition of Curtis, 17 Aug. 1865, SCCP, 1865, PAPEI

43 See Dodd to Hodgson, 29 Aug. 1865, Accession 2514/10, PAPEI.

44 Dodd to W.H. Pope, 15 Aug. 1865, printed in PEI, Assembly, *Journal*, 1866, app. G; deposition of Curtis, 17 Aug. 1865, SCCP, 1865, PAPEI

45 Hodgson to Cardwell, 16 Aug. 1865 (despatch no. 67), printed in PEI, Assembly, *Journal*, 1866, app. G. See, for example, PEI, Assembly, *Journal*, 1838, 39; 1840, app. L, 105, 106; 1843, 69; 1844, 61; 1845, 45.

46 Notice of W.H. Pope, dated 21 Aug. 1865, in *Royal Gazette*, 23 Aug. 1865

47 See Marjorie L. Hyndman Coffin, 'Joseph Wightman,' *DCB*, vol. 11, 922–3.

48 Ian Ross Robertson, ed., *The Prince Edward Island Land Commission of 1860* (Fredericton, 1988), 130

49 PEI, Executive Council Minutes, 1, 14 Aug. 1865; C. DesBrisay to Wightman, 2 Aug. 1865, and Wightman to C. DesBrisay, 5 Aug. 1865, in *Royal Gazette*, 16 Aug. 1865

50 PEI, Executive Council Minutes, 1 Aug. 1865

51 Balderston to C. DesBrisay, 17 Aug. 1865, Accession 2514/10, PAPEI

52 PEI, Executive Council Minutes, 29 Aug., 19 Sept. 1865. For Campbell's perspective, see PEI, Assembly, *Debates*, 1874, 67, 71, 128. He would become a justice of the peace once again in the late 1870s, appointed by the government of Louis Henry Davies; see W.W. Sullivan and Campbell in PEI, Assembly, *Debates*, 1877, 212–13.

53 See J.D. Gordon and D. Laird, eds., *Abstract of the Proceedings before the Land Commissioners' Court, Held During the Summer of 1860, to Inquire into the Differences Relative to the Rights of Landowners and Tenants in Prince Edward Island* (Charlottetown, 1862), Proceedings, 40, 64, 71–77.

54 See Robertson, ed., *The Prince Edward Island Land Commission of 1860*, Glossary of Names, 210–11.

55 Bronwen Douglas and Bruce W. Hodgins, 'George Nicol Gordon,' *DCB*, vol. 9 (Toronto, 1976), 325–6; Bruce W. Hodgins, 'James Douglas Gordon,' *DCB*, vol. 10 (Toronto, 1972), 308

56 PEI, Executive Council Minutes, 14 Aug., 19 Sept. 1865, 1 July 1867; Gordon to C. DesBrisay, 30 Aug. 1865, Accession 2514/10, PAPEI; memorial of Gordon, 10 Jan. 1866, CO 226/102, 33–41

57 Letter of Cornelius Richard O'Leary to the Editor of *Patriot*, dated 30 Jan. 1866, in *Herald*, 7 March 1866

58 See Stewart to Dundas, 18 April 1866, in PEI, Assembly, *Journal*, 1866, app. Z;

'General Order,' dated 30 Aug. 1865, over the name of Stewart, in *Royal Gazette*, 30 Aug. 1865.

59 See PEI, Executive Council Minutes, 14 Aug., 19 Sept. 1865; affidavit of McKenna, dated 5 Aug. 1865, Accession 2514/10, PAPEI.

60 Calculation based on data in *Islander*, 4 Jan. 1867

61 Extract from the Minutes of the Board of Education (printed), 27 July 1865, Accession 2514/9, PAPEI

62 John McNeill, Secretary of the Board of Education, to the Clerk of the Executive Council, 5 Oct. 1865, Accession 2514/9, PAPEI

63 See M. Brook Taylor, 'John Longworth,' *DCB*, vol. 11, 528–9; Ian Ross Robertson, 'Edward Palmer,' *DCB*, vol. 11, 664–70; Andrew Robb, 'Thomas Heath Haviland [Jr],' *DCB*, vol. 12 (Toronto, 1990), 415–18. It appears that this committee may have been established on 2 September; see PEI, Executive Council Minutes, 19 Sept. 1865.

64 *Royal Gazette*, 20 Sept. 1865

65 PEI, Executive Council Minutes, 7 Oct. 1865

66 Palmer to C. DesBrisay, 16 Oct. 1865, Accession 2514/9, PAPEI

67 PEI, Executive Council Minutes, 13 Nov. 1865

68 Palmer to C. DesBrisay, 2 Dec. 1865, Accession 2514/9, PAPEI; PEI, Executive Council Minutes, 19 Dec. 1865

69 J. McNeill to C. DesBrisay, 2 Dec. 1865, Accession 2514/9, PAPEI

70 J. McNeill to C. DesBrisay, 7 Feb. 1866, in *Islander*, 23 Feb. 1866; PEI, Executive Council Minutes, 20 Feb. 1866

71 *Royal Gazette*, 21 Feb. 1866; *Islander*, 23 Feb. 1866; *Examiner*, 26 Feb. 1866; *Herald*, 28 Feb. 1866

72 *Herald*, 28 Feb. 1866

73 *Islander*, 13 July 1866

74 Hodgson to MacDonnell, 1 Aug. 1865, printed in PEI, Assembly, *Journal*, 1866, app. G

75 Dundas to Cardwell, 25 Aug. 1865, confidential, CO 226/101, 665

76 *Islander*, 14 July 1865

77 See *Islander*, 17 Aug. 1866.

78 Palmer to C. DesBrisay, 16 Oct. 1865, Accession 2514/9, PAPEI (emphasis added)

79 See Tom Garvin, 'Defenders, Ribbonmen and Others: Underground Political Networks in Pre-Famine Ireland,' *Past and Present*, no. 96 (Aug. 1982), 137; and M.R. Beames, 'The Ribbon Societies: Lower-Class Nationalism in Pre-Famine Ireland,' *Past and Present*, no. 97 (Nov. 1982), 134–5.

80 Kevin B. Nowlan, 'Conclusion,' in T. Desmond Williams, ed., *Secret Societies in Ireland* (Dublin, 1973), 180

81 Paul E.W. Roberts, 'Caravats and Shanavests: Whiteboyism and Faction Fighting in East Munster, 1802–11,' in Samuel Clark and James S. Donnelly Jr, eds., *Irish Peasants: Violence and Political Unrest 1780–1914* (Madison, 1983), 64, 66

82 Gale E. Christianson, 'Secret Societies and Agrarian Violence in Ireland, 1790–1840,' *Agricultural History*, 46/3 (July 1972), 372–3, 376–9

83 Roberts, 'Caravats and Shanavests,' 79n60 (emphasis added); 66; also see 82–6.

84 Roberts, 'Caravats and Shanavests,' 67; also see 66, 87–90, 93. The Caravats, as the representatives of the poor, were more numerous than the Shanavests, who had to rely on the middle class and their allies for recruits. Therefore, the Caravats usually prevailed at fair-ground fights; they also bore the brunt of the punishment meted out by the special commission. The two sides distinguished themselves by sporting, respectively, cravats and old waistcoats, emblems identified with the peculiarities of dress their archetypal leaders favoured. In fact their names derived from these distinctions of dress, and as alternate names they had, respectively, Cravats and Old Waistcoats. For a fascinating analysis of the symbolism embedded in the sartorial emblems, see 68–73.

85 Note, for example, the initial reaction of the Roman Catholic church in North America to the Knights of Labor, founded in 1869. In the view of the founding leadership of this union movement of working people, which G.S. Kealey describes as 'the major labour reform organization of the late 19th century,' secrecy with respect to membership was vital because of the hostility of employers, which could result in firing or blacklisting. See 'Knights of Labor,' *The Canadian Encyclopedia*, 2nd edition (Edmonton, 1988), vol. 2, 1144.

86 'Preface,' in Williams, ed., *Secret Societies in Ireland*, ix. On the origins and popular legitimacy of such bodies, see Maureen Wall, 'The Age of the Penal Laws (1691–1778),' in T.W. Moody and F.X. Martin, eds., *The Course of Irish History* (Cork, 1967), 228–9.

87 See Garvin, 'Defenders, Ribbonmen and Others,' 148; Beames, 'The Ribbon Societies,' 139; Donal McCartney, 'The Churches and Secret Societies,' in Williams, ed., *Secret Societies in Ireland*, 68–78.

88 Nowlan, 'Conclusion,' in Williams, ed., *Secret Societies in Ireland*, 180

89 R.V. Comerford, 'Ireland 1850–70: Post-Famine and Mid-Victorian,' in W.E. Vaughan, ed., *A New History of Ireland*, vol. 5: *Ireland under the Union, I (1801–70)* (Oxford, 1989), 390

90 Dickieson to the Editor of the *Islander*, dated 23 Aug. 1866, in *Islander*, 31 Aug. 1866. Also see Benjamin Davies and W.S. McNeill in PEI, Assembly, *Debates*, 1868, 18, 19.

91 Rowe to the Editor of the *Herald*, dated 1 Jan. 1867 [*sic*; 1868], in *Herald*, 22 Jan. 1868
92 Cardwell, memorandum dated 12 Aug. 1865, CO 226/101, 439. See E.D. Steele, *Irish Land and British Politics: Tenant-Right and Nationality 1865–1870* (Cambridge, 1974), 68–70; J.C. Brady, 'Legal Developments, 1801–79,' in Vaughan, ed., *A New History of Ireland*, vol. 5, 457–63.
93 *Dictionary of National Biography*, vol. 17, 120
94 Rogers, memorandum dated 13 Aug. 1865, CO 226/101, 440
95 Elliot, memorandum dated 1 March 1866, CO 226/102, 29
96 Disregarding the rescue attempt on 26 July 1865
97 Morrison to the attorney general and the solicitor general, 5 Aug. 1865, Colonial Secretary's Letterbooks, PAPEI
98 Dundas to Cardwell, 25 Aug. 1865, confidential, CO 226/101, 666–7
99 Palmer to C. DesBrisay, 16 Oct. 1865, Accession 2514/9, PAPEI
100 PEI, Executive Council Minutes, 7 Oct. 1865
101 PEI, Executive Council Minutes, 13 Nov. 1865
102 The title of the article is given according to 'The Queen *v.* John Ross,' Indictment for Libel, marked 'No Bill,' PEI, SCCP, 1866, PAPEI, although the Executive Council Minutes cited a slightly different version of the title, namely, 'Courts of Law and Tenant Union'; PEI, Executive Council Minutes, 13 Nov. 1865.
103 PEI, Executive Council Minutes, 12 Dec. 1865

9: Military Desertion and the Tenant League

1 *Islander*, 3 Nov. 1865
2 Printed in PEI, House of Assembly, *Journal*, 1866, app. G
3 *Islander*, 22 Sept., 13 Oct. 1865; *Royal Gazette*, 6, 13, 20 Sept. 1865; PEI, Assembly, *Journal*, 1866, app. Q
4 Donald Graves, '"Tommy" in Canada,' *Horizon Canada*, 2/20 (July 1985), 466
5 For a complete list of the seventeen, with descriptions, see *Royal Gazette*, 20 Sept. 1865.
6 J.H. Francklyn to Robert Hodgson, 4 Aug. 1865, printed in PEI, Assembly, *Journal*, 1866, app. G
7 Carol M. Whitfield, *Tommy Atkins: The British Soldier in Canada, 1759–1870* (Ottawa, 1981), app. C, table 1, 143. Whitfield has made her calculations on the assumption that 'the true desertion rate' was the sum of the men reported 'absent without leave' and the men reported as having deserted; see 137–8. The reference work by David J. Bercuson and J.L. Granatstein, *Dictionary of Canadian Military History* (Toronto, 1992), contains no entry for desertion or for military discipline in general.

8 Consequently, the 'strangers' became natural recruits for the Shanavests, the group militantly opposed to the Caravats. See Paul E.W. Roberts, 'Caravats and Shanavests: Whiteboyism and Faction Fighting in East Munster, 1802–11,' in Samuel Clark and James S. Donnelly Jr, eds., *Irish Peasants: Violence and Political Unrest 1780–1914* (Madison, 1983), 83–4, 90.

9 Whitfield, *Tommy Atkins*, 61. Also see 60, 62, 67, app. C, 140; Peter Burroughs, 'Tackling Army Desertion in British North America,' *Canadian Historical Review*, 61/1 (March 1980), 32–7, 48, 50–2, 57.

10 Note the editorial entitled 'Desertions' in *Islander*, 8 Sept. 1865; and Kennedy Wells, *The Fishery of Prince Edward Island* (Charlottetown, 1986), 108–9, 113–25.

11 See Whitfield, *Tommy Atkins*, 62.

12 Campbell to Earl Grey, 2 Jan. 1850, in PEI, Assembly, *Journal*, 1852, app. S; Bannerman to Sir John Pakington, 9 April 1852, in PEI, Assembly, *Journal*, 1853, app. J. Historians have tended to cite Campbell uncritically, and also comments made by Sir John Harvey (when he was lieutenant governor of Nova Scotia), who was relying on him, although Bannerman explicitly contradicted his predecessor. See, for example, Burroughs, 'Tackling Army Desertion in British North America,' 33, 42–3; Whitfield, *Tommy Atkins*, app. C, 140, and 216n9. One military historian who is an exception to this generalization is J. Mackay Hitsman, 'Military Defenders of Prince Edward Island 1775–1864,' *Annual Report of the Canadian Historical Association*, 1964, 33. No researcher has investigated thoroughly the question of pre-1854 desertion on the Island.

13 *Royal Gazette*, 30 Aug. 1865

14 For the relevant provisions of the statute, see *Royal Gazette*, 20 Sept. 1865.

15 *Islander*, 8 Sept. 1865

16 *Royal Gazette*, 20 Sept. 1865

17 Whitfield, *Tommy Atkins*, 46; also see 45, 200n28.

18 Private Thomas Goodman, age twenty-two, from Newry, Ireland, was one of the deserters never captured; see *Royal Gazette*, 16 Aug. 1865.

19 *Royal Gazette*, 20 Sept. 1865

20 *Royal Gazette*, 20 Sept. 1865

21 *Examiner*, 18 Sept. 1865

22 *Royal Gazette*, 20 Sept. 1865. Brennan would apply for a reward for apprehending a deserter, but the Executive Council rejected the application because there was no evidence that Glynn had been convicted of desertion. There was a precedent for such a rejection. In an earlier instance, Deputy Sheriff James Curtis had applied for £15 for apprehension of a deserter; 'The Board declined to allow this account – it not being certified by the Officer in

command of the Detachment.' PEI, Executive Council Minutes, 6 Feb. 1866, 2 Sept. 1865.

23 *Royal Gazette,* 20 Sept. 1865

24 *Examiner,* 18 Sept. 1865

25 *Herald,* 4 Oct. 1865

26 PEI, Executive Council Minutes, 7 Oct. 1865. Depending on the source, her name is given as 'Halloran' or 'O'Halloran.'

27 *Islander,* 27 Oct. 1865

28 *Examiner,* 23 Oct. 1865; *Herald,* 25 Oct. 1865; *Islander,* 27 Oct. 1865

29 There is no doubt that this oath relates to the prosecution of Jane O'Halloran or Halloran, for the appended statement by the justice of the peace refers to it explicitly.

30 John Balderston in PEI, Legislative Council, *Debates and Proceedings,* 1874, 61; also see 67.

31 PEI, Executive Council Minutes, 17 Oct. 1865

32 J.W. Morrison to Colonel Langley, 28 Dec. 1865, Colonial Secretary's Letter-books, PAPEI

33 PEI, Executive Council Minutes, 6 Feb. 1866. The delay in authorizing payment in these cases may have been owing to a desire to await the outcome of courts martial, and to receive official confirmation that the persons apprehended had been found guilty of desertion.

34 *Islander,* 3 Nov. 1865. The newspaper report spells the name of the soldier-witness Macgory–Macgrory inconsistently.

35 *Royal Gazette,* 30 Aug. 1865

36 *Islander,* 3 Nov. 1865

37 Thomas Moffatt, age twenty-five, a native of Ireland, had been reported as a deserter by 17 August; see *Royal Gazette,* 23 Aug. 1865. He was never captured.

38 *Islander,* 3 Nov. 1865

39 Dr Brian Cuthbertson of PANS on 28 July 1981

40 See Whitfield, *Tommy Atkins,* 59, 62; Burroughs, 'Tackling Army Desertion in British North America,' 51–4.

41 See Burroughs, 'Tackling Army Desertion in British North America,' 43–8; Whitfield, *Tommy Atkins,* 59–101, app. C, 137–47, app. D, 148–69.

42 Edward Lagard (?) to T.F. Elliot, 25 Sept. 1865, CO 226/101, 666, microfilm, PAC

43 *Herald,* 4, 11 Oct. 1865. Whatever else happened to the captured deserters, at least for a period they were confined in the Queens County Jail because of lack of a military jail on Prince Edward Island. See PEI, Executive Council Minutes, 6 March 1865; Morrison to Thomas Pethick and four others, 2 Oct. 1865, Colonial Secretary's Letterbooks, PAPEI.

44 *Royal Gazette*, 6 Sept. 1865. Whitfield, *Tommy Atkins*, 61 notes that it was not unusual for a soldier who had been punished previously for attempted desertion to try again.

45 Whitfield, *Tommy Atkins*, 63

46 Hodgson to Edward Cardwell, 30 Aug. 1865, despatch no. 71, printed in PEI, Assembly, *Journal*, 1866, app. G; PEI, Assembly, *Journal*, 1866, app. Q; PEI, Executive Council Minutes, 7, 17 Oct. 1865

47 Hodgson to Cardwell, 30 Aug. 1865, despatch no. 71, printed in PEI, Assembly, *Journal*, 1866, app. G

48 PEI, Assembly, *Journal*, 1866, app. Q

49 Burroughs, 'Tackling Army Desertion in British North America,' 66. One writer who has noted the hiring of watchmen on Prince Edward Island has misunderstood its significance entirely. He portrays them as being hired to ward off 'the threat of civil disobedience by tenant farmers protesting the Island's unresolved land question,' and has the soldiers arriving on the Island in order to back up the watchmen. See Greg Marquis, 'Enforcing the Law: The Charlottetown Police Force,' in Douglas Baldwin and Thomas Spira, eds., *Gaslights Epidemics and Vagabond Cows: Charlottetown in the Victorian Era* (Charlottetown, 1988), 91–2.

50 Burroughs, 'Tackling Army Desertion in British North America,' 41

51 PEI, Assembly, *Journal*, 1866, app. Q; *Royal Gazette*, 16 Aug., 20 Sept. 1865; Whitfield, *Tommy Atkins*, 61; Burroughs, 'Tackling Army Desertion in British North America,' 35

52 PEI, Executive Council Minutes, 19 Sept., 7, 17 Oct., 12 Dec. 1865, 3 April 1866

53 Hodgson to Cardwell, 30 Aug. 1865, despatches nos. 70, 71, printed in PEI, Assembly, *Journal*, 1866, app. G; PEI, Assembly, *Journal*, 1866, app. C; also see PEI, Executive Council Minutes, 29 Aug. 1865. It is on the powder magazine erected at the Victoria Barracks, Charlottetown – a peculiar choice, perhaps – that the Historic Sites and Monuments Board of Canada has placed a plaque entitled 'The Land Tenure Question.'

54 Whitfield, *Tommy Atkins*, app. C, 141; table 1, 143

55 *Royal Gazette*, 20 Sept. 1865. Even if subtraction is made for the three who were captured, the annualized rate was 93.6 per cent.

10: Resistance, Outrage, Escalation, and Coercion

1 Accession 2514/10, PAPEI

2 Joint deposition of James Curtis and Jonathan Collings, 22 Sept. 1865, PEI, SCCP, 1865, PAPEI

3 In *Islander*, 10 Nov. 1865

4 Donald Graves, '"Tommy" in Canada,' *Horizon Canada*, 2/20 (July 1985), 465

5 J. Mackay Hitsman, 'Military Defenders of Prince Edward Island 1775–1864,' *Annual Report of the Canadian Historical Association*, 1964, 31–2

6 See Edward Cardwell to Robert Hodgson, 23 Aug. 1865, printed in PEI, House of Assembly, *Journal*, 1866, app. G; Earl of Carnarvon to George Dundas, 16 Aug. 1866, CO 226/101, 503, microfilm, PAC; memorandum of T.F. Elliot, 25 Sept. 1866, CO 226/102, 308–10; Carol M. Whitfield, *Tommy Atkins: The British Soldier in Canada, 1759–1870* (Ottawa, 1981), 192n43.

7 Cardwell to R. Hodgson, 25 Sept. 1865; R. Hodgson to Cardwell, 24 Oct. 1865, both printed in PEI, Assembly, *Journal*, 1866, app. G

8 Thomas Tydd to Assistant to the Major General, 9 Aug. 1865, CO 226/101, 660; J.H. Francklyn to Secretary of State for War, 11 Aug. 1865, CO 226/101, 661–2; Undersecretary of State for War to Elliot, 10 Nov. 1865, CO 226/101, 654–5; Douglas Galton to Undersecretary of State, Colonial Affairs, 12 Jan. 1866, CO 226/102, 486; Elliot to Galton, 16 Jan. 1866, CO 226/101, 656–8; Cardwell to Dundas, 20 Jan. 1866, CO 226/101, 658–9; Dundas to Cardwell, 8 Feb. 1866, CO 226/102, 58–9; PEI, Executive Council Minutes, 6 March, 6 July 1866; Public Accounts for the financial year ending 31 January 1867, in PEI, Assembly, *Journal*, 1867, app. B. The total for this expense was £72 18s. currency.

9 Charles Hastings Doyle to the Secretary of State for War, 21 Nov. 1865, CO 226/102, 490–1; ?? [name illegible] at War Office to Undersecretary of State, Colonial Office, 27 March 1866, CO 226/102, 488–9; Cardwell to Dundas, 9 April 1866, CO 226/102, 492–3; Dundas to Cardwell, 2 May 1866, CO 226/102, 136–7; PEI, Executive Council Minutes, 15 May 1866.

10 Galton to Undersecretary of State, Colonial Office, 1 Dec. 1866, CO 226/102, 509–10; Carnarvon to Dundas 15 Dec. 1866, CO 226/102, 511–12; Dundas to Carnarvon, 4 Feb. 1867, CO 226/103, 64–5

11 For a particularly flagrant example, see Carnarvon to Dundas, 16 Aug. 1866, CO 226/101, 503–5.

12 See Cardwell to R. Hodgson, 23 Aug., 16 Sept. 1865; R. Hodgson to Cardwell, 25 Oct. 1865; Cardwell to Dundas, 23 Dec. 1865; Dundas to Cardwell, 8 Feb. 1866; Extracts from Minutes of the Executive Council, 17 Oct. 1865, 6 Feb. 1866 (misdated as 5 Feb. 1866), all printed in PEI, Assembly, *Journal*, 1866, app. G; Carnarvon to Dundas, 3 Oct. 1866, CO 226/102, 311–13, and 4 March 1867, CO 226/102, 370–1; Dundas to Carnarvon, 7 Nov. 1866, CO 226/102, 368–9; Dundas to Duke of Buckingham and Chandos, 17 June 1867, CO 226/103, 240–3; Buckingham to Dundas, 10 Aug. 1867, CO 226/103, 563–4; PEI, Assembly, *Journal*, 1868, app. A, AA. Despite considerable sympathy among the clerks at the Colonial Office for the position of the Island government,

there was still a tendency to let expediency override equity if extraction of the soldiers from the colony required that pressure be applied to the local government. See memoranda of Elliot, 6 Nov. 1865, CO 226/101, 493, and 25 Sept. 1866, CO 226/102, 310; Arthur Blackwood, 28 Feb. 1866, CO 226/102, 54; and the précis on the issue of military expenses prepared by Elliot, 3 Aug. 1866, CO 226/101, 498–502.

13 PEI, Executive Council Minutes, 16 April 1867

14 Sir Charles FitzRoy, 'Return of Township Lands in Prince Edward Island, shewing their Extent, population, (from the Census of 1833), Names of the principal Proprietors, the general Tenure of the Settlers, and the causes which have tended to retard their improvement,' 3 Oct. 1837, in CO 226/54, 269; the 1861 census in PEI, Assembly, *Journal*, 1862, app. A; Edward J. Hodgson in P.S. Macgowan, ed., *Report of Proceedings before the Commissioners Appointed under the Provisions of 'The Land Purchase Act, 1875'* (Charlottetown, 1876), 286.

15 Sir George F. Seymour, 'Memo Prince Edwd Island 1840,' Misc. Lot 13 Papers, Seymour of Ragley Papers, CR 114A/565, WCRO

16 See, for example, *Royal Gazette*, 22 June 1847; PEI, Assembly, *Journal*, 1853, 75, 103, 105–6, 109, 112–15, app. B–b; Ian Ross Robertson, ed., *The Prince Edward Island Land Commission of 1860* (Fredericton, 1988), 39, 53, 209.

17 B.H. Cumberland to Cardwell, 16 June 1864, CO 226/100, 538–9

18 Joint deposition of Curtis and Collings, 22 Sept. 1865, SCCP, 1865, PAPEI

19 Thomas W. Dodd to W.H. Pope, 23 Sept. 1865, CO 226/101, 461; 'The Queen at the prosecution of James Curtis and Jonathan Collings *v*. James Gorveatt *et al*.,' indictment for riot, assault, and conspiracy, SCCP, 1865, PAPEI

20 Joint deposition of Curtis and Collings, 22 Sept. 1865, SCCP, 1865, PAPEI. Also see Dodd to W.H. Pope, 23 Sept. 1865, CO 226/101, 461–2; *Examiner*, 25 Sept. 1865 (reprinted from *Patriot*, 23 Sept. 1865); *Herald*, 27 Sept. 1865.

21 Dodd to W.H. Pope, 23 Sept. 1865, CO 226/101, 462

22 Lady Georgiana Fane to Cardwell, 18 Oct. 1865, CO 226/101, 686–8

23 *Islander*, 29 Sept. 1865

24 W.E. Vaughan, *Sin, Sheep and Scotsmen: John George Adair and the Derryveagh Evictions, 1861* (Belfast, 1983), 29

25 John E. Archer, 'Under Cover of Night: Arson and Animal Maiming,' in G.E. Mingay, ed., *The Unquiet Countryside* (London, 1989), 76; also see John E. Archer, *By a Flash and a Scare: Incendiarism, Animal Maiming, and Poaching in East Anglia, 1815–1870* (Oxford, 1990), 6, 202, 214, 254.

26 See Stanley H. Palmer, *Police and Protest in England and Ireland 1780–1850* (Cambridge, 1988), 52.

27 See Rusty Bittermann, 'Women and the Escheat Movement: The Politics of

Everyday Life on Prince Edward Island,' in Janet Guildford and Suzanne Morton, eds., *Separate Spheres: Women's Worlds in the Nineteenth-Century Maritimes* (Fredericton, 1994), 26, 32.

28 John E. Archer, '"A Fiendish Outrage?" A Study of Animal Maiming in East Anglia: 1830–1870,' *Agricultural History Review*, 33/2 (1985), 147–8, 154; also see 156; Archer, *By a Flash and a Scare*, 6–7, 198–221, 255–6; Archer, 'Under Cover of Night,' 75–7; David J.V. Jones, *Crime in Nineteenth-Century Wales* (Cardiff, 1992), 133.

29 Archer, *By a Flash and a Scare*, 221; also see 200, 216–17; Archer, 'Under Cover of Night,' 76. Animals were not understood as having 'rights,' and therefore maiming would be perceived as a crime against the owner and his property, not against the animal, who, although a living creature, was not conceptualized as a rights-bearing individual.

30 Archer, 'Under Cover of Night,' 75; also see 76; Archer, '"A Fiendish Outrage"?,' 147. The address of Curtis appears in the joint deposition of Curtis and Collings, 5 Oct. 1865, SCCP, 1865, PAPEI; also see letter of 'A Trouter' to the Editor of *Herald*, dated 10 Sept. 1865, in *Herald*, 27 Sept. 1865.

31 Jones, *Crime in Nineteenth-Century Wales*, 133

32 Archer, *By a Flash and a Scare*, 255–6

33 Dodd to W.H. Pope, 23 Sept. 1865, CO 226/101, 462

34 Dodd to R. Hodgson, 29 Aug. 1865, Accession 2514/10, PAPEI

35 Bernard McKenna to the Administrator in Council, 4 Sept. 1865, in Accession 2514/10, PAPEI; PEI, Executive Council Minutes, 19 Sept. 1865. There appears to be no surviving evidence concerning investigation of McKenna's complaint against Farquharson, which had been referred to the attorney general – 'and if ... he shall be deemed to have been guilty of a criminal offence, – that he be prosecuted accordingly.' Judging from an item in PEI, Executive Council Minutes, 17 April 1866, Farquharson became involved in the extended search for Samuel Fletcher, at least to the extent of claiming expenses for his efforts. The Council referred the claim to a committee.

36 Fane to Cardwell, 18 Oct. 1865, CO 226/101, 687

37 Blackwood, memorandum dated 23 Oct. 1865, CO 226/101, 690; also see Seymour to Blackwood, 6 Oct. 1865, Seymour of Ragley Papers, CR114A/508/12, Letterbooks, WCRO

38 Sir Frederic Rogers, memorandum dated 24 Oct. 1865, CO 226/101, 690. A bracketed passage in the same memorandum was remarkable for the impatience Rogers showed with the proprietors: 'If troops are kept there the *fair* thing w^d [would] be to pay for them by a rate upon profits of land' (original emphasis). This was a rare demonstration of exasperation on the part of Rogers, a bureaucrat who normally did not go beyond consideration of the legal,

constitutional, and administrative aspects of a situation placed before him in the line of duty. Also see memorandum from Elliot to Rogers, 28 Feb. 1868, CO 226/104, 21.

39 David J.V. Jones, *Before Rebecca: Popular Protests in Wales 1793–1835* (London, 1973), 200

40 Private collection of T.W. Stewart, Ottawa. Deborah Stewart, 'Robert Bruce Stewart: A Study of Resident Proprietorship on Prince Edward Island' (research paper, University of Toronto, June 1984), app. N, and Deborah Stewart, 'Robert Bruce Stewart and the Land Question,' *Island Magazine*, no. 21 (Spring–Summer 1987), 7 both include photocopies of the note; the passage under the latter copy erroneously gives the year as 1856. R.B. Stewart's claim to own almost eighty thousand acres is contained in Stewart to Cardwell, July 1865, CO 226/101, 683.

41 The following report, unconfirmed, from the *Islander* newspaper of 13 Oct. 1865 concerning an unnamed participant in the Lot 22 riot of 2 Oct. 1865, who had apparently given information to justices of the peace, falls under the heading of forcible recruitment rather than a threatening letter, as usually understood, in the study of landlord-tenant relations: 'He is a blacksmith, and the posting upon his forge [of] a notice to the effect that if he did not join the League his house would be burnt, induced him to avow himself a Tenant Leaguer.'

42 David J.V. Jones, 'Rural Crime and Protest,' in G.E. Mingay, ed., *The Victorian Countryside* (London, 1981), vol. 2, 573; also see 575.

43 Fane to Cardwell, 15 Nov. 1865, CO 226/101, 692

44 Stewart, 'Robert Bruce Stewart and the Land Question,' 11, 7; 'Robert Bruce Stewart: A Study of Resident Proprietorship on Prince Edward Island,' 111n177

45 Eric Richards, 'How Tame Were the Highlanders during the Clearances?,' *Scottish Studies*, 17/1 (1973), 42; Vaughan, *Sin, Sheep and Scotsmen*, 30. The threatening letter was more common even in England than in Scotland; see the evidence in Palmer, *Police and Protest in England and Ireland 1780–1850*, 52, 386.

46 Ms. Mairin Nic Dhiarmada of the Celtic Studies Program at St Michael's College, University of Toronto, 15 Dec. 1994. Dr G. Edward MacDonald, curator of history at the Prince Edward Island Museum and Heritage Foundation, has pointed out that if the text of this note were Irish Gaelic, it would be the only known Prince Edward Island example of writing in that language.

47 Note by Black, 16 Feb. 1994, enclosed in Michael Kennedy to MacDonald, May 1994 (photocopies in MacDonald to author, 7 Sept. 1994)

48 Professor Ronald S. Blair, Political Science, Scarborough College, University

of Toronto, 2 Nov. 1993 and 2 Aug. 1995. He is not persuaded by Black's interpretation of the text of the note.

49 Henry Jones Cundall Diaries, vol. 5, entry for 27 Sept. 1865, PEIM

50 R. Hodgson to Cardwell, 27 Sept. 1865, printed in PEI, Assembly, *Journal,* 1866, app. G; J.W. Morrison to Edward Palmer, 27 Sept. 1865, Colonial Secretary's Letterbooks, PAPEI

51 Today Fulham is part of London, and although Sulivan's residence, Broom House, no longer stands, more than a century later the visitor to Fulham would encounter Sulivan Hall, Sulivan Court, Sulivan Road, Broomhouse Lane, Broomhouse Road, Sulivan Elementary School, and so on.

52 Donald Montgomery in PEI, Legislative Council, *Debates and Proceedings,* 1864, 85

53 Calculations based on PEI, Assembly, *Journal,* 1862, app. A

54 Kenneth Bourne, ed., *The Letters of the Third Viscount Palmerston to Laurence and Elizabeth Sulivan 1804–1863* (London, 1979), Introduction, 27; also see Kenneth Bourne, *Palmerston: The Early Years, 1784–1841* (London, 1982), 32

55 On the office of secretary-at-war, see Whitfield, *Tommy Atkins,* 15–19.

56 Copy of will of Laurence Sulivan, 19 July 1865, box 1, bundle S, Sulivan Family Papers, DD/476, Archives Department, Shepherds Bush Library, Uxbridge Road, London, England. Also see Bourne, *Palmerston: The Early Years, passim,* but especially 108–9, 126–31; Bourne, ed., *Letters of the Third Viscount Palmerston to Laurence and Elizabeth Sulivan,* 18–21, 123.

57 E.D. Steele, *Irish Land and British Politics: Tenant-Right and Nationality 1865–1870* (Cambridge, 1974), 45; also see 44; Steele, 'Ireland and the Empire in the 1860s. Imperial Precedents for Gladstone's First Irish Land Act,' *Historical Journal,* 11/1 (1968), 64–5; Steele, *Palmerston and Liberalism, 1855–1865* (Cambridge, 1991), 355; Jasper Ridley, *Lord Palmerston* (London, 1970), 512–14, 585

58 *Vindicator,* 23 March 1864. The *Vindicator* misspelled his name as 'Lawrence Sullivan.'

59 Louis Henry Davies in Macgowan, ed., *Proceedings of Land Commissioners of 1875–76,* 152

60 See A.B. McKenzie in PEI, Legislative Council, *Debates,* 1877, 17.

61 J. Simpson Sr in Macgowan, ed., *Proceedings of Land Commissioners of 1875–76,* 116

62 George W. DeBlois in Macgowan, ed., *Proceedings of Land Commissioners of 1875–76,* 80

63 Robertson, ed., *The Prince Edward Island Land Commission of 1860,* 46; also see 45, 47–9; for more information on the state of indebtedness of Lot 22 tenants, see J.D. Gordon and D. Laird, eds., *Abstract of the Proceedings before the Land Commissioners' Court, Held During the Summer of 1860, to Inquire into the Differ-*

ences Relative to the Rights of Landowners and Tenants in Prince Edward Island (Charlottetown, 1862), Proceedings, 243. Peter McGuigan has published two articles relating to Lot 22. In 'From Wexford and Monaghan: The Lot 22 Irish,' *Abegweit Review*, 5/1 (1985), 61–96 he has dealt with the early years of Irish settlement; in 'Tenants and Troopers: The Hazel Grove Road, 1865–68,' *Island Magazine*, no. 32 (Fall–Winter 1992), 22–8 he has covered some of the events in this chapter concerning Hazel Grove, Lot 22.

64 Michael McGuigan in Macgowan, ed., *Proceedings of Land Commissioners of 1875–76*, 118–19

65 PEI, Assembly, *Journal*, 1862, app. A. P. McGuigan refers to this assessment as excessively favourable; 'From Wexford and Monaghan,' 71.

66 Deposition of Curtis, 5 Oct. 1865, SCCP, 1865, PAPEI

67 The account of this incident is drawn from the following sources: depositions of Curtis and Collings, 5 Oct. 1865; examination and evidence of Andrew Cranston, Richard Bagnall, Montague C. Irving, Zechariah Mayhew, Curtis, and Collings, 14 Oct. 1865, all in SCCP, 1865, PAPEI; examination and evidence of Curtis, Cranston, and William Harris (Corney), 8 Jan. 1866, in SCCP, 1866, PAPEI; and Extracts from James Horsfield Peters to R. Hodgson, 3 Oct. 1865, as printed in PEI, Assembly, *Journal*, 1866, app. G; *Islander*, 13 Oct. 1865. Sources of verbatim quotations are identified individually by means of endnotes.

68 Deposition of Curtis, 5 Oct. 1865, SCCP, 1865, PAPEI

69 Examination and evidence of Cranston, 14 Oct. 1865, SCCP, 1865, PAPEI

70 Examination and evidence of Irving, 14 Oct. 1865, SCCP, 1865, PAPEI

71 Deposition of Curtis, 5 Oct. 1865, SCCP, 1865, PAPEI

72 Examination and evidence of Cranston, 14 Oct. 1865, SCCP, 1865, PAPEI

73 Deposition of Collings, 5 Oct. 1865, SCCP, 1865, PAPEI

74 *Islander*, 13 Oct. 1865

75 Examination and evidence of Cranston, 14 Oct. 1865, SCCP, 1865, PAPEI

76 Deposition of Collings, 5 Oct. 1865, SCCP, 1865, PAPEI

77 Examination and evidence of Bagnall, 14 Oct. 1865, SCCP, 1865, PAPEI

78 Examination and evidence of Curtis, Cranston, and William Harris (Corney), 8 Jan. 1866, in SCCP, 1866, PAPEI

79 See David Williams, *A History of Modern Wales* (London, 1950), 208; David Williams, *The Rebecca Riots: A Study in Agrarian Discontent* (Cardiff, 1955), 187–8, 199, 207; Jones, *Crime in Nineteenth-Century Wales*, 10–11.

80 Examination and evidence of William Harris (Corney), 8 Jan. 1866, in SCCP, 1866, PAPEI.

81 Confession of William Harris (Corney), 6 Sept. 1866, SCCP, 1866, PAPEI. Early in 1867 Chief Justice Hodgson sentenced him to two months' hard labour in

prison after a guilty plea to a charge of theft; see *Islander*, 18 Jan. 1867; *Herald*, 6 Feb. 1867. Of course, there may have been more than one reason for Harris to disguise himself on 2 Oct. 1865.

82 An item in PEI, Executive Council Minutes, 15 May 1866 suggests that he may also have been captured separately from the others charged as a result of the 2 Oct. 1865 incident. Collings and others were to receive £6 'for arresting William Harris Corney under Warrant – including horses & Sleighs hire.'

83 Extracts from Peters to R. Hodgson, 3 Oct. 1865, printed in PEI, Assembly, *Journal*, 1866, app. G

84 Peters to T.H. Haviland Sr, 17 July 1843, in 226/65, 216; see Ian Ross Robertson, 'James Horsfield Peters,' *DCB*, vol. 12 (Toronto, 1990), 838–42.

85 Peters to Sir John Pakington, 24 May 1852, CO 226/80, 626–7

86 Extracts from Peters to R. Hodgson, 3 Oct. 1865, printed in PEI, Assembly, *Journal*, 1866, app. G

87 Dodd to R. Hodgson, 6 Oct. 1865, printed in PEI, Assembly, *Journal*, 1866, app. G

88 W.H. Pope to William Swabey, 6 Oct. 1865, Colonial Secretary's Letterbooks, PAPEI. In fact, whatever Hodgson's desire may have been, the only persons actually in command of the soldiers would be their officers. A magistrate could request a military officer to order his men to intervene, but only the officer could issue the order – a point explained to me by Dr Brian Cuthbertson of PANS on 28 July 1981.

89 *Islander*, 13 Oct. 1865; *Patriot*, 7 Oct. 1865 gave the number of bailiffs or special constables as eight, and although the *Patriot* itself does not survive, this information was reprinted in *Herald*, 11 Oct. 1865.

90 W.H. Pope to Swabey, 6 Oct. 1865, Colonial Secretary's Letterbooks, PAPEI

91 See Swabey to Edward Whelan, dated 30 Dec. 1865, in *Examiner*, 8 Jan. 1866; PEI, Executive Council Minutes, 13 Nov. 1865. Predictably, Swabey's performance of his duty evoked bitter criticism; see, for example, *Herald*, 17 Jan. 1866. The sheriff received £32 13s. 4d., and financial information presented to the House of Assembly on 24 April 1867 indicated that he was paid, in addition, £6 for 'capture of rioters at Hazel Grove.' See PEI, Executive Council Minutes, 2 June 1866; PEI, Assembly, *Journal*, 1867, app. C.

92 *Islander*, 13 Oct. 1865

93 R. Hodgson to Cardwell, 25 Oct. 1865, printed in PEI, Assembly, *Journal*, 1866, app. G; also see R. Hodgson to Cardwell, 11 Oct. 1865, printed in PEI, Assembly, *Journal*, 1866, app. G. P. McGuigan, 'Tenants and Troopers,' 27 has dealt with the expedition, and added some colourful details, drawn from Margaret Ruth Bagnall, 'The Red Fox,' in Mary Brehaut, ed., *Historic Sidelights of Prince Edward Island* (Charlottetown, 1956), 37–47. 'The Red Fox' is a

retelling of a story based on the sending of troops to Hazel Grove, Lot 22; unfortunately for historical accuracy, Bagnall has soldiers combing Hazel Grove for Tenant Leaguers in 1862, three years before they arrived in the colony.

94 *Herald*, 11 Oct. 1865

95 R. Hodgson to Cardwell, 25 Oct. 1865, printed in PEI, Assembly, *Journal*, 1866, app. G

96 Richards, 'How Tame Were the Highlanders during the Clearances?,' 40. On 39–40, Richards constructs a four-stage composite picture of a typical incident of Highland resistance to landlord authority. Also see Richards, 'Patterns of Highland Discontent, 1790–1860,' in Roland Quinault and John Stevenson, eds., *Popular Protest and Public Order: Six Studies in British History 1790–1920* (London, 1974), 78, 92–3, 95, 105; Charles W.J. Withers, *Gaelic Scotland: The Transformation of a Culture Region* (London, 1988), 355–91, especially table 6.2 on 359–62; T.M. Devine, *Clanship to Crofters' War: The Social Transformation of the Scottish Highlands* (Manchester, 1994), 208–27.

97 P. McGuigan, 'Tenants and Troopers,' 26 states that the 15th was 'a less "Irish" regiment' than the 16th, but gives no indication of the basis for this assertion. He implies strongly that the changing of regiments accounted for the stoppage of desertion; in fact, the last deserter was reported in *Royal Gazette, 20 Sept. 1865, more than five weeks before the change.*

98 W.H. Pope to Swabey, 31 Oct. 1865, Colonial Secretary's Letterbooks, PAPEI; R. Hodgson to Cardwell, 22 Nov. 1865, printed in PEI, Assembly, *Journal*, 1866, app. G

99 PEI, Executive Council Minutes, 31 July, 20 Nov. 1866

100 R. Hodgson to Cardwell, 22 Nov. 1865, printed in PEI, Assembly, *Journal*, 1866, app. G

101 Letter of R.P. Haythorne to the Editor of *Islander*, dated 22 Nov. 1865, in *Islander*, 24 Nov. 1865

102 Haythorne in PEI, Legislative Council, *Debates*, 1867, 60

103 *Examiner*, 28 Oct. 1867

104 R. Hodgson to Cardwell, 22 Nov. 1865, printed in PEI, Assembly, *Journal*, 1866, app. G

105 Peters to Haviland, 17 July 1843, in 226/65, 216

106 DeBlois in Macgowan, ed., *Proceedings of Land Commissioners of 1875–76*, 95. DeBlois's experience as a land agent on Prince Edward Island stretched back at least to 1853; see Elinor Vass, 'George Wastie DeBlois,' DCB, vol. 11 (Toronto, 1982), 241.

107 Fragments of Family History of John McEachern, Diary (hereafter cited as McEachern Diary), undated entry at the commencement of 1870; entries for

19, 20, 21 Oct. 1865, microfilm, PAPEI. Concerning McEachern, see David Weale, 'The Emigrant,' *Island Magazine*, no. 16 (Fall–Winter 1984), 15–22 and no. 17 (Summer 1985), 3–11. Although the part of the collection entitled 'Sundry Accounts' gives no figures for the 1860s, between 1841 and 1851 McEachern, who had arrived from Scotland at age twenty in 1830, had been as much as 11.8 years in arrears.

108 McEachern Diary, 1 Jan. 1866, PAPEI

109 This is not to deny, of course, that in various elections there may have been a relationship between demands to pay up back-rent and how a tenant voted, especially if the landlord or land agent was a candidate. Toleration of continuing arrears could be one of those non-contractual privileges of which sociologist Samuel Clark has written in the context of the Irish land question.

110 Richards, 'How Tame Were the Highlanders during the Clearances?,' 43

111 Morrison to St. George M. Nugent, 6 Dec. 1865, Colonial Secretary's Letter-books, PAPEI. Morrison closed the communication by explaining on the administrator's behalf that it was only on 2 Dec. that Hodgson received the 20 Nov. letter, addressed to W.H. Pope, the colonial secretary, and that Pope had left for England between 30 Oct. and 20 Nov. without telling Hodgson that he had not replied to the earlier letter. The Halifax military authorities may have concluded that Pope was ignoring their protest. According to *Islander*, 24 Nov. 1865, Pope had departed on the previous day.

112 *Islander*, 8 Dec. 1865

11: Collapse and the Courts

1 *Islander*, 2 Feb. 1866

2 *Islander*, 3 Nov. 1865

3 See T.E. MacNutt, 'Aid to the Civil Power in PEI in Times Past' (unpublished typescript, 1937; rev. 1961), Accession 2825/103a, PAPEI.

4 *Examiner*, 25 Sept. 1865 (original emphasis); also see 18 Sept., 9 Oct. 1865.

5 *Herald*, 15 Nov. 1865. Provision for assistance to distressed tenants had been contemplated under Article 11; see *Examiner*, 9 Oct. 1865.

6 In 1986 a single issue of *Ross's Weekly* for 1866 came to light.

7 *Examiner*, 16 July 1866

8 *Examiner*, 28 Jan. 1880; this reference comes courtesy of genealogist George F. Sanborn Jr, 12 Aug. 1981.

9 Ewen MacNeill, 'McNeill of Barra on P.E. Island: Ancestors and Descendants of Alexander McNeill' (privately circulated, 1982), unpaginated; copy at PAPEI

10 PEI, Supreme Court Minutes, 1, 3 May 1866, PAPEI; also see *Lady Louisa Wood and Maria Matilda Fanning v. Samuel Sabine*, affidavit of debt, dated 4 March 1865, PEI, SCCP, 1865, PAPEI.

11 *Herald*, 19 Sept. 1866

12 *Herald*, 10 Oct. 1866

13 *Examiner*, 11 Feb. 1867; also see *Richard Joseph Clark v. George F. Adams*, affidavit of debt, dated 6 Nov. 1866, PEI, SCCP, 1866, PAPEI.

14 George F. Adams, assignment of lease, 4 March 1868, Land Title Documents, Leases and Related Documents, Lease No. 155, Lot 50, PAPEI

15 *Herald*, 22 Aug., 19 Sept. 1866

16 Letter of John Roach Bourke to the Editor of *Examiner*, dated 25 Sept. 1865, in *Examiner*, 2 Oct. 1865

17 Letter of Bourke to the Editor of *Islander*, dated 26 Aug. 1865, in *Islander*, 1 Sept. 1865. The offending article from *Ross's Weekly*, 24 Aug. 1865 will be found reprinted in full, along with Bourke's response in the *Islander*, in *Examiner*, 4 Sept. 1865; Whelan expressed the view that the person who wrote the article in *Ross's Weekly* knew that it was false when he wrote it, and wrote it in order to deceive.

18 Bourke to the Editor of *Examiner*, dated 25 Sept. 1865, in *Examiner*, 2 Oct. 1865

19 Francis Seymour to Sir George F. Seymour, 10 Oct. 1859, Correspondence, Seymour of Ragley Papers, CR 114A/526/10, WCRO

20 Bourke to the Editor of *Examiner*, dated 25 Sept. 1865, in *Examiner*, 2 Oct. 1865; also see letter of John Currie, JP, C.L. Lane, and William Inman, JP (unaddressed), dated 25 Sept. 1865, in *Examiner*, 2 Oct. 1865.

21 Bourke to the Editor of *Islander*, dated 26 Aug. 1865, in *Islander*, 1 Sept. 1865

22 Notice of Bourke, dated 23 Oct. 1865, in *Islander*, 27 Oct. 1865

23 Calculation based on the 1861 census in PEI, House of Assembly, *Journal*, 1862, app. A

24 *Herald*, 25 Oct. 1865

25 *Herald*, 29 Nov. 1865. The peculiar use of the term 'legal' recalls the example of such groups as the Caravats of Ireland, who liked to refer to their 'laws.'

26 Henry Jones Cundall Diaries, vol. 5, entries for 19, 24, 25 Oct. 1865, PEIM

27 Letter of James Sullivan to the Editor of *Herald*, dated 28 Nov. 1865, in *Herald*, 6 Dec. 1865 (original emphasis)

28 Letter of Cornelius Richard O'Leary to the Editor of *Patriot*, dated 30 Jan. 1866, in *Herald*, 7 March 1866; the *Patriot* of 6 Jan. 1866 does not survive.

29 'A Plain Man' to 'Mr. Editor,' n.d., in *Examiner*, 11 Sept. 1865. Peter McGuigan, 'Tenants and Troopers: The Hazel Grove Road, 1865–68,' *Island Magazine*, no. 32 (Fall–Winter 1992), 23 asserts that the leaders 'were commit-

ted to passive resistance,' and implies that the mass membership got out of control (23, 25). In 1868 in the assembly Benjamin Davies did state that 'after a time, it [the Tenant League] was joined by large bodies of men over whom the leaders had no control,' but examination of Adams's rhetoric in 1864 suggests that the militant actions of the members, which went well beyond passive resistance, were consistent with a reasonable reading of his message. See PEI, House of Assembly, *Debates and Proceedings*, 1868, 15.

30 *Examiner*, 2 Oct. 1865

31 *Herald*, 11 Oct. 1865; see *Examiner*, 9 Oct. 1865, for Whelan's dismissive response to the item, which would have appeared in print elsewhere already – in either *Ross's Weekly* or the *Patriot*, or both. He had previously noted the attempt to have Tenant Leaguers stop their subscriptions in *Examiner*, 21 Aug. 1865.

32 *Examiner*, 4 Sept. 1865; see letter of Sullivan to the Editor of *Herald*, dated 28 Nov. 1865, in *Herald*, 6 Dec. 1865.

33 George Dundas to Edward Cardwell, 25 Aug. 1865, confidential, CO 226/101, 665, microfilm, PAC

34 See, for example, *Islander*, 29 Dec. 1865.

35 Haythorne to the Editor of *Islander*, dated 15 Nov. 1865, in *Islander*, 17 Nov. 1865

36 *Herald*, 8 Nov. 1865

37 *Examiner*, 2 Oct. 1865

38 *Examiner*, 4 Dec. 1865

39 For the resolution, see *Examiner*, 18 Dec. 1865.

40 PEI, Executive Council Minutes, 19 Dec. 1865

41 See examination and evidence of Andrew Cranston *et al.*, 14 Oct. 1865, PEI, SCCP, 1865, PAPEI; PEI, Supreme Court Minutes, 31 Oct., 1 Nov. 1865, PAPEI; *Herald*, 8 Nov. 1865.

42 *Examiner*, 15 Jan. 1866

43 PEI, *Statutes*, 24 Vic., c. 10

44 Davies in PEI, Assembly, *Debates*, 1867, 27

45 *Examiner*, 15 Jan. 1866

46 *Examiner*, 22 Jan. 1866

47 PEI, Supreme Court Minutes, 10 Jan. 1866, PAPEI; 'The Queen at the prosecution of Joseph Doucette *v.* James Curtis,' bill of indictment for assault, SCCP, 1866, PAPEI

48 PEI, Supreme Court Minutes, 20 Jan. 1866, PAPEI; *Herald*, 17 Jan. 1866

49 'The Queen *v.* John Ross,' bill of indictment, SCCP, 1866, PAPEI

50 *Patriot*, 20 Jan. 1866, reprinted in *Examiner*, 22 Jan. 1866

51 'The Queen at the prosecution of Alexander Campbell, Alexander Russell,

and Donald Stewart *v.* James Curtis and George Swan,' indictment for assault and false imprisonment, SCCP, 1866, PAPEI; PEI, Supreme Court Minutes, 17 Jan. 1866, PAPEI. There is room for doubt about the date because eight days *before* 5 October the *Herald* newspaper published an account of Curtis waylaying and detaining three men with the same names that appear in the presentment, reported that they had sworn to their story before justices of the peace, and stated that the matter would come before the Supreme Court in January 1866. If 5 October is the correct date of the offences alleged in the presentment, the other possibility is that there were two incidents involving the same cast of characters and fitting the same general description – although the details which are known are remarkably similar. See letter of 'A Trouter' to the Editor of *Herald*, dated 10 Sept. 1865, in *Herald*, 27 Sept. 1865.

52 'The Queen at the prosecution of Elizabeth Devine *v.* James Curtis,' indictment for assault, SCCP, 1866, PAPEI; PEI, Supreme Court Minutes, 17 Jan. 1866, PAPEI

53 'The Queen at the prosecution of Bernard McKenna *v.* Francis McKenna, Hugh McKenna, and Patrick McKenna,' indictment, marked 'True Bill,' January 1866, SCCP, 1866, PAPEI; affidavit of Francis Kelly, dated 23 Jan. 1866, SCCP, 1866, PAPEI

54 *Examiner,* 15 Jan. 1866

55 See PEI, Assembly, *Debates,* 1868, 72; *Herald,* 6 Feb. 1867.

56 Francis L. Haszard and A. Bannerman Warburton, eds., *Reports of Cases Determined in the Supreme Court, Court of Chancery, and Vice Admiralty Court of Prince Edward Island .... [1850–1874],* vol. 1 (Charlottetown, 1885), 263–6

57 Michael J.W. Finley to author, 20 March 1978; Mr Finley, a member of the Saskatchewan bar, kindly provided me with an extensive critical commentary on the quality of Peters's written legal judgments.

58 PEI, Supreme Court Minutes, 26 June 1866, PAPEI

59 PEI, Assembly, *Debates,* 1867, 85

60 *Examiner,* 8 Jan. 1866

61 There were two other accused, Patrick Murphy and George Rackem, but they were not brought to trial because they could not be found. See PEI, Supreme Court Minutes, 13 Jan. 1866, PAPEI.

62 *Islander,* 26 Jan. 1866

63 *Islander,* 26 Jan. 1866

64 PEI, Supreme Court Minutes, 16, 17, 24 Jan. 1866, PAPEI. The other two McKennas charged in the assault on Bernard McKenna were found not guilty; see PEI, Supreme Court Minutes, 22 Jan. 1866, PAPEI. Later in 1866 Peters would sentence one John Pickering, convicted of assault on a consta-

ble, to eight months; the fact that Pickering was indicted for 'rescue & assault' may explain the long sentence. It is not apparent from the available evidence whether there was a Tenant League connection. PEI, Supreme Court Minutes, 2, 3 Oct. 1866, PAPEI.

65 *Islander*, 2 Feb. 1866 (emphasis added)

66 PEI, Assembly, *Debates*, 1867, 27

67 See SCCP, 1866, PAPEI.

68 See, for example, letter of John Lawson to the Editor of *Islander*, dated 4 July 1866, in *Islander*, 6 July 1866; *Herald*, 8 Aug. 1866; letter of Robert Stewart to the Editor of *Islander*, dated 10 Aug. 1866, in *Islander*, 17 Aug. 1866; letter of Dickieson, dated 23 Aug. 1866, in *Islander*, 31 Aug. 1866; W.S. McNeill and Davies in PEI, Assembly, *Debates*, 1874, 270–1, 394; *Island Argus*, 9 Nov. 1875.

69 *Herald*, 31 Jan. 1866; *Examiner*, 5 Feb., 26 March 1866

70 See *Examiner*, 30 April, 6 Aug. 1866; *Herald*, 8 Aug. 1866; George-Antoine Bellecourt to C.F. Cazeau, 12 Aug. 1866, series 310 CN II: 99, Archives de l'Archevêché de Québec, microfilm, Confederation Centre Library, Charlottetown. A typewritten copy of the Bellecourt letter (incorrectly dated as 2 Aug. 1866) appears as an appendix to Cécile Gallant, 'L'engagement social de Georges[sic]-Antoine Belcourt, curé de Rustico, 1859–1869,' *Les cahiers de la Société Historique Acadienne*, 11/4 (Dec. 1980), 338. The chronology in James Michael Reardon, *George Anthony Belcourt, Pioneer Catholic Missionary of the Northwest, 1803–1874: His Life and Times* (St Paul, 1955), 178–9 is fanciful: he has Bellecourt going to Charlottetown, interviewing Dundas, and returning the three prisoners to their homes – all on Christmas Eve. The explanation of the circumstances that led to the charges against the three is also inaccurate.

71 Letter of Stewart to the Editor of *Islander*, dated 10 Aug. 1866, in *Islander*, 17 Aug. 1866

72 Letter of Dickieson to the Editor of *Islander*, dated 23 Aug. 1866, in *Islander*, 31 Aug. 1866

73 J.W. Morrison to John Binns, 24 April 1866, Colonial Secretary's Letterbooks, PAPEI

74 PEI, *Statutes*, 28 Vic., c. 8 s. 1; also see PEI, *Statutes*, 23 Vic., c. 12.

75 'Presentment against the careless way Oaths are administered,' 30 June 1866, SCCP, 1866, PAPEI (original emphasis)

76 PEI, Supreme Court Minutes, 30 June 1866, PAPEI

77 Maps in *Illustrated Historical Atlas of the Province of Prince Edward Island* (Belleville, Ont., 1972 [orig. publ. 1880]), 51 (detail), 59, published by J.H. Meacham and Co., show Binns and Laird occupying premises directly across the road from each other.

78 PEI, Executive Council Minutes, 17, 31 July 1866
79 *Osborn's Concise Law Dictionary*, 6th edition (London, 1976), 212
80 PEI, Supreme Court Minutes, 26 June 1866, PAPEI
81 PEI, Assembly, *Debates*, 1868, 72
82 *Royal Gazette*, 9 May 1866
83 *Examiner*, 23 April 1866; *Islander*, 20 April 1866
84 Dundas to Earl of Carnarvon, 7 Sept. 1866, CO 226/102, 307
85 Memorandum of Arthur Blackwood, 24 Sept. 1866, in CO 226/102, 308
86 Memorandum of T.F. Elliot, 25 Sept. 1866, CO 226/102, 310 (original emphasis)
87 See Carnarvon to Dundas, 3 Oct. 1866, CO 226/102, 311–13; Douglas Galton to Sir Frederic Rogers, 31 July 1867, CO 226/103, 560–1; Dundas to Duke of Buckingham and Chandos, 1 July, 7 Oct. 1868, CO 226/104, 394–6, 481–2; PEI, Assembly, *Journal*, 1869, app. G; Haythorne in PEI, Legislative Council, *Debates and Proceedings*, 1870, 55.
88 Charles Hastings Doyle to the Quarter Master General to the Forces, Horse Guards, 8 April 1867, CO 226/103, 533
89 Doyle to the Secretary of State for War, 21 Nov. 1865, CO 226/102, 490
90 Doyle to Dundas, 18 March 1867, CO 226/103, 131–2; Doyle to the Quarter Master General to the Forces, Horse Guards, 8, 10 April 1867, CO 226/103, 532–3, 541–3. Doyle wrote his letter of 10 April in a rather caustic tone, in response to a plea by the Island government (a new Liberal administration led by George Coles) for retention of the soldiers. That Minute of the Executive Council, dated 2 April, also provoked Dundas, for the government implied that 'oppressive conduct' on the part of proprietors had caused the disorders of 1865. Dundas submitted a memorandum to the Council, in effect challenging the councillors on this point, and they issued a clarification of sorts. Having obtained a degree of satisfaction, Dundas then wrote to London expressing his own wish that the troops be retained. See PEI, Executive Council Minutes, 2, 3 April, 19 June 1867; Dundas to Buckingham, 3 April 1867, CO 226/103, 129–30; *Islander*, 17 May 1867.
91 PEI, Assembly, *Journal*, 1875, app. E
92 *Examiner*, 30 April 1866
93 No precise figure for the acreage of the Sulivan estate in 1866 is available, and I am using the number of 66,937, which is a minimum, for it is the amount which would be purchased under a compulsory land purchase act passed nine years later. See P.S. Macgowan, ed., *Report of Proceedings before the Commissioners Appointed under the Provisions of 'The Land Purchase Act, 1875'* (Charlottetown, 1876), 669.
94 *Islander*, 4 May 1866. The writer appears to have been someone other than

W.H. Pope, who was absent several months on a trade delegation to the southern hemisphere, and did not return until 9 May. See *Examiner*, 11 May 1866.

95 See PEI, Executive Council Minutes, 3 April 1866; *Islander*, 6 April 1866.

96 The statistical calculations are based on various sources, all of which have been cited elsewhere in this study.

12: The Impact and Significance of the Tenant League

1 PEI, House of Assembly, *Debates and Proceedings*, 1871, 176

2 *Herald*, 9 Jan. 1867

3 Fragments of Family History of John McEachern, Diary, undated entry at commencement of 1870, 19 Dec. 1866; also see 11 Feb. 1863; microfilm, PAPEI. In calculating the number who voted in 1863, it is necessary to total the votes and divide by two, since at that one election each voter could choose two candidates.

4 *Islander*, 14 Dec. 1866

5 PEI, Legislative Council, *Debates and Proceedings*, 1867, 41

6 Calculations based on data in *Islander*, 4 Jan. 1867

7 PEI, Legislative Council, *Debates*, 1874, 166

8 *Islander*, 14 Dec. 1866

9 On the latter point, see Benjamin Davies in PEI, Assembly, *Debates*, 1867, 8.

10 PEI, Assembly, *Debates*, 1867, 22, 148

11 Calculations based on *Examiner*, 11 March 1867

12 Edward Palmer to Joseph Howe (draft), 8 March 1867, item 256, Palmer Family Papers, PAPEI

13 George Dundas to Duke of Buckingham and Chandos, 3 June 1868, CO 226/104, 333–4, microfilm, PAC. Pope's political support on Lot 27 appears to have been weak from the beginning of his political career, for in 1857, when first elected in a by-election, 606 to 507, he lost the Lot 27 poll, 134 to 87. Thus he won only 39.4 per cent of the votes cast at that poll, while winning 58.2 per cent of the votes in the remainder of the electoral district. Calculations are based on electoral data in *Examiner*, 8 June 1857.

14 PEI, Assembly, *Debates*, 1872 (2nd session), 117

15 PEI, Assembly, *Debates*, 1872 (2nd session), 167, 169

16 *Islander*, 7 Dec. 1866

17 *Examiner*, 11 March 1867

18 *Islander*, 5 April 1867; PEI, Legislative Council, *Debates*, 1867, 41; PEI, Assembly, *Debates*, 1868, 40

19 *Islander*, 15 Feb. 1867

20 Letter of 'Fair Play' to the Editor of the *Islander*, dated 27 March 1867, in *Islander*, 5 April 1867

21 *Examiner*, 22 April 1867

22 *Examiner*, 11 March 1867

23 *Islander*, 26 April 1867

24 See Ian Ross Robertson, 'Edward Whelan,' *DCB*, vol. 9 (Toronto, 1976), 832–3; 'Edward Reilly,' *DCB*, vol. 10 (Toronto, 1972), 612–13. Although no longer an assemblyman, Whelan remained queen's printer until his death; then Reilly assumed the post.

25 PEI, Assembly, *Debates*, 1867, 25, 27, 29–30

26 Henry Jones Cundall to W.C.A. Williams, 8 May 1867, Henry Jones Cundall Letterbook, PAPEI

27 See PEI, Legislative Council, *Debates*, 1867, 41; 1870, 130; Jean Layton MacKay, 'William Warren Lord,' *DCB*, vol. 11 (Toronto, 1982), 531–3.

28 PEI, Executive Council Minutes, 16 April 1867; *Royal Gazette*, 22 May 1867

29 See PEI, Assembly, *Journal*, 1867, 13; Davies in PEI, Assembly, *Debates*, 1874, 74.

30 PEI, Executive Council Minutes, 30 April 1867; Robert Hodgson to Edward Cardwell, 16 Aug. 1865, CO 226/101, 380–8, reprinted in PEI, Assembly, *Journal*, 1866, app. G

31 *Examiner*, 6 May 1867

32 PEI, Assembly, *Journal*, 1867, 56

33 PEI, Assembly, *Debates*, 1867, 82–4

34 Dundas to Buckingham, 17 May 1867, CO 226/103, 169

35 See *Examiner*, 13 May 1867; PEI, Executive Council Minutes, 11 May 1867.

36 Dundas to Buckingham, 22 May 1867, confidential, CO 226/103, 181–5

37 See *Royal Gazette*, 12 June 1867.

38 PEI, Assembly, *Debates*, 1868, 2

39 PEI, Assembly, *Debates*, 1868, 2

40 See Ian Ross Robertson, 'Political Realignment in Pre-Confederation Prince Edward Island, 1863–1870,' *Acadiensis*, 15/1 (Autumn 1985), 35–58.

41 See Francis W.P. Bolger, *Prince Edward Island and Confederation, 1863–1873* (Charlottetown, 1964), 8–9, 61, 135–7, 166–8; and Bolger's chapters in Francis W.P. Bolger, ed., *Canada's Smallest Province: A History of P.E.I.* (Charlottetown, 1973), 180–1, 230. The latter volume is purportedly a general history of the Island; Bolger does not mention the Cunard purchase in either book, although there can be no doubt about its relevance to the state of the land question.

42 Daniel Cobb Harvey, 'Confederation in Prince Edward Island,' *Canadian Historical Review*, 14/2 (June 1933), 155; also see 151.

43 Andrew Hill Clark, *Three Centuries and the Island: A Historical Geography of Settlement and Agriculture in Prince Edward Island, Canada* (Toronto, 1959), 93

44 For example, see Coles in PEI, Assembly, *Debates*, 1866, 9, 13.

45 In 1860 the numbers of leaseholders occupying land owned by Sir Samuel and Edward Cunard, respectively, had been 971 and 376; see PEI, Assembly, *Journal*, 1863, app. A. Sixteen Cunard tenants had purchased under the Fifteen Years Purchase Bill of 1864; and in the following year (1865) the estate had ceased issuing new leases. See Dundas, 'Statistics respecting Land Tenure,' 20 May 1868, CO 226/104, 311.

46 See Hensley to Dundas, 20 Sept. 1867, printed in PEI, Assembly, *Journal*, 1868, app. A; Cundall to Captain [John] Orlebar, 24 Sept. 1867, Cundall Letterbook, PAPEI; *Journal* (Summerside), 14 Nov. 1867.

47 For a list, see PEI, Assembly, *Journal*, 1875, app. E.

48 PEI, Legislative Council, *Debates*, 1871, 21. Haviland was not oblivious to pecuniary considerations, for in 1874 he would state that without receiving a price of 9s. per acre – rather than the old limit of 7s. 6d. (or 5s. sterling) – he would not have sold. See PEI, Assembly, *Debates*, 1874, 313.

49 PEI, Assembly, *Debates*, 1871, 369. In 1874 Emanuel McEachen, a Conservative assemblyman, would note that fellow-assemblyman A.E.C. Holland, first elected in 1873, was a proprietor; see PEI, Assembly, *Debates*, 1874, 87. It appears that Holland and his wife owned slightly under five thousand acres spread over four townships and three counties. See P.S. Macgowan, ed., *Report of Proceedings before the Commissioners Appointed under the Provisions of 'The Land Purchase Act, 1875'* (Charlottetown, 1876), 633–6, 666.

50 PEI, Assembly, *Debates*, 1856, 10

51 See, for example, the remarks of George Beer, who represented eastern Queens County, regarding the Tracadie estate and the Fanning estate on Lot 50, in PEI, Legislative Council, *Debates*, 1867, 35–6; and of Henry Beer regarding the Tracadie estate, which was in his assembly district of 3rd Queens, in PEI, Assembly, *Debates*, 1874, 51.

52 PEI, Assembly, *Debates*, 1871, 176. Others had already spoken in the legislature, crediting the purchase to the league; see George W. Howlan in PEI, Assembly, *Debates*, 1869, 210; W.S. McNeill in PEI, Assembly, *Debates*, 1870, 246–7.

53 See Ian Ross Robertson, 'James Colledge Pope,' DCB, vol. 11, 700–1.

54 Hodgson to Sir John A. Macdonald, 26 Sept. 1873, Sir John A. Macdonald Papers, vol. 119, p. 48275, PAC

55 *Examiner*, 21 Oct. 1867

56 Phyllis R. Blakeley, 'Sir Samuel Cunard,' DCB, vol. 9, 184

57 See G. Beer and J. Balderston in PEI, Legislative Council, *Debates*, 1867, 35, 41; Haythorne in PEI, Legislative Council, *Debates*, 1868, 8.

58 *Examiner*, 19 Aug. 1867

59 PEI, Legislative Council, *Debates*, 1874, 44–5; also see Joseph Wightman in PEI, Legislative Council, *Debates*, 1875, 216; William Welsh in PEI, Assembly, *Debates*, 1876, 67.

60 See John Francis Maguire, *The Irish in America* (New York, 1969 [orig. publ. 1868]), 31, 44; *Dictionary of National Biography*, vol. 12, 774–5; E.D. Steele, *Palmerston and Liberalism, 1855–1865* (Cambridge, 1991), 326; *Islander*, 19 Oct. 1866; *Examiner*, 22, 29 Oct. 1866.

61 E.D. Steele, *Irish Land and British Politics: Tenant-Right and Nationality 1865–1870* (Cambridge, 1974), 62–3; also see 186–7. On the Irish Tenant League, see F.S.L. Lyons, *Ireland since the Famine* (Glasgow, 1973 ed.), 115–22; for a characterization of Maguire as a politician, see J.H. Whyte, *The Independent Irish Party, 1850–9* (Oxford, 1958), 139–41. Although Maguire may not have been aware of the Island's Tenant League, he had taken some interest in the colony over the years, for in 1863 he had written a letter to *The Times*, and also an editorial in the Cork *Examiner*, on the subject of Orangeism in Prince Edward Island; see *Vindicator*, 7 Aug. 1863.

62 Maguire, *The Irish in America*, 42, 43. The reference is to Cade's rebellion, an agrarian revolt in the south of England in 1450, crushed by force. For response in the legislature to Maguire's book, see Haythorne in PEI, Legislative Council, *Debates*, 1868, 8, 115, 162–3.

63 See Steele, *Irish Land and British Politics*, 124–6, 193–4.

64 Charles Mackay, 'A Week in Prince Edward Island,' *Fortnightly Review*, 5/26 (1 June 1866), 143; also see *Dictionary of National Biography*, vol. 12, 564–5.

65 Maguire, *The Irish in America*, 40

66 There is negative commentary on Mackay's article in *Herald*, 8 Aug. 1866.

67 Mackay, 'A Week in Prince Edward Island,' 155, 147. Also see James Hunter, *A Dance Called America: The Scottish Highlands, the United States, and Canada* (Edinburgh, 1994), 89–90, 131–3.

68 Sir Andrew Macphail, *The Master's Wife* (Charlottetown, 1994 [orig. publ. 1939]), 18

69 Historian Sung Bok Kim, writing of neo-feudal tenurial relationships in colonial New York, refers to a 'psychology of yeomanry' whose obverse was 'hatred of the elements of leaseholding – landlords and rent.' See Kim, *Landlord and Tenant in Colonial New York: Manorial Society, 1664–1775* (Chapel Hill, 1978), 130.

70 PEI, Assembly, *Journal*, 1862, app. A

71 J.E. Lattimer, *Economic Survey of Prince Edward Island with Particular Emphasis on Agricultural Needs* (Charlottetown, 1944), 18; also see chart on the same page.

72 *Examiner*, 13 Nov. 1865

73 Maguire, *The Irish in America*, 29

74 See *Examiner*, 30 Oct. 1865; Palmer in PEI, Legislative Council, *Debates*, 1866, 11; *Journal* (Summerside), 4 July 1867; Davies in PEI, Assembly, *Debates*, 1874, 57.

75 Mackay, 'A Week in Prince Edward Island,' 148

76 Scott W. See, 'Polling Crowds and Patronage: New Brunswick's "Fighting Elections" of 1842–3,' *Canadian Historical Review*, 72/2 (June 1991), 155–6

77 Eric Richards, *A History of the Highland Clearances*, vol. 1: *Agrarian Transformation and the Evictions 1746–1886* (London, 1982), 446. Also see his 'Problems on the Cromartie Estate, 1851–3,' *Scottish Historical Review*, 52/2 (Oct. 1973), 157; T.M. Devine, *Clanship to Crofters' War: The Social Transformation of the Scottish Highlands* (Manchester, 1994), 210.

78 See David Maldwyn Ellis, *Landlords and Farmers in the Hudson-Mohawk Region 1790–1850* (New York, 1967 ed. [orig. publ. 1946]), 258, 264–6; Henry Christman, *Tin Horns and Calico: A Decisive Episode in the Emergence of Democracy* (Cornwallville, NY, 1975 ed. [orig. publ. 1945]), 177–83, 200.

79 There are no data on the extent of turnover as settlers gave up, were evicted, or were encouraged to vacate, but anecdotal evidence suggests it was a quite common occurrence.

80 Laura Maynard, a student at Scarborough College, University of Toronto, made this point to me on 29 Jan. 1991.

81 See Robertson, 'Political Realignment in Pre-Confederation Prince Edward Island, 1863–1870,' 56–7; Robertson, 'James Colledge Pope,' 701–2. Also note the suggestive comments of J.M. Bumsted, '"The Only Island There Is": The Writing of Prince Edward Island History,' in Verner Smitheram, David Milne, and Satadal Dasgupta, eds., *The Garden Transformed: Prince Edward Island, 1945–1980* (Charlottetown, 1982), 29.

82 Dundas to Cardwell, 25 Aug. 1865, confidential, CO 226/101, 665

83 A point made to me by historian Kenneth McNaught on 30 May 1995.

84 Charles Townshend, *Political Violence in Ireland: Government and Resistance since 1848* (Oxford, 1983), 126

85 See Colin Read, *The Rebellion of 1837 in Upper Canada*, Canadian Historical Association Booklet no. 46 (Ottawa, 1988), 1, 11–24; Colin Read and Ronald J. Stagg, 'Introduction' to *The Rebellion of 1837 in Upper Canada: A Collection of Documents* (Toronto, 1985), xcviii–xcix.

86 Interview with Roy Dickieson, 27 Aug. 1986

87 Ernest R. Forbes makes the point about such regional bias for a later period in his powerful article, 'In Search of a Post-Confederation Maritime Historiography, 1900–1967,' *Acadiensis*, 8/1 (Autumn 1978), 3–21.

Appendices

1 PEI, Legislative Council, *Journal*, 1865, 36
2 PEI, Executive Council Minutes, 9 Jan. 1868, appendix
3 P.S. Macgowan, ed., *Report of Proceedings before the Commissioners Appointed under the Provisions of 'The Land Purchase Act, 1875'* (Charlottetown, 1876), 49, 255
4 Saving = £1,380 (£1,146 [Montgomery's estimate of fifteen years' rent − £4,000] + £384 [arrears] − £150 [expenses])
5 Saving = £1,772 (£1,146 [Montgomery's estimate of fifteen years' rent − £4,000] + £776 [arrears] − £150 [expenses])

Historiographical Background

Over the years, researchers have given little attention to the Tenant League or, more formally, the Tenant Union of Prince Edward Island. Political scientist Frank MacKinnon incorporated a brief account of its context and activities in his classic study, *The Government of Prince Edward Island*, published in 1951, but did not attempt to assess its significance.[1] Eight years later, historical geographer Andrew Hill Clark, in his highly original *Three Centuries and the Island*, made more of a judgment, although the effect appeared to be to dismiss it as of minimal importance.[2] In the 1960s, with the approach of the centennial of Confederation, there was an outpouring of historical publications on the movement for union of the British North American colonies, and because the league flourished in precisely the same years, some writers took note of it. Historian Peter B. Waite included information on the confrontations between leaguers and Island authorities in *The Life and Times of Confederation 1864–1867*, and even reported the desertion of soldiers sent to deter the organization.[3] But Francis W.P. Bolger, also an historian, and one specializing in the history of the Island, gave the league scant notice in *Prince Edward Island and Confederation, 1863–1873*, and conveyed less appreciation of its nature than MacKinnon, Clark, or Waite.[4] In more recent years historian Andrew Robb has referred to the Tenant League in a brief and general article on violence in Prince Edward Island history, and historian and genealogist Peter McGuigan has dealt with a sequence of events relating to the league in a particular area where it flourished.[5]

There has also been a stream of historical writing concerning Prince Edward Island which has not been based on systematic research. Authors in this tradition who have mentioned the league in their writings include H.M. Anderson, George E. Hart, Lorne C. Callbeck, Errol Sharpe, and Douglas Baldwin. In sympathy, they range from the totally enthusiastic (Sharpe) to the seemingly bemused (Callbeck). These authors have not added anything reliable to our knowledge or understanding of the league, and occasionally their statements have been simply inaccurate, for example, insisting that there was no violence (Anderson).[6] In addition, there has been what could be described as a pamphleteering tradition, and at one end of that spectrum there is an eccentric and choleric little tract, anonymously authored, published in 1881, which attacks the Tenant League in the course of linking Fenianism, the Irish Land League, and communism. At the other end is a published address delivered by a farm activist in 1987, making reference to dramatic events in the history of the league for didactic purposes.[7]

I first became aware of the Tenant League when doing research for my Master of Arts thesis at McGill University in the late 1960s on 'Religion, Politics, and Education in Prince Edward Island, from 1856 to 1877.' It was clear from the primary sources that the league had played a major role in the history of the colony during the middle years of the Confederation decade. My work began to touch closely on the league with articles on such figures as George Coles, William Henry Pope, and Edward Whelan for volumes of the *Dictionary of Canadian Biography* which were published during the 1970s. Coles and Whelan were prominent Reformers who strongly opposed the league; they had emerged in local politics in the 1840s, had governed in the 1850s, and appeared to constitute a Reform establishment by the time the league arose. Pope, a Conservative, was particularly furious at the reluctance of traditional Conservative partisans to distance themselves from the organization. The emergence of the league had, in effect, driven a wedge between the élite and the mass of supporters in each political party. It became increasingly apparent in the course of this research, particularly on the brilliant journalist Whelan, who was a close and perceptive observer of the league's rise and fall, that there was an important part of the history of popular movements in pre-Confederation British North America to be uncovered.

At a conference of researchers specializing in the histories of Ireland and Scotland held in Dublin in 1976, I delivered a general paper entitled 'Highlanders, Irishmen, and the Land Question in Nineteenth-Century Prince Edward Island,' of which approximately four pages concern the history of the league from formation to demise. Published the following year, it remains, despite its brevity, the most comprehensive treatment of the movement available before the appearance of this book.[8] In the 1980s I published three pieces which dealt with specific aspects of the history of the league: an article in the scholarly periodical *Acadiensis* focusing on its political impact; the introduction to an abridged edition of the proceedings and report of a royal commission, the rejection of whose recommendations was a factor precipitating the formation of the league; and an article in *The Island Magazine* on the curious endeavour to apprehend a rank and file Tenant Leaguer in his own district through the ancient device of the *posse comitatus*.[9]

NOTES

1 Frank MacKinnon, *The Government of Prince Edward Island* (Toronto, 1951), 107, 119
2 Andrew Hill Clark, *Three Centuries and the Island: A Historical Geography of Settlement and Agriculture in Prince Edward Island, Canada* (Toronto, 1959), 93
3 Peter B. Waite, *The Life and Times of Confederation 1864–1867: Politics, Newspapers, and the Union of British North America* (Toronto, 1962 ed.), 181–2, 188–91
4 Francis W.P. Bolger, *Prince Edward Island and Confederation, 1863–1873* (Charlottetown, 1964), 8–9, 61, 135–7, 166–8; also see Bolger's chapters in Francis W.P. Bolger, ed., *Canada's Smallest Province: A History of P.E.I.* (Charlottetown, 1973), 180–1.
5 Andrew Robb, 'Rioting in Nineteenth Century P.E.I.,' *Canadian Frontier*, 2 (1979), 32–3; Peter McGuigan, 'Tenants and Troopers: The Hazel Grove Road, 1865–68,' *Island Magazine*, no. 32 (Fall–Winter 1992), 22–8. Also see Peter McGuigan, 'The Lot 61 Irish: Settlement and Stabilization,' *Abegweit Review*, 6/1 (Spring 1988), 42–5. Robb cites no sources in his article.
6 G.U. Hay, *The History of Canada to which has been added a sketch of the History of Prince Edward Island by H.M. Anderson* (Toronto, 1905), 118; George E. Hart, *The Story of Old Abegweit: A Sketch of Prince Edward Island History* ([Charlottetown?], [1935?]), 55; Lorne C. Callbeck, *The Cradle of Confederation: A Brief History of Prince Edward Island from Its Discovery in 1534 to the Present Time* (Fredericton, 1964), 175, 177–82, 203–4; Errol Sharpe, *A People's History of Prince Edward Island* (Toronto, 1976), 107–11; Douglas Baldwin, *Land of the Red Soil: A Popular History of Prince Edward Island* (Charlottetown, 1990), 90–1. In his text for grade 6 students, Baldwin also mentions the Tenant League; see *Abegweit: Land of the Red Soil* (Charlottetown, 1985), 194–5. Professor Kenneth MacKinnon of Saint Mary's University kindly brought the Hay-Anderson book to my attention.
7 'A Native of P.E. Island,' *Fenianism, Irish Land Leagueism, and Communism* (n.p., 1881), 15–17; Reg Phelan, 'P.E.I. Land Struggle,' *Cooper Review*, no. 3 (1988), 30–1
8 It was published in L.M. Cullen and T.C. Smout, eds., *Comparative Aspects of Scottish and Irish Economic and Social History 1600–1900* (Edinburgh, [1977]), 227-40; the relevant pages are 231–4. It is most accessible in J.M. Bumsted, ed., *Interpreting Canada's Past*, vol. 1: *Pre-Confederation* (Toronto, 2nd ed.,

1993), 515–32, where it appears under a new title chosen by the editor: 'Ethnicity and Rural Discontent.'
9 'Political Realignment in Pre-Confederation Prince Edward Island, 1863–1870,' *Acadiensis*, 15/1 (Autumn 1985), 35–58; 'Introduction' to Ian Ross Robertson, ed., *The Prince Edward Island Land Commission of 1860* (Fredericton, 1988); 'The *Posse Comitatus* Incident of 1865,' *Island Magazine*, no. 24 (Fall–Winter 1988), 3–10

Bibliography

A. Contemporary Sources

MANUSCRIPT MATERIAL

Accession 2514/9 & 10, PAPEI
Archives de l'Archevêché de Québec, série 310 CN II, microfilm, Confederation
   Centre Library, Charlottetown
Colonial Office, 226 series, microfilm, PAC

Colonial Secretary's Letterbooks, PAPEI
Fragments of Family History of John McEachern, microfilm, PAPEI
Government House Letterbooks, PAPEI
Headquarters Office Papers for Nova Scotia, 1783–1907, PANS
Henry Jones Cundall Diaries, PEIM
Henry Jones Cundall Letterbooks, PAPEI
Joseph Howe Papers, PAC
Palmer Family Papers, PAPEI
Palmerston Letterbooks, Add. MSS. 48579, British Library, London
Palmerston Papers, MS 62, LB/1. University of Southampton Library, Southampton, England
PEI, Conveyance Registers, Land Registry Records, PAPEI
PEI, Executive Council Minutes, microfilm, PAPEI
PEI, Land Title Documents, PAPEI
PEI, Supreme Court Case Papers, PAPEI
PEI, Supreme Court Minute Books, PAPEI
Pope Scrapbook, PAPEI
Seymour of Ragley Papers, CR114A, WCRO
Sir John A. Macdonald Papers, PAC
Sulivan Family Papers, DD/476, Archives Department, Shepherds Bush Library, Uxbridge Road, London, England
William Frederick Rennie Papers, Newfoundland Historical Society, St John's

GOVERNMENT PUBLICATIONS: GENERAL

British Parliamentary Papers: Correspondence and Papers Relating to Canada 1854–58, vol. 21. Shannon, Ireland: Irish University Press, 1970
Census of 1861 for Prince Edward Island. PEI, House of Assembly, *Journal*, 1862, appendix A
Census of 1871 for Prince Edward Island. PEI, House of Assembly, *Journal*, 1872, appendix I
PEI, House of Assembly, *Debates and Proceedings*
PEI, House of Assembly, *Journal*
PEI, Legislative Council, *Debates and Proceedings*
PEI, Legislative Council, *Journal*
PEI, *Statutes*
War Office, 1/563, microfilm, PAC

GOVERNMENT PUBLICATIONS: COMMISSIONS

Gordon, J.D., and D. Laird, eds. *Abstract of the Proceedings before the Land Commis-*

sioners' Court, Held During the Summer of 1860, to Inquire into the Differences Relative to the Rights of Landowners and Tenants in Prince Edward Island. Charlottetown, 1862

Macgowan, P.S., ed. *Report of Proceedings before the Commissioners Appointed under the Provisions of 'The Land Purchase Act, 1875.'* Charlottetown, 1876

PRINTED PRIMARY SOURCES, NON-GOVERNMENT: GENERAL

Baglole, Harry, comp. and ed. *The Land Question: A Study Kit of Primary Documents.* Charlottetown: PEI Department of Education, 1975

Bell, Marilyn, ed. 'Mr. Mann's Island: The Journal of an Absentee Proprietor, 1840.' *Island Magazine*, no. 33 (Spring–Summer 1993), 17–24

Bourne, Kenneth, ed. *The Letters of the Third Viscount Palmerston to Laurence and Elizabeth Sulivan 1804–1863.* London: Royal Historical Society, 1979

Grand Orange Lodge of Prince Edward Island Records, in private possession

Haszard, Francis L., and A. Bannerman Warburton, eds. *Reports of Cases Determined in the Supreme Court, Court of Chancery, and Vice Admiralty Court of Prince Edward Island ... [1850–1874]*, vol. 1. Charlottetown, 1885

Robertson, Ian Ross, ed. *The Prince Edward Island Land Commission of 1860.* Fredericton: Acadiensis Press, 1988

PRINTED PRIMARY SOURCES, NON-GOVERNMENT: NEWSPAPERS CONSULTED

Examiner (Charlottetown)
Herald (Charlottetown)
Islander (Charlottetown)
Journal (Summerside)
Monitor (Charlottetown)
Patriot (Charlottetown)
Protestant and Evangelical Witness (Charlottetown)
Ross's Weekly (Charlottetown)
Royal Gazette (Charlottetown)
Semi-Weekly Advertiser (Charlottetown)
Vindicator (Charlottetown)

OTHER PUBLICATIONS, MORE OR LESS CONTEMPORARY

[Anonymous]. 'Historical Sketch of the Province of Prince Edward Island.' In *Illustrated Historical Atlas of the Province of Prince Edward Island*, 3–12. Belleville, Ont.: Mika, 1972 (orig. publ. 1880)

Bagster, C. Birch. *The Progress and Prospects of Prince Edward Island, written during the leisure of a visit in 1861. A sketch intended to supply information upon which enquiring emigrants may rely, and actual settlers adopt as the basis of a wider knowledge of their beautiful island home.* Charlottetown, 1861

'Celt.' 'Chips.' *Prince Edward Island Magazine*, 2/2 (April 1900), 51–2

Cotton, W.L. 'The Press in Prince Edward Island.' In *Past and Present of Prince Edward Island ...* , edited by D.A. MacKinnon and A.B. Warburton, 112–21. Charlottetown, [1906]

Hay, G.U. *The History of Canada to which has been added a sketch of the History of Prince Edward Island by H.M. Anderson.* Toronto: Copp Clark, 1905

Lake, D.J. *Topographical Map of Prince Edward Island in the Gulf of St. Lawrence. From actual Surveys and the late Coast Survey of Capt. H.W. Bayfield.* Saint John, 1863

Le Page, John. *The Island Minstrel: A Collection of Some of the Poetical Writings of John Le Page ...* vol. 2. Charlottetown, 1867

Mackay, Charles. 'A Week in Prince Edward Island.' *Fortnightly Review*, 5/26 (1 June 1866), 143–57

Maguire, John Francis. *The Irish in America.* New York: Arno Press, 1969 (orig. publ. 1868)

'A Native of P.E. Island.' *Fenianism, Irish Land Leagueism, and Communism.* N.p., 1881

'Rambler.' 'The Tenant League Articles.' *Prince Edward Island Magazine*, 2/5 (July 1900), 162

Ross, John. *Reminiscences from the Life of John Ross, Formerly Publisher of 'Ross's Weekly,' the Avowed Champion of the Tenant Union Organization, also Advocate of the Prince Edward Island Railway, and Now Manufacturer and Proprietor of the Celebrated 'Magic Healer' Salve.* Charlottetown, 1892

– 'A Page from the History of P.E. Island.' *Prince Edward Island Magazine*, 1/8 (Oct. 1899), 276–8

– 'The Land Question of Prince Edward Island – Continued.' *Prince Edward Island Magazine*, 1/9 (Nov. 1899), 323–5

– 'Tenant League Proceedings – Continued.' *Prince Edward Island Magazine*, 1/10 (Dec. 1899), 356–9

– 'Tenant League Results.' *Prince Edward Island Magazine*, 1/12 (Feb. 1900), 432–5

– 'Tenant League Results.' *Prince Edward Island Magazine*, 2/1 (March 1900), 22–5

B. Selected Later Works

PUBLISHED STUDIES

Acorn, Milton. *The Island Means Minago.* Toronto: NC Press, 1975

Archer, John E. '"A Fiendish Outrage"? A Study of Animal Maiming in East Anglia: 1830–1870.' *Agricultural History Review*, 33/2 (1985), 147–57
– 'Under Cover of Night: Arson and Animal Maiming.' In *The Unquiet Countryside*, edited by G.E. Mingay, 65–79. London: Routledge, 1989
– *By a Flash and a Scare: Incendiarism, Animal Maiming, and Poaching in East Anglia, 1815–1870*. Oxford: Clarendon Press, 1990
Baglole, Harry. 'William Cooper.' In *Dictionary of Canadian Biography*, vol. 9, 155–8. Toronto: University of Toronto Press, 1976
– 'William Cooper of Sailor's Hope.' *Island Magazine*, no. 7 (Fall–Winter 1979), 3–11
– 'Walter Patterson.' In *Dictionary of Canadian Biography*, vol. 4, 605–11. Toronto: University of Toronto Press, 1979
– 'Drawing Lots.' *Horizon Canada*, 6/55 (March 1986), 1297–1303
– 'The Legacy of William Cooper: Protecting the Island Homeland.' *Cooper Review*, no. 3 (1988), 16–20
Baglole, Harry, ed. *Exploring Island History: A Guide to the Historical Resources of Prince Edward Island*. Belfast, PEI: Ragweed, 1977
Bagnall, Mrs Frank. 'Road Houses and Taverns.' In *Historic Highlights of Prince Edward Island*, edited by Mary Brehaut, 52–8. Charlottetown: Historical Society of Prince Edward Island, 1955
Bagnall, Margaret Ruth. 'The Red Fox.' In *Historic Sidelights of Prince Edward Island*, edited by Mary Brehaut, 37–47. Charlottetown: Historical Society of Prince Edward Island, 1956
Baldwin, Douglas. *Land of the Red Soil: A Popular History of Prince Edward Island*. Charlottetown: Ragweed, 1990
Beames, M.R. 'The Ribbon Societies: Lower-Class Nationalism in Pre-Famine Ireland,' *Past and Present*, no. 97 (Nov. 1982), 128–43
Beloff, Max. *Public Order and Popular Disturbances 1660–1714*. London: Frank Cass and Co., 1963 ed. (orig. publ. 1938)
Bew, Paul. *Land and the National Question in Ireland 1858–82*. Dublin: Gill and Macmillan, 1978
Bittermann, [K.] Rusty. 'Agrarian Protest and Cultural Transfer: Irish Immigrants and the Escheat Movement on Prince Edward Island.' In *The Irish in Atlantic Canada, 1780–1900*, edited by Thomas P. Power, 96–106. Fredericton: New Ireland Press, 1991
– 'Women and the Escheat Movement: The Politics of Everyday Life on Prince Edward Island.' In *Separate Spheres: Women's Worlds in the Nineteenth-Century Maritimes*, edited by Janet Guildford and Suzanne Morton, 23–38. Fredericton: Acadiensis Press, 1994

Blakeley, Phyllis R. 'Sir Samuel Cunard.' In *Dictionary of Canadian Biography*, vol. 9, 172–86. Toronto: University of Toronto Press, 1976

Bolger, Francis W.P. *Prince Edward Island and Confederation, 1863–1873.* Charlottetown: St Dunstan's University Press, 1964

Bolger, Francis, W.P., ed. *Canada's Smallest Province: A History of P.E.I.* Charlottetown: Prince Edward Island 1973 Centennial Commission, 1973

Bourne, Kenneth. *Palmerston: The Early Years, 1784–1841.* London: Allen Lane, 1982

Brady, J.C. 'Legal Developments, 1801–79.' In *A New History of Ireland, vol. 5: Ireland under the Union, I (1801–70)*, edited by W.E. Vaughan, 451–81. Oxford: Clarendon Press, 1989

Buckner, Phillip A. 'Charles Douglass Smith.' In *Dictionary of Canadian Biography*, vol. 8, 823–8. Toronto: University of Toronto Press, 1985

– 'The Colonial Office and British North America, 1801–50.' In *Dictionary of Canadian Biography*, vol. 8, xxiii–xxxvii. Toronto: University of Toronto Press, 1985

Bumsted, J.M. 'Sir James Montgomery and Prince Edward Island, 1767–1803.' *Acadiensis*, 7/2 (Spring 1978), 76–102

– 'Lord Selkirk of Prince Edward Island.' *Island Magazine*, no. 5 (Fall–Winter 1978), 3–8

– 'Captain John MacDonald and the Island.' *Island Magazine*, no. 6 (Spring–Summer 1979), 15–20

– 'The Origins of the Land Question on Prince Edward Island, 1767–1805.' *Acadiensis*, 11/1 (Autumn 1981), 43–56

– '"The Only Island There Is": The Writing of Prince Edward Island History.' In *The Garden Transformed: Prince Edward Island, 1945–1980*, edited by Verner Smitheram, David Milne, and Satadal Dasgupta, 11–38. Charlottetown: Ragweed, 1982

– 'The Patterson Regime and the Impact of the American Revolution on the Island of St. John, 1775–1786.' *Acadiensis*, 13/1 (Autumn 1983), 47–67

– *Land, Settlement, and Politics on Eighteenth-Century Prince Edward Island.* Kingston and Montreal: McGill-Queen's University Press, 1987

Burroughs, Peter. 'Tackling Army Desertion in British North America.' *Canadian Historical Review*, 61/1 (March 1980), 28–68

Callbeck, Lorne C. *The Cradle of Confederation: A Brief History of Prince Edward Island from Its Discovery in 1534 to the Present Time.* Fredericton: Brunswick Press, 1964

'CFH' [Catherine F. Horan]. 'William Epps Cormack.' In *Encyclopedia of Newfoundland and Labrador*, vol. 1, 536. St John's: Newfoundland Book Publishers (1967), 1981

Charlesworth, Andrew, ed. *An Atlas of Rural Protest in Britain 1548–1900*. London: Croom Helm, 1983

Christianson, Gale E. 'Secret Societies and Agrarian Violence in Ireland, 1790–1840.' *Agricultural History*, 46/3 (July 1972), 369–84

Christman, Henry. *Tin Horns and Calico: A Decisive Episode in the Emergence of Democracy*. Cornwallville, NY: Hope Farm Press, 1975 ed. (orig. publ. 1945)

Clark, Andrew Hill. *Three Centuries and the Island: A Historical Geography of Settlement and Agriculture in Prince Edward Island, Canada*. Toronto: University of Toronto Press, 1959

Clark, Samuel. *Social Origins of the Irish Land War*. Princeton: Princteon University Press, 1979

– 'Landlord Domination in Nineteenth-Century Ireland.' In *UNESCO Yearbook on Peace and Conflict Studies 1986*, 5–29. Paris: UNESCO, 1988

Coffin, Marjorie L. Hyndman. 'Joseph Wightman.' In *Dictionary of Canadian Biography*, vol. 11, 922–3. Toronto: University of Toronto Press, 1982

Comerford, R.V. 'Ireland 1850–70: Post-Famine and Mid-Victorian.' In *A New History of Ireland*, vol. 5: *Ireland under the Union, I (1801–70)*, edited by W.E. Vaughan, 372–95. Oxford: Clarendon Press, 1989

Crawford, W.H. 'Landlord-Tenant Relations in Ulster 1609–1820.' *Irish Economic and Social History*, 2 (1975), 5–21

Cross, Michael S. '"The Laws Are Like Cobwebs": Popular Resistance to Authority in Mid-Nineteenth Century British North America.' In *Law in a Colonial Society: The Nova Scotia Experience*, edited by Peter B. Waite, Sandra Oxner, and Thomas Barnes, 103–23. Toronto: Carswell Co., 1984

Cullen, Mary K. 'Charlottetown Market Houses: 1813–1958.' *Island Magazine*, 6 (Spring–Summer 1979), 27–32.

Davies, James C. 'Toward a Theory of Revolution.' *American Sociological Review*, 27/1 (Feb. 1962), 5–19

de Jong, Nicolas J., and Marven E. Moore. *Shipbuilding on Prince Edward Island: Enterprise in a Maritime Setting 1787–1920*. Hull: Canadian Museum of Civilization, 1994

Devine, T.M. *The Great Highland Famine: Hunger, Emigration and the Scottish Highlands in the Nineteenth Century*. Edinburgh: John Donald, 1988

– *Clanship to Crofters' War: The Social Transformation of the Scottish Highlands*. Manchester: Manchester University Press, 1994

– *The Transformation of Rural Scotland: Social Change and the Agrarian Economy, 1660–1815*. Edinburgh: Edinburgh University Press, 1994

Donnelly, James S., Jr. *The Land and the People of Nineteenth-Century Cork: The Rural Economy and the Land Question*. London: Routledge and Kegan Paul, 1975

– 'Landlords and Tenants.' In *A New History of Ireland*, vol. 5: *Ireland under the*

Union, I (1801–70), edited by W.E. Vaughan, 332–49. Oxford: Clarendon Press, 1989

Douglas, Bronwen, and Bruce W. Hodgins. 'George Nicol Gordon.' In *Dictionary of Canadian Biography*, vol. 9, 325–6. Toronto: University of Toronto Press, 1976

Ellis, David Maldwyn. *Landlords and Farmers in the Hudson-Mohawk Region 1790–1850*. New York: Octagon Books, 1967 ed. (orig. publ. 1946)

Foster, R.F. *Modern Ireland, 1600–1972*. Harmondsworth, England: Penguin, 1989 ed.

Fraser, W. Hamish. 'Patterns of Protest.' In *People and Society in Scotland I: 1760–1830*, edited by T.M. Devine and Rosalind Mitchison, 268–91. Edinburgh: John Donald, 1988

Gallant, Cécile. 'L'engagement social de Georges[sic]-Antoine Belcourt, curé de Rustico, 1859–1869.' *Les cahiers de la Société Historique Acadienne*, 11/4 (Dec. 1980), 316–39

Garvin, Tom. 'Defenders, Ribbonmen and Others: Underground Political Networks in Pre-Famine Ireland.' *Past and Present*, no. 96 (Aug. 1982), 133–55

Glen, William M. 'The Making of a Gentleman? The Curious Life of W.W. Irving.' *Island Magazine*, no. 30 (Fall–Winter 1991), 3–8

Graves, Donald. '"Tommy" in Canada.' *Horizon Canada*, 2/20 (July 1985), 464–9

Hart, George E. *The Story of Old Abegweit: A Sketch of Prince Edward Island History*. [Charlottetown?]: n.p., [1935?]

Harvey, Daniel Cobb. 'Dishing the Reformers.' *Transactions* of the Royal Society of Canada, 3rd ser., vol. 25, sect. 2 (1931), 37–44

– 'Confederation in Prince Edward Island.' *Canadian Historical Review*, 14/2 (June 1933), 143–60

Hitsman, J. Mackay. 'Military Defenders of Prince Edward Island 1775–1864.' *Annual Report of the Canadian Historical Association*, 1964, 25–36

Hodgins, Bruce W. 'James Douglas Gordon.' In *Dictionary of Canadian Biography*, vol. 10, 308. Toronto: University of Toronto Press, 1972

Holman, H.T. 'William Douse.' In *Dictionary of Canadian Biography*, vol. 9, 222–3. Toronto: University of Toronto Press, 1976

– 'James Bardin Palmer.' In *Dictionary of Canadian Biography*, vol. 6, 565–9. Toronto: University of Toronto Press, 1987

Hornby, Jim. *Black Islanders: Prince Edward Island's Historical Black Community*. Charlottetown: Institute of Island Studies, 1991

Hunter, James. *The Making of the Crofting Community*. Edinburgh: John Donald, 1976

– *Scottish Highlanders: A People and Their Place*. Edinburgh: Mainstream, 1992

– *A Dance Called America: The Scottish Highlands, the United States, and Canada*. Edinburgh: Mainstream, 1994

Jones, David J.V. *Before Rebecca: Popular Protests in Wales 1793–1835*. London: Allen Lane, 1973

– 'Rural Crime and Protest.' In *The Victorian Countryside*, vol. 2, edited by G.E. Mingay, 566–79. London: Routledge and Kegan Paul, 1981

– *Crime in Nineteenth-Century Wales*. Cardiff: University of Wales Press, 1992

Jordan, Donald E., Jr. *Land and Popular Politics in Ireland: County Mayo from the the Plantation to the Land War*. Cambridge: Cambridge University Press, 1994

Kealey, Gregory S. 'Knights of Labor.' *The Canadian Encyclopedia*, 2nd edition, vol. 2, 1144. Edmonton: Hurtig, 1988

Kealey, Gregory S., and Bryan D. Palmer. *Dreaming of What Might Be: The Knights of Labor in Ontario, 1880–1900*. Toronto: New Hogtown Press, 1987 ed. (orig. publ. 1982)

Kim, Sung Bok. *Landlord and Tenant in Colonial New York: Manorial Society, 1664–1775*. Chapel Hill: University of North Carolina Press, 1978

Lattimer, J.E. *Economic Survey of Prince Edward Island with Particular Emphasis on Agricultural Needs*. Charlottetown: Department of Reconstruction, 1944

Lyons, F.S.L. *Ireland since the Famine*. Glasgow: Collins, 1973 ed.

MacDonald, G. Edward. 'Joseph Hensley.' In *Dictionary of Canadian Biography*, vol. 12, 425–7. Toronto: University of Toronto Press, 1990

MacKay, Jean Layton. 'William Warren Lord.' In *Dictionary of Canadian Biography*, vol. 11, 531–3. Toronto: University of Toronto Press, 1982

MacKinnon, Frank. *The Government of Prince Edward Island*. Toronto: University of Toronto Press, 1951

MacKinnon, Wayne E. *The Life of the Party: A History of the Liberal Party in Prince Edward Island*. Summerside: Williams and Crue, 1973

Macphail, Sir Andrew. *The Master's Wife*. Charlottetown: Institute of Island Studies, 1994 (orig. publ. 1939)

Marlow, Joyce. *Captain Boycott and the Irish*. London: André Deutsch, 1973

Marquis, Greg. 'Enforcing the Law: The Charlottetown Police Force.' In *Gaslights Epidemics and Vagabond Cows: Charlottetown in the Victorian Era*, edited by Douglas Baldwin and Thomas Spira, 86–102. Charlottetown: Ragweed, 1988

Mather, F.C. *Public Order in the Age of the Chartists*. Manchester: Manchester University Press, 1959

McCartney, Donal. 'The Churches and Secret Societies.' In *Secret Societies in Ireland*, edited by T. Desmond Williams, 68–78. Dublin: Gill and Macmillan, 1973

McDonald, Ronald H. 'Sir Charles Hastings Doyle.' In *Dictionary of Canadian Biography*, vol. 11, 278–81. Toronto: University of Toronto Press, 1982

McGuigan, Peter. 'From Wexford and Monaghan: The Lot 22 Irish.' *Abegweit Review*, 5/1 (Winter 1985), 61–96

– 'The Lot 61 Irish: Settlement and Stabilization.' *Abegweit Review*, 6/1 (Spring 1988), 33–63
– 'Tenants and Troopers: The Hazel Grove Road, 1865–68.' *Island Magazine*, no. 32 (Fall–Winter 1992), 22–8
Moody, T.W. 'Fenianism, Home Rule and the Land War (1850–91).' In *The Course of Irish History*, edited by T.W. Moody and F.X. Martin, 275–93. Cork: Mercier, 1967
– *Davitt and Irish Revolution 1846–82*. Oxford: Clarendon Press, 1981.
Morrison, A.L. *My Island Pictures: The Story of Prince Edward Island*. Charlottetown: Ragweed, 1980
Morrow, Marianne. 'The Builder: Isaac Smith and Early Island Architecture.' *Island Magazine*, no. 18 (Fall–Winter 1985), 17–23
Nowlan, Kevin B. 'Conclusion.' In *Secret Societies in Ireland*, edited by T. Desmond Williams, 180–9. Dublin: Gill and Macmillan, 1973
Orr, Willie. *Deer Forests, Landlords and Crofters: The West Highlands in Victorian and Edwardian Times*. Edinburgh: John Donald, 1982
Palmer, Stanley H. *Police and Protest in England and Ireland 1780–1850*. Cambridge: Cambridge University Press, 1988
Phelan, Reg. 'P.E.I. Land Struggle.' *Cooper Review*, no. 3 (1988), 21–33
Pigot, F.L. 'John MacDonald of Glenaladale.' *Dictionary of Canadian Biography*, vol. 5, 514–17. Toronto: University of Toronto Press, 1983
Reardon, James Michael. *George Anthony Belcourt, Pioneer Catholic Missionary of the Northwest, 1803–1874: His Life and Times*. St Paul: North Central Publishing, 1955
Richards, Eric. 'Problems on the Cromartie Estate, 1851–3.' *Scottish Historical Review*, 52/2 (Oct. 1973), 149–64
– 'How Tame Were the Highlanders during the Clearances?' *Scottish Studies*, 17/1 (1973), 35–50
– *The Leviathan of Wealth: The Sutherland Fortune in the Industrial Revolution*. London: Routledge and Kegan Paul, 1973
– 'Patterns of Highland Discontent, 1790–1860.' In *Popular Protest and Public Order: Six Studies in British History 1790–1920*, edited by Roland Quinault and John Stevenson, 75–114. London: Allen and Unwin, 1974
– *A History of the Highland Clearances*, vol. 1: *Agrarian Transformation and the Evictions 1746–1886*. London: Croom Helm, 1982
– *A History of the Highland Clearances*, vol. 2: *Emigration, Protest, Reasons*. London: Croom Helm, 1985
Ridley, Jasper. *Lord Palmerston*. London: Constable, 1970
Robb, Andrew. 'Rioting in Nineteenth Century P.E.I.' *Canadian Frontier*, 2 (1979), 30–4

- 'Robert Poore Haythorne.' In *Dictionary of Canadian Biography*, vol. 12, 418–19. Toronto: University of Toronto Press, 1990
- 'Thomas Heath Haviland [Jr].' In *Dictionary of Canadian Biography*, vol. 12, 415–18. Toronto: University of Toronto Press, 1990
- Roberts, Paul E.W. 'Caravats and Shanavests: Whiteboyism and Faction Fighting in East Munster, 1802–11.' In *Irish Peasants: Violence and Political Unrest 1780–1914*, edited by Samuel Clark and James S. Donnelly Jr, 64–101. Madison: University of Wisconsin Press, 1983
- Robertson, Ian Ross. 'Edward Reilly.' In *Dictionary of Canadian Biography*, vol. 10, 612-13. Toronto: University of Toronto Press, 1972
- 'George Coles.' In *Dictionary of Canadian Biography*, vol. 10, 182–8. Toronto: University of Toronto Press, 1972
- 'George Dundas.' In *Dictionary of Canadian Biography*, vol. 10, 264–5. Toronto: University of Toronto Press, 1972
- '[Rev.] John McDonald.' In *Dictionary of Canadian Biography*, vol. 10, 460–2. Toronto: University of Toronto Press, 1972
- 'Sir Robert Hodgson.' In *Dictionary of Canadian Biography*, vol. 10, 352–3. Toronto: University of Toronto Press, 1972
- 'William Henry Pope.' In *Dictionary of Canadian Biography*, vol. 10, 593–9. Toronto: University of Toronto Press, 1972
- 'Edward Whelan.' In *Dictionary of Canadian Biography*, vol. 9, 828–35. Toronto: University of Toronto Press, 1976
- 'Highlanders, Irishmen, and the Land Question in Nineteenth-Century Prince Edward Island.' In *Comparative Aspects of Scottish and Irish Economic and Social History 1600–1900*, edited by L.M. Cullen and T.C. Smout, 227–40. Edinburgh: John Donald, [1977]
- 'Party Politics and Religious Controversialism in Prince Edward Island from 1860 to 1863.' *Acadiensis*, 7/2 (Spring 1978), 29–59
- 'Edward Palmer.' In *Dictionary of Canadian Biography*, vol. 11, 664–70. Toronto: University of Toronto Press, 1982
- 'James Colledge Pope.' In *Dictionary of Canadian Biography*, vol. 11, 699–705. Toronto: University of Toronto Press, 1982
- 'Political Realignment in Pre-Confederation Prince Edward Island, 1863–1870.' *Acadiensis*, 15/1 (Autumn 1985), 35–58
- 'Donald McDonald.' In *Dictionary of Canadian Biography*, vol. 8, 530–3. Toronto: University of Toronto Press, 1985
- 'The *Posse Comitatus* Incident of 1865.' *Island Magazine*, no. 24 (Fall–Winter 1988), 3–10
- 'Reform, Literacy, and the Lease: The Prince Edward Island Free Education Act of 1852.' *Acadiensis*, 20/1 (Autumn 1990), 52–71

– 'James Horsfield Peters.' In *Dictionary of Canadian Biography*, vol. 12, 838–42. Toronto: University of Toronto Press, 1990
– 'The 1850s: Maturity and Reform.' In *The Atlantic Region to Confederation: A History*, edited by Phillip A. Buckner and John G. Reid, 333–59. Toronto and Fredericton: University of Toronto Press and Acadiensis Press, 1994
– 'The Maritime Colonies, 1784 to Confederation.' In *Canadian History: A Reader's Guide*, vol. 1: *Beginnings to Confederation*, edited by M. Brook Taylor, 237–79. Toronto: University of Toronto Press, 1994
Rogers, Irene. *Charlottetown: The Life in Its Buildings*. Charlottetown: Prince Edward Island Museum and Heritage Foundation, 1983
Sager, Eric W., and Lewis R. Fischer. 'Patterns of Investment in the Shipping Industries of Atlantic Canada, 1820–1900.' *Acadiensis*, 9/1 (Autumn 1979), 19–43
See, Scott W. 'Polling Crowds and Patronage: New Brunswick's "Fighting Elections" of 1842–3.' *Canadian Historical Review*, 72/2 (June 1991), 127–56
– '"Mickeys and Demons" vs. "Bigots and Boobies": The Woodstock Riot of 1847.' *Acadiensis*, 21/1 (Autumn 1991), 110–31
– *Riots in New Brunswick: Orange Nativism and Social Violence in the 1840s*. Toronto: University of Toronto Press, 1993
Sharpe, Errol. *A People's History of Prince Edward Island*. Toronto: Steel Rail, 1976
Smout, T.C. 'The Highland Clearances.' *Scottish International*, 5/2 (Feb. 1972), 13–16
Steele, E.D. 'Ireland and the Empire in the 1860s. Imperial Precedents for Gladstone's First Irish Land Act.' *Historical Journal*, 11/1 (1968), 64–83
– *Irish Land and British Politics: Tenant-Right and Nationality 1865–1870*. Cambridge: Cambridge University Press, 1974
– *Palmerston and Liberalism, 1855–1865*. Cambridge: Cambridge University Press, 1991
Stewart, Deborah J. 'Robert Bruce Stewart and the Land Question.' *Island Magazine*, no. 21 (Spring–Summer 1987), 3–11
Story, G.M. 'William Eppes Cormack.' In *Dictionary of Canadian Biography*, vol. 9, 158–62. Toronto: University of Toronto Press, 1976
Taylor, M. Brook. 'John Longworth.' In *Dictionary of Canadian Biography*, vol. 11, 528–9. Toronto: University of Toronto Press, 1982
– 'Charles Worrell.' In *Dictionary of Canadian Biography*, vol. 8, 953–5. Toronto: University of Toronto Press, 1985
– 'William Johnston.' In *Dictionary of Canadian Biography*, vol. 6, 359–61. Toronto: University of Toronto Press, 1987
Townshend, Charles. *Political Violence in Ireland: Government and Resistance since 1848*. Oxford: Clarendon Press, 1983

Vass, Elinor Bernice. 'George Wastie DeBlois.' In *Dictionary of Canadian Biography*, vol. 11, 241–2. Toronto: University of Toronto Press, 1982

Vaughan, W.E. 'Landlord and Tenant Relations in Ireland between the Famine and the Land War, 1850–78.' In *Comparative Aspects of Scottish and Irish Economic and Social History 1600–1900*, edited by L.M. Cullen and T.C. Smout, 216–26. Edinburgh: John Donald, [1977]

– *Sin, Sheep and Scotsmen: John George Adair and the Derryveagh Evictions, 1861.* Belfast: Appletree Press, 1983

– *Landlords and Tenants in Ireland 1848–1904.* [Dublin]: Economic and Social History Society of Ireland, 1984

– *Landlords and Tenants in Mid-Victorian Ireland.* Oxford: Clarendon Press, 1994

Vincent, Thomas B. 'John LePage.' In *Dictionary of Canadian Biography*, vol. 11, 511–12. Toronto: University of Toronto Press, 1982

Waite, Peter B. *The Life and Times of Confederation 1864–1867: Politics, Newspapers, and the Union of British North America.* Toronto: University of Toronto Press, 1962 ed.

Wall, Maureen. 'The Age of the Penal Laws (1691–1778).' In *The Course of Irish History*, edited by T.W. Moody and F.X. Martin, 217–31. Cork: Mercier, 1967

Weale, David. 'The Gloomy Forest.' *Island Magazine*, no. 13 (Spring–Summer 1983), 8–13

– 'The Emigrant.' *Island Magazine*, no. 16 (Fall–Winter 1984), 15–22, and no. 17 (Summer 1985), 3–11

Weale, David, and Harry Baglole. *The Island and Confederation: The End of an Era.* Summerside: Williams and Crue, 1973

Webb, Sidney, and Beatrice Webb. *The Parish and the County.* London: Frank Cass and Co., 1963 ed. (orig. publ. 1906)

Webber, David. *A Thousand Young Men: The Colonial Volunteer Militia of Prince Edward Island 1775–1874.* Charlottetown: Prince Edward Island Museum and Heritage Foundation, 1990

Wells, Kennedy. *The Fishery of Prince Edward Island.* Charlottetown: Ragweed, 1986

Whitfield, Carol M. *Tommy Atkins: The British Soldier in Canada, 1759–1870.* Ottawa: Parks Canada, 1981

Whyte, J.H. *The Independent Irish Party, 1850–9.* Oxford: Oxford University Press, 1958

Williams, David. *A History of Modern Wales.* London: John Murray, 1950

– *The Rebecca Riots: A Study in Agrarian Discontent.* Cardiff: University of Wales Press, 1955

Williams, T. Desmond. 'Preface.' In *Secret Societies in Ireland*, edited by T. Desmond Williams, 1–12. Dublin: Gill and Macmillan, 1973

Winstanley, Michael J. *Ireland and the Land Question 1800–1922*. London: Methuen, 1984
Withers, Charles W.J. *Gaelic Scotland: The Transformation of a Culture Region*. London: Routledge, 1988

UNPUBLISHED STUDIES

Baglole, Harry. 'A Reassessment of the Role of Absentee Proprietors in Prince Edward Island History.' Research paper, Memorial University of Newfoundland, Dec. 1970, PAPEI
Bittermann, [K.] Rusty. 'Escheat!: Rural Protest on Prince Edward Island, 1832–1842.' PHD thesis, University of New Brunswick, 1991
MacNutt, T.E. 'Aid to the Civil Power in PEI in Times Past.' Unpublished typescript, 1937; rev. 1961, PAPEI
MacNeill, Ewen. 'McNeill of Barra on P.E. Island: Ancestors and Descendants of Alexander McNeill.' Privately circulated, 1982, PAPEI
Robertson, Ian Ross. 'Religion, Politics, and Education in Prince Edward Island, from 1856 to 1877.' MA thesis, McGill University, 1968
Stewart, Deborah J. 'Robert Bruce Stewart: A Study of Resident Proprietorship on Prince Edward Island.' Research paper, University of Toronto, June 1984

WORKS OF REFERENCE

Bercuson, David J., and J.L. Granatstein. *Dictionary of Canadian Military History*. Toronto: Oxford University Press, 1992
The Concise Oxford Dictionary, 7th edition. Oxford: Clarendon Press, 1982
Dictionary of National Biography
Osborn's Concise Law Dictionary, 6th edition. Edited by John Burke. London: Sweet and Maxwell, 1976
Pratt, T.K., ed. *Dictionary of Prince Edward Island English*. Toronto: University of Toronto Press, 1988
Rayburn, Alan. *Geographical Names of Prince Edward Island*. Ottawa: Department of Energy, Mines and Resources, 1973

C. Other Sources

INTERVIEWS AND EXPERT CONSULTATIONS

Anderson, Robert C. 12 Aug. 1981
Blair, Ronald S. 2 Nov. 1993, and 2 Aug. 1995

Cuthbertson, Brian. 28 July 1981
Dickieson, Herb. 20 Nov. 1992, and 4 Sept. 1994
Dickieson, Roy. 27 Aug. 1986, 4 Sept. 1994, and 21 April 1995
MacNeill, Ewen. 26 July and 1 Sept. 1995
Nic Dhiarmada, Mairin. 15 Dec. 1994
Pope, Maurice. 18 May 1968
Raywalt, James K. 26 Nov. 1989
Sanborn, George F., Jr. 12 Aug. 1981
Shipley, Allan Q. 1 Sept. 1992

LETTERS

Arsenault, Georges, to author. 2 May 1995
Blanchard, J. Wilmer, to author. 5 Dec. 1994
Finley, Michael J.W., to author. 20 March 1978
MacDonald, G. Edward, to author. 7 Sept. 1994
Raywalt, James K., to author. 8 July 1990
Sanborn, George F., Jr, to author. 23 Feb. 1995

ADDITIONAL

CBC television, 'Prime Time News.' 18 June 1993
Note by Black, Ronald. 16 Feb. 1994

Illustration Credits

Index